SOCIETY FOR EXPERIMENTAL BIOLOGY
SEMINAR SERIES · 37

TECHNIQUES IN
COMPARATIVE RESPIRATORY PHYSIOLOGY

SOCIETY FOR EXPERIMENTAL BIOLOGY SEMINAR SERIES

A series of multi-author volumes developed from seminars held by the Society for Experimental Biology. Each volume serves not only as an introductory review of a specific topic, but also introduces the reader to experimental evidence to support the theories and principles discussed, and points the way to new research.

TECHNIQUES IN COMPARATIVE RESPIRATORY PHYSIOLOGY

An experimental approach

Edited by

C. R. BRIDGES

Institute for Zoology, University of Düsseldorf, FRG

and

P. J. BUTLER

School of Biological Sciences
University of Birmingham, UK

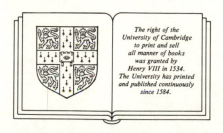

The right of the
University of Cambridge
to print and sell
all manner of books
was granted by
Henry VIII in 1534.
The University has printed
and published continuously
since 1584.

CAMBRIDGE UNIVERSITY PRESS

Cambridge
New York   Port Chester
Melbourne   Sydney

Published by the Press Syndicate of the University of Cambridge
The Pitt Building, Trumpington Street, Cambridge CB2 1RP
40 West 20th Street, New York NY 10011, USA
10 Stamford Road, Oakleigh, Melbourne 3166, Australia

First published 1989

Printed in Great Britain at the University Press, Cambridge

British Library cataloguing in publication data

Techniques in comparative respiratory
physiology – an experimental approach.
1. Animals. Respiration
I. Bridges, C. R.   II. Butler, P. J.
III. Series
591.1′2

Library of Congress cataloguing in publication data

Techniques in comparative respiratory physiology: an experimental
approach/edited by C. R. Bridges & P. J. Butler.
p.   cm. (Seminar Series/Society for Experimental Biology: 37)
Papers presented at the York meeting of the Respiration Group of the Society of
Experimental Biology, March 31–April 3, 1987.
Includes index.
ISBN 0 521 34568 5
1. Respiratory organs – Physiology – Research – Methodology – Congresses.
2. Physiology, Comparative – Research – Methodology – Congresses.
I. Bridges, C. R.   II. Butler, P. J.
III. Society for experimental Biology (Great Britain). Respiration Group.
IV. Series: Seminar series (Society for Experimental Biology (Great Britain)): 37.
[DNLM: 1. Physiology, Comparative – methods – congresses. 2. Respiration –
congresses. 3. Respiratory System – physiology – congresses. W1 SO8539 v.
37/WF 102 T255 1987]
QP121.T43 1989   DNLM/DLC   88–25609

ISBN 0 521 34568 5

Kjell
VIKING & PHYSIOLOGIST

(Reproduced with permission of B. Linzen & Springer Verlag, Heidelberg)

CONTENTS

Contents

CONTRIBUTORS

Bryant, D. M.
Department of Biology, University of Stirling, Stirling, Scotland, UK

Butler, P. J.
School of Biological Sciences, Birmingham, B15 2TT, UK

Dejours, P.
Laboratoire de Physiology Respiratoire du C.N.R.S., 23 rue Becquerel, 67087 Strasbourg, France

Elliott, C. J. H.
Department of Biology, University of York, York, YO1 5DD, UK

Farrel, A. P.
Biology Department, Simon Fraser University, Burnaby, British Columbia, Canada

Forstner, H.
Abteilung Zoophysiology, University of Innsbruck, Technikerstraße 25, A-6020 Innsbruck, Austria

Gäde, G.
Institut für Zoologie IV, Universität Düsseldorf, D-4000, Düsseldorf, FRG

Gnaiger, E.
Institut für Zoologie, Abteilung Zoophysiologie, Universität Innsbruck, Peter Mayr Straße 1a, A-6020 Innsbruck, Austria

Grieshaber M. K.
Institut für Zoologie IV, Universität Düsseldorf, D 4000, Düsseldorf, FRG

Heisler, N. G.
Max Planck Inst. für Exp. Mediz., Abteilung Physiologie, Hermann Rein Str 3, D-3400 Göttingen, FRG

Hughes, G. M.
Research Unit for Comparative Animal Respiration, The University, Woodland Road, Bristol BS8 1UG, UK

Kaufmann, R.
Abteilung Zoophysiology, University of Innsbruck, Technikerstraße 25, A-6020 Innsbruck, Austria

Perry, S. F.
Department of Biology, Univ. of Ottawa, Ottawa, Ontario, Canada

Shick, J. M.
Department of Zoology, Murray Hall, University of Maine, Orono, Maine 04460–0146, USA

Tatner, P.
Department of Biology, University of Stirling, Stirling, Scotland, UK

Taylor, E. W.
School of Biological Sciences, University of Birmingham, Birmingham, B15 2TT, UK

Truchot, J.-P.
Laboratoire de Neurobiologie, et Physiologie Comparee, Place du Docteur Betrand Peyneau, 33120 Arcachon, France

Weber, R. E.
Zoophysiology Laboratory, Institute for Zoology and Zoophysiology, University of Aarhus, DK-8000 Aarhus, Denmark

Wells, R. M. G.
Department of Zoology, University of Auckland, Private Bag, Auckland, New Zealand

West, N. H.
Department of Physiology, University of Saskatchewan, Saskatoon S7N 0WO, Canada

Widdows, J.
Institute for Marine Environmental Research, The Hoe, Plymouth, PL13 DH, UK

Wieser, W.
Abteilung Zoophysiology, University of Innsbruck, Technikerstraße 25, A-6020 Innsbruck, Austria

Woakes, A. J.
School of Biological Sciences, PO Box 363, Birmingham, B15 2TT, UK

ABBREVIATIONS, SYMBOLS AND UNITS

The following list is based on quantities and units defined by the System International d'Unités (SI). However, certain units lying outside the SI are maintained because they are unambiguously defined and still of common use in Respiratory Physiology: Torr (1 Torr = 133.3 Pa); litre (1 L = 1 dm^3); minute (1 min = 60 sec). The writing of symbols follows the rules set out by the Atlantic City Convention (Pappenheimer, 1950). Lower-case or small capital subsidiary symbols indicate the medium or location to which the main symbol refers, whereas subscripts are used to specify the chemical species:

m medium	I inspired
g gas	E expired
w water	A alveolar
b blood	L pulmonary
a arterial	G gill
v venous	D dead space
v̄ mixed venous	T tidal
c capillary	B barometric
t tissular	

Description	Symbol	Unit
quantity of substance	M	mol
quantity of substance per unit time ($= M \cdot t^{-1}$)	$\dot{M}$	mol·sec^{-1} or mol·min^{-1}
oxygen consumption	e.g. $\dot{M}_{O_2}$	
oxygen uptake through the gill	$\dot{M}_{G_{O_2}}$	
gas pressure (may be partial pressure)	P	Torr (or Pa)
barometric pressure	e.g. P_B	
carbon dioxide partial pressure in expired medium	$P_{E_{CO_2}}$	
oxygen partial pressure in inspired medium	$P_{I_{O_2}}$	

Description	Symbol	Unit
oxygen partial pressure in arterial blood	Pa_{O_2}	
concentration of gas $(= M \cdot V^{-1})$	C	$mol \cdot dm^{-3}$ or $mol \cdot L^{-1a}$
oxygen concentration in water	e.g. Cw_{O_2}	
oxygen concentration in inspired medium	Ci_{O_2}	
concentration of dissolved oxygen	$C_{O_2 diss}$	
concentration of oxyhemoglobin	C_{HbO_2}	
total oxygen concentration (dissolved plus combined to the respiratory pigment)	e.g. $C_{O_2 tot}$	
saturation (ratio of the amount of oxygen actually in combined form to the maximum amount of combined oxygen)	S	
oxygen saturation in mixed venous blood	e.g. $S\bar{v}_{O_2}$	
volume fraction of a gas in dry gas phase	F	
volume fraction of carbon dioxide in alveolar gas	e.g. Fa_{CO_2}	
respiratory exchange ratio $(= \dot{M}_{CO_2}/\dot{M}_{O_2})$	RE	
respiratory exchange ratio between arterial and mixed venous blood	e.g. $Ra, \bar{v}$	
thermodynamic constant of ideal gas	R	$J \cdot mol^{-1} \cdot K^{-1}$
volume	V	dm^3 or L
tidal volume	e.g. V_T	
stroke volume	V_S	
lung volume	V_L	
volume per unit time or volume flow rate $(V \cdot t^{-1})$	$\dot{V}$	$dm^3 \cdot min^{-1}$ or $L \cdot min^{-1}$
water flow rate	e.g. $\dot{V}w$	
gas flow rate (e.g. $\dot{V}air$)	$\dot{V}g$	

[a] $mol \cdot L^{-1}$ (or $mmol \cdot L^{-1}$) is preferred to M (or mM) which is in fact a quantity and not a concentration. Molar gas concentrations in a gas phase should state if total gas volume is in BTPS or ATPS conditions, or any other conditions. Volume fraction F of the gas in dry gas phase can be used instead of molar concentration provided that temperature and barometric pressure are stated, so that the molar concentration can be calculated. Ambiguous concentration units as ppm should be avoided. Concentration of a non gaseous chemical species is generally given in brackets, e.g. $[HCO_3^-]$.

Description	*Symbol*	*Unit*
blood flow rate	$\dot{V}b$	
rate of inspired air or water flow	$\dot{V}I$	
solubility coefficient of a gas $(M \cdot V^{-1} \cdot p^{-1})$	α	$mol \cdot L^{-1} \cdot Torr^{-1}$ (Pa^{-1})
oxygen solubility coefficient in blood	αb_{O_2}	
capacitance coefficient of a gas $(\Delta C \cdot \Delta P^{-1})^b$	β	$mol \cdot L^{-1} \cdot Torr^{-1}$ (Pa^{-1})
carbon dioxide capacitance coefficient - in water	e.g. βw_{CO_2}	
coefficient of diffusion	D	$m^2 \cdot sec^{-1}$
Krogh's constant of diffusion of a gas $(= D \cdot \alpha)$	K	$mol \cdot m^{-1} \cdot sec^{-1} \cdot$ $Torr^{-1}$ (Pa^{-1})
conductance of a gas species, or gas transfer rate per unit partial pressure difference $(= \dot{M} \cdot \Delta P^{-1})$	G	$mol \cdot sec^{-1} \cdot Torr^{-1}$ (Pa^{-1})
conductance between inspired and alveolar gas	e.g. $G_{I,A}$	
lung diffusing capacity (= conductance) for oxygen	$D_{L_{O_2}}$	
frequency	f	sec^{-1} or min^{-1}
heart frequency	e.g. f_H	
respiratory frequency	f_R	
partial pressure of oxygen at which Hb is 50% oxygenated $(S_{O_2} = 0.5)$	P_{50}	Torr (Pa)
body mass	B	kg or g
mass-specific oxygen consumption rate	e.g. $\dot{M}_{O_2} \cdot B^{-1}$	
Isotope standard, Standard Mean Ocean Water	SMOW	
Isotope standard, Standard Light Antarctic Precipitation	SLAP	
catabolic heat flux	$_k\dot{Q}$	μW
total heat flux synonymous with total rate of heat dissipation	$_t\dot{Q}$	μW
weight specific heat flux	$_t\dot{q}$	$mW \cdot g^{-1}$
oxygen flux synonymous with rate of oxygen consumption	N_{O_2}	$nmol \cdot O_2 \cdot s^{-1}$
weight specific oxygen flux	$\dot{n}_{O_2}$	$nmol \; O_2 \cdot s^{-1} \cdot g^{-1}$

[b] Takes into account both dissolved and reversibly bound forms of a gas. May also be used for a gas in a gas phase.

Description	Symbol	Unit
theoretical oxycaloric equivalent	$\Delta_k H_{O_2}$	kJ (mol O_2)$^{-1}$
measured calorimetric-respirometric ratio: CR ratio	$\Delta_t Q_{O_2}$	kM (mol O_2)$^{-1}$
respiratory combustion equivalent	$\Delta_c H_{O_2}$	kJ (mol O_2)$^{-1}$
respiratory combustion equivalent for ammonia	$\Delta_c H_{NH_3}$	kJ (mol NH_3)$^{-1}$
dry weight	$_d W$	g
ash-free (organic) dry weight	$_{af} W$	g
caloric content of organic matter synonymous with enthalpy of combustion	$\Delta_c h$	kJ (g $_{af}W$)$^{-1}$
mass fraction of organic carbon	w_C	g C (g $_{af}W$)$^{-1}$
mass fraction of organic nitrogen	w_N	g N (g $_{af}W$)$^{-1}$
respiratory quotient	RQ	mol CO_2/mol O_2
nitrogen quotient	NQ	mol N/mol O_2
exponential time constant	τ	s
polarographic oxygen sensor	POS	
oxygen reduction ratio	R_{O_2}	

PREFACE

This volume in the seminar series marks a decade in the history of the Respiration Group of the Society of Experimental Biology, which staged its first official meeting in January 1977 at the University of Birmingham. From this first meeting, with 20 papers and 3 posters, the group has expanded to its present size and at the York meeting, 31 March–3 April 1987, 46 papers and 37 posters were presented. The group has become a truly international one with 45 per cent of the 490 poster/papers presented over the last ten years coming from mainland Europe and 17 per cent from North America. Thirty-eight per cent of the papers have involved work with invertebrates and of the remaining papers on vertebrates 68 per cent have been concerned with work on fish.

It is therefore not surprising that in the last decade there has been a dramatic increase in both the breadth and scope of comparative respiratory physiology. This has been due both to the development of new technology and to the application, in the comparative field, of many techniques evolved for medical purposes.

It is evident, with the evolution of high technology, that no one scientist can keep abreast of all the new methods available in the field of comparative respiratory physiology. The present volume, which originates from review lectures given at the York meeting, provides a working basis for those scientists who wish to use some of these techniques. Particular attention is given to possible drawbacks and problems that may arise. Special emphasis has been placed on 'an experimental approach', where possible, and the authors have tried to convey some of the 'tips and tricks' associated with each method. Our thanks are due to the contributors to the volume, to Dr C. Elliott the local secretary at the meeting, to the chairmen of the various sessions and also to the numerous referees who gave up their time in order to make this volume successful.

On a sadder note, just prior to the meeting on 4 March Kjell Johansen died at the age of 54. Kjell had agreed to participate in the writing of this present volume and had been a supporter of the Respiration Group from its inception. His enthusiasm for comparative physiology and his encouragement of others in the field will be greatly missed.

On behalf of the Respiration Group we should therefore like to dedicate this volume to the memory of Kjell Johansen, Viking and Physiologist.

C. R. Bridges and P. J. Butler
Editors for the Society of Experimental Biology

Respiratory systems – theory and morphology

J. P. TRUCHOT and P. DEJOURS

Comparative respiratory physiology – quantities, dimensions and units

Introduction

Comparative Respiratory Physiology deals with gas exchange and transfer in animals, i.e. the inward movement of oxygen molecules from the ambient medium to the cells, as well as the outward movement of carbon dioxide molecules along the reverse path. These tasks are of course performed by every aerobically-fuelled organism, in widely differing ambient media and with various structural and functional organizations. The aim of Comparative Respiratory Physiology is to assess and compare the performance and efficiency, as well as the adaptability, of such diverse gas exchange systems.

Continuous improvement of instrumentation and techniques is promoting a more and more quantitative approach to gas exchange physiology. However, in respiratory physiology, as in many areas of science, quantitative data can attain a highly significant value and serve true comparative purposes only if they are expressed within a comprehensive corpus of concepts, quantities and units. Ideally, it would be desirable to describe gas exchange with the same concepts and quantities whatever the medium – air, water or blood – in which the gases are transferred, and whatever the physical process – diffusion or convection – accomplishing the transfer. Much progress has been realized toward this goal during the last decades, leading to better descriptions and easier comparisons of gas transfer processes. This took several forms, in a rather intermingled manner. First of all, a system physiology framework has been generalized that considers any organism as a gas exchange system made up of several, serially-disposed compartments through which gas movements can be described by appropriate transfer equations. An adequate system of basic physical quantities required to describe gas exchange has also been refined, together with a rationalization of the writing of symbols used to specify these quantities. Finally, new concepts of high comparative value, such as those of capacitance and conductance, have been introduced.

This chapter presents a summary of these advances, to form a common basis for communication and understanding of new results that hopefully will continue to arise, due to extending techniques.

General remarks on quantities, symbols and units

Simple physical processes govern gas transfer, and respiratory physiology is thus based largely on the elementary physics of gases. In an effort to arrive at a uniform scientific language on an interdisciplinary basis, the Système International d'Unités (SI) (1985) has introduced a dimensionally coherent set of physical quantities and units as well as rules and recommendations for their use and for the writing of corresponding symbols. This now worldwide-adopted system should also be used in respiratory physiology. However, the SI itself acknowledges that certain units lying outside its rules may be maintained, either for their practical importance or by force of habit. Such is the case for some units of common use in respiratory physiology – for example the Torr for partial pressure – whose usage will probably continue for a long time without any particular problem as long as they are unambiguously defined and easily convertible into SI units.

The widely accepted principles set out by the Atlantic City Convention (Pappenheimer, 1950) are to be used for the writing of symbols in Respiratory Physiology. In this system, main quantities are written with single large capital letters, lower case or small capital subsidiary symbols specify the medium or location to which the main symbol refers, whereas subscripts designate the chemical species. Examples for the application of these principles and a list of the major quantities in common use can be found in the list of Abbreviations.

Physical quantities in the description of gas exchange
Basic quantities

To describe gas exchange, the following basic quantities are required (symbols in parentheses):

> *amount of substance* (gas); (M); SI unit: mole (mol)
> *volume* (gas or liquid); (V); SI unit: cubic meter (m^3); common unit: liter (L or dm^3)
> *partial pressure*; (P); SI unit: Pascal (Pa); common unit: Torr (Torr), 1 Torr = 133.3 Pa
> *time*; (t); SI unit: second (s; but best abbreviated sec); common unit: minute (min) or hour (h)

The introduction of the SI concept of *amount of substance (as)* represents a major clarification in Comparative Respiratory Physiology. Considering the amount of a gas species as dimensionally equivalent to an *as* in mole units instead of a standard volume makes it possible to distinguish unambiguously between the amount of gas exchanged and the volume of medium – air, water or blood – that carries it. In addition, this usage greatly simplifies handling respiratory data when stoichiometric relationships are being dealt with, for example in metabolism.

Derived quantities

Transfer rate and volume flow rate. The transfer rate of a gas species x (uptake, output, or net flux):

$$\dot{M}_x = dM_x/dt \tag{1}$$

has the dimension of $as \cdot time^{-1}$ (unit: $mol \cdot sec^{-1}$) and thus can be distinguished clearly from the volume flow rate of the carrier medium y:

$$\dot{V}_y = dV_y/dt \tag{2}$$

which has the dimension of $volume \cdot time^{-1}$. To be physiologically relevant, volume flow rates should always be expressed in physical conditions of biological meaning, i.e., for gas phase volumes, in BTPS conditions, that is at body temperature and pressure, and saturated with water vapor.

Concentration. The concentration of a gas species x has in all media the dimension of $as \cdot volume^{-1}$ (units: $mol \cdot m^{-3}$ or $mol \cdot L^{-1}$):

$$C_x = M_x \cdot V^{-1} \tag{3}$$

This is obviously true even if the amount of gas is expressed in volume STPD (standard temperature and pressure, dry), but, for reasons given above, mole units are to be preferred. Again, when the gas concentration in a gas phase is expressed in this way, the volume must be biologically relevant, i.e. in BTPS conditions, meaning that for a given molar amount of gas x in a gas phase, the concentration depends on temperature and total pressure. By contrast, the fractional concentration, F_x, usually defined as a volume fraction in a dry gas phase, V_x/V_{tot} (with both volumes expressed in the same conditions), but having also the meaning of a mole fraction (as long as the gases can be considered perfect), is adimensional and independent of temperature and total pressure. This last mode of expressing gas concentration in a gas phase has practical value since F_x is directly determined by volumetric absorption analysis. As pointed out by Dejours (1974; 1981), the ideal gas law allows easy conversion of C_x to F_x and vice versa. Since:

$$P_x V = M_x RT \tag{4}$$
$$C_x = M_x/V = (1/RT) \cdot P_x = (1/RT) \cdot (P_B - P^T_{H_2O}) \cdot F_x \tag{5}$$

in which P_B is the barometric (total) pressure and $P_{H_2O}^T$ the water partial pressure at temperature T. Values of this conversion factor for a range of temperatures and barometric pressures have been tabulated by Dejours (1981). At sea level, the ideal gas law may be considered valid for most practical purposes in Respiratory Physiology. At high gas pressure in hyperbaric experiments, significant deviation from the ideal, may, however, require appropriate correction terms (see Imbert, Dejours & Hildwein, 1982).

Capacitance coefficient. The capacitance coefficient β_x of a medium for a gas species x (Piiper, Dejours, Haab & Rahn, 1971) is defined as the increment of gas concentration C_x per increment of partial pressure P_x:

$$\beta_x = \Delta C_x / \Delta P_x \tag{6}$$

and has thus the same dimension as a solubility coefficient, i.e. *as·volume*$^{-1}$ *·pressure*$^{-1}$; units: $mol·m^{-3}·Pa^{-1}$. Its physical meaning can also be reduced to the inverse of the molar free energy of the gas.

The capacitance coefficient is a general concept of high comparative value because it applies equally to a gas physically dissolved or reversibly bound in a liquid, and to a gas in a gas phase. Provided that it is defined in molar units of concentration, it allows for easy comparison of the abilities of external and internal media to carry gases. In a gas phase, according to the gas law (equations (4) and (5)), it equals $1/RT$, at least at low or moderate total pressure, and so has the same value for any gas. In a liquid phase in which the gas exists only in dissolved form, it is equivalent to the solubility coefficient α_x. When the gas is present in reversibly bound form (oxygen combined to a respiratory pigment; carbon dioxide in the form of carbonates or carbamate compounds), the capacitance coefficient varies as a function of the gas partial pressure, since C_x is usually a non-linear function of P_x.

Gas transfer equations and the concept of conductance

Respiratory gases move through the organism by the physical processes of diffusion and convection that can be described by simple transfer equations.

Transport by convection

Simple considerations of mass balance indicate that the amount of gas substance M_x taken up or released in the steady state by a tissue, organ or organism through which a carrier medium flows, equals the product of the volume of medium V_m times the difference of concentration ($Cin_x - Cout_x$) in the ingoing and outgoing media:

Table 1. *Steady state equations describing convective oxygen transfer in water, blood and gas. Similar equations can be written for carbon dioxide transfer*

Convection by	Transfer rate $(\text{mol}\cdot\text{sec}^{-1})$	=	Flow rate $(\text{L}\cdot\text{sec}^{-1})$	×	Concentration difference $(\text{mol}\cdot\text{L}^{-1})$	
water	$\dot{M}_{O_2}$	=	$\dot{V}w$	×	$(C_{I_{O_2}} - C_{E_{O_2}})$	(9)
blood	$\dot{M}_{O_2}$	=	$\dot{V}b$	×	$(C_{a_{O_2}} - C_{v_{O_2}})$	(10)
gas	$\dot{M}_{O_2}$	=	$^a\dot{V}g$	×	$^a(C_{I_{O_2}} - C_{E_{O_2}})$	(11)

Conductance

Convection by	Transfer rate $(\text{mol}\cdot\text{sec}^{-1})$	=	Flow rate $(\text{L}\cdot\text{sec}^{-1})$	×	Capacitance coefficient $(\text{mol}\cdot\text{L}^{-1}\cdot\text{Torr}^{-1})$	×	Pressure difference (Torr)	
water	$\dot{M}_{O_2}$	=	$\dot{V}w$	×	$^b\beta w_{O_2}$	×	$(P_{I_{O_2}} - P_{E_{O_2}})$	(15)
blood	$\dot{M}_{O_2}$	=	$\dot{V}b$	×	$^c\beta b_{O_2}$	×	$(P_{a_{O_2}} - P_{v_{O_2}})$	(16)
gas	$\dot{M}_{O_2}$	=	$^a\dot{V}g$	×	$^d\beta g_{O_2}$	×	$(P_{I_{O_2}} - P_{E_{O_2}})$	(17)

[a] volumes in a gas phase are to be expressed in BTPS conditions.
[b] oxygen solubility in water; for carbon dioxide, may be higher than solubility in carbonated waters.
[c] usually a function of P_{O_2} when the blood contains a respiratory pigment.
[d] $\beta g = 1/RT = 0.01603 \times (1/T)$ mol·L$_{\text{BTPS}}^{-1}$·Torr^{-1} (T in K)

$$M_x = Vm \cdot (Cin_x - Cout_x) \tag{7}$$

Dividing by time to express transfer rate and volume flow rate, this translates into the general form of the Fick principle for convection:

$$\dot{M}_x = \dot{V}m \cdot (Cin_x - Cout_x) \tag{8}$$

Convection media may be blood, water or air and examples of applications of the Fick equation are given in Table 1 (equations (9), (10), (11)). In these equations again, the gas transfer rate $\dot{M}_x$ is best expressed in *as·time*$^{-1}$ ($\text{mol}\cdot\text{sec}^{-1}$ for example), the volume flow rate $\dot{V}m$ in *volume·time*$^{-1}$ ($\text{L}\cdot\text{sec}^{-1}$) and the concentrations C_x in *as·volume*$^{-1}$ ($\text{mol}\cdot\text{L}^{-1}$), volume units being expressed in BTPS conditions when convection in a gas phase is considered. In this form, Fick-type equations are of general application, dimensionally valid and physiologically relevant.

If air is the carrier medium and if fractional concentrations F_x are used instead of molar concentrations C_x, a conversion term as defined in equation (5) should be added in order to express $\dot{M}_x$ in mole units:

$$\dot{M}_x = \dot{V}_{\text{BTPS}} \cdot (1/RT) \cdot (P_B - P_{H_2O}^T) \cdot (Fin_x - Fout_x) \tag{12}$$

Note however, that equations (11) and (12) are in most cases only approximate when applied to O_2 and CO_2 exchanges in the lung. Because the O_2 uptake rate often differs from the CO_2 output rate, the respiratory exchange ratio $RE = \dot{M}_{CO_2}/\dot{M}_{O_2}$ is not equal to one, and the volume flow rates of inspired and expired gas. $\dot{V}I$ and $\dot{V}E$ are different if expressed both in BTPS conditions. Thus, equation (11) for example can be exactly written as:

$$\dot{M}_{O_2} = \dot{V}I \cdot C_{IO_2} - \dot{V}E \cdot C_{EO_2} \tag{13}$$

with $\dot{V}E = \dot{V}I \cdot (F_{IN_2}/F_{EN_2})$; F_{IN_2} and F_{EN_2} being the volume fractions of nitrogen in inspired and expired gases, respectively (Rahn & Fenn, 1955; Otis, 1964).

This complication does not hold, however, when the convection medium is water or blood and, even if approximate, the general form of the Fick principle (equation (8)) remains very useful for comparative purposes. If the capacitance coefficient β_x is introduced, equation (8) can be written in terms of partial pressure instead of concentration difference:

$$\dot{M}_x = \dot{V}m \cdot \beta_x \cdot (Pin_x - Pout_x) \tag{14}$$

When applied to a gas phase in which $\beta_x = 1/RT$, equation (14) is no more than a particular form of the gas law with the dimension of time added (Dejours, 1974). Table 1 gives examples of the application of this general equation to convective gas transport by water, blood and gaseous media.

Transport by diffusion

The Fick first Law of diffusion states that the steady state rate of transfer of gas x, $\dot{M}_x$, along a diffusion path of cross-sectional area A and thickness E is directly proportional to the concentration gradient $\Delta C_x/E$:

$$\dot{M}_x = D_x \cdot (A/E) \cdot \Delta C_x \tag{18}$$

D_x is the diffusion coefficient (or diffusitivity) which is dependent on the particular gas and on the diffusion medium.

For a system of two homogenous, that is perfectly mixed, compartments separated by a diffusion barrier, this relationship is valid only if the same phase is present on both sides, i.e. if ΔC_x is directly proportional to ΔP_x. Using the capacitance coefficient $\beta_x = \Delta C_x/\Delta P_x$, one can write:

$$\dot{M}_x = D_x \cdot \beta_x \cdot (A/E) \cdot \Delta P_x \tag{19}$$

In this form, the Fick equation also validly describes the steady state diffusion of a gas between two compartments containing different media, for example air and blood, D_x and β_x being respectively the diffusion coefficient and the capacitance coefficient of the material of the diffusion barrier. In many cases, β_x can be equated to the solubility coefficient α_x. However, for media in which gases are chemically bound (haemoglobin or

Table 2. *Steady state equations describing diffusive oxygen transfer in water, gas, lung and tissue. Similar equations can be written for carbon dioxide transfer*

Diffusion in	Transfer rate $(mol \cdot sec^{-1})$	=	Transport $(L \cdot sec^{-1})$	×	Capacitance coefficient $(mol \cdot L^{-1} \cdot Torr^{-1})$	×	Pressure difference (Torr)	
					Conductance			
water	$\dot{M}_{O_2}$	=	$Dw_{O_2} \cdot (A/E)$	×	βw_{O_2}	×	ΔP_{O_2}	(20)
gas	$\dot{M}_{O_2}$	=	$Dg_{O_2} \cdot (A/E)$	×	$^a\beta g_{O_2}$	×	ΔP_{O_2}	(21)
lung	$\dot{M}_{O_2}$	=			$^bDL_{O_2}$	×	ΔP_{O_2}	(22)
tissue	$\dot{M}_{O_2}$	=			$^bDt_{O_2}$	×	ΔP_{O_2}	(23)

[a] $mol \cdot L_{BTPS}^{-1} \cdot Torr^{-1}$ in a gas phase (see also Table 1)
[b] DL_{O_2} and Dt_{O_2} refer to diffusing capacity of lung and tissue, respectively.

myoglobin solutions for O_2, bicarbonate solutions for CO_2), facilitated diffusion can occur, gas movement taking place in both free and bound forms. Description of diffusive gas transfer by equation 19 is then no longer accurate and more refined analysis may be required (e.g. Fletcher, 1980). Various applications of the Fick equation for diffusive gas transfer are given in Table 2 (eqs 20, 21, 22, 23). It can be seen that the term $D_x \cdot (A/E)$ has the same dimension, *volume·time*$^{-1}$, as the volume flow rate of medium $\dot{V}m$, in the equation describing convective gas transfer (equation 14).

Table 3 shows the dimension and meaning of the various parameters that can be used to describe gas transfer by diffusion. Because numerical values for all four terms D_x, β_x, A and E are not always known, they can be grouped in various ways, depending on what can be measured.

D_x, the diffusion coefficient of physicists, is the transfer rate per unit surface area, under one unit of concentration gradient, $\Delta C_x/E$. Its physical dimension reduces to *length*$^2 \cdot time^{-1}$.

The ratio D_x/E is the permeability coefficient, i.e. the transfer rate per unit surface area, under one unit concentration difference ΔC_x. Its dimension is that of a velocity, *length·time*$^{-1}$.

The product $D_x \cdot \beta_x = K_x$ is called the Krogh's constant of diffusion. It is the transfer rate per unit surface area, under one unit of partial pressure gradient $\Delta P_x/E$. Its physical dimension reduces to *as·mass*$^{-1} \cdot time$.

The concept of conductance

Equations 14 and 19 for convective and diffusive gas transfer take the same form, similar to Ohm's law in electricity:

Table 3. *Dimensions and examples of units for various parameters describing diffusive gas transfer*

Names	Diffusion coefficient Diffusivity constant	Permeability Permeability coefficient	Krogh's constant Permeation coefficient	Diffusive conductance Diffusing capacity Transfer factor
Symbols	D_x	D_x/E	$D_x \cdot \beta_x$	$(A/E) \cdot D_x \cdot \beta_x$
Meaning	$\dfrac{\text{transfer rate of } as}{\text{surface area} \times \text{C. gradient}}$	$\dfrac{\text{transfer rate of } as}{\text{surface area} \times \text{C. difference}}$	$\dfrac{\text{transfer rate of } as}{\text{surface area} \times \text{P. gradient}}$	$\dfrac{\text{transfer rate of } as}{\text{P. difference}}$
Examples of units	$\text{m}^2 \cdot \text{sec}^{-1}$	$\text{m} \cdot \text{sec}^{-1}$	$\text{mol} \cdot \text{sec}^{-1} \cdot \text{m}^{-1} \cdot \text{Torr}^{-1}$	$\text{mol} \cdot \text{sec}^{-1} \cdot \text{Torr}^{-1}$
Physical dimension	$length^2 \cdot time^{-1}$	$length \cdot time^{-1}$	$as \cdot mass^{-1} \cdot time$	$as \cdot mass^{-1} \cdot time \cdot length$

From Piiper *et al.*, 1971.

$$\dot{M}_x = G_x \cdot \Delta P_x \tag{24}$$

$\dot{M}_x$ and ΔP_x being analogous, but of course not dimensionally equivalent, to current intensity and potential difference, respectively. G_x has the meaning of a *conductance*, or of the inverse of a resistance. It is the gas transfer rate in the system under one unit of partial pressure difference, i.e. *as·time^{-1} ·partial pressure^{-1}*. In terms of SI base units, its physical dimension can be shown to reduce to *as·mass^{-1}·time·length*.

According to the transfer process at work, the conductance term takes two different forms:
(i) the convective conductance (equation 14) is the product of a volume flow rate times the capacitance coefficient of the carrier medium:

$$Gconv_x = \dot{V}m \cdot \beta_x$$

(ii) the diffusive conductance (equation 19):

$$Gdiff_x = D_x \cdot \beta_x \cdot (A/E) = K_x \cdot (A/E)$$

takes into account diffusive properties of the particular medium and gas (D, β_x, K_x) and a geometrical factor (A/E). Other names for the diffusive conductance are the diffusing capacity, or the transfer factor.

In the steady state, each transfer step along the gas exchange system, either diffusive or convective, is accompanied by an obligate decrease of gas partial pressure. This pressure difference is required for transfer, either as the driving force for diffusion or as the condition for a concentration change in a convective transport process. The concept of conductance obviously has a potent comparative usefulness in that it affords the possibility of analyzing and comparing, within a common framework and in terms of partial pressure differences: (i) the efficiency of gas transfer by different processes, in different media and in organisms with various structural organizations; (ii) the adaptive response to changing environmental or physiological conditions in the various transfer steps of gas exchange systems.

Some applications of the concept of conductance
Transport by combination of convection and diffusion
Diffusive and convective processes are not always easy to isolate in an experimental analysis of gas transfer because they are often intermingled in such a way that the relevant parameters cannot be measured. When diffusion takes place between two moving streams of fluid, for example between water and blood across the gill epithelium of fish, an estimate of the mean driving partial pressure difference is difficult to obtain for many reasons (see for example, Piiper, Scheid, Perry and Hughes, 1986). Since

the conductance concept (equation 24) allows the description of gas transfer with the same equations, quantities and units whatever the process at work, convective and diffusive conductances can validly be lumped together, according to what can be conveniently measured. For O_2 transfer in the mammalian lung for example, one can define a conductance between mean alveolar gas and mixed arterial blood:

$$G_{A,aO_2} = \dot{M}_{O_2}/(P_{AO_2} - P_{aO_2}) \tag{25}$$

which can easily be quantified. Obviously, this is a composite term describing various factors accounting for the alveolar-arterial partial pressure difference, including the diffusive conductance of the lung epithelium itself but also regional inequalities of the ventilation/blood flow ratio, $\dot{V}_A/\dot{V}b$, venous admixture, etc.

In terms of resistance to gas transfer, $1/G$ (i.e. the partial pressure difference required for one unit transfer rate), any gas exchange system can in this way be resolved into a sum of experimentally quantifiable terms, for example:

$$(1/G_{I,v}) = (1/G_{I,A}) + (1/G_{A,a}) + (1/G_{a,v}) \tag{26}$$

Such a decomposition of the total conductance (or total resistance to gas transfer) proves very useful to analyze the adaptive responses of the gas exchange system to environmental and physiological challenges, at a number of serially-disposed, easily defined transfer steps (see Dejours, 1981). Note, however, that such an analysis is valid only for animals having a single, well-differentiated breathing apparatus, one-way circulation and systemic capillarization. The respiratory organization in many zoological groups may be much more complex than simplified models. In addition, being based on conductive properties of the system, the approach just described applies only to ideal steady state conditions. Cyclical changes over a short period such as those occurring during a respiratory cycle usually deviate little from a true steady state, but the analysis of transitions having a longer period should include additional capacitive terms describing the time course of changes of the gas stores in the body (see for example Piiper, 1982).

Conductance ratios and the nature of the limitations to gas transfer

There are two major sources of limitations to gas transfer: (1) The constraints resulting from the properties of gases in the media which carry them or in which they diffuse; (2) The constraints resulting from the structure and particular design of the gas exchangers in which gas transfer simultaneously involves both convective and diffusive processes. Identifying the limiting process is of interest to predict how the efficiency of transfer can be increased by physiological responses.

These limitations can be conveniently analyzed using conductance ratios, either for the two gases O_2 and CO_2 (e.g. G_{O_2}/G_{CO_2}) in the same transfer process, or for one gas (O_2 or CO_2) in different transfer processes (e.g. $(Gdiff/Gconv)_{O_2}$).

According to the basic transfer equation (equation 24), the gas partial pressure difference required to effect a given transfer rate depends on the corresponding conductance that is itself determined in part by the properties of the media in which the gases move. As these properties may be different for O_2 and for CO_2, simultaneous transfer of the two gases may require different partial pressure differences. This is conveniently analyzed in expressing the steady state respiratory exchange ratio, $RE = \dot{M}_{CO_2}/\dot{M}_{O_2}$, in terms of equation (24):

$$RE = (\dot{M}_{CO_2}/\dot{M}_{O_2}) = (G_{CO_2}/G_{O_2}) \cdot (\Delta P_{CO_2}/\Delta P_{O_2}) \tag{27}$$

or

$$\Delta P_{CO_2} = RE \cdot (G_{O_2}/G_{CO_2}) \cdot \Delta P_{O_2} \tag{28}$$

an equation that can be plotted on an $O_2 \cdot CO_2$ diagram for convective and diffusive transfers in the air, water or any medium (Rahn & Fenn, 1955; Rahn, 1966; Rahn, Wangensteen & Fahri, 1971, Dejours, 1981). In this plot, the slope of the so-called R lines (or R isopleths), $\Delta P_{CO_2}/\Delta P_{O_2}$, equals $-RE \cdot (G_{O_2}/G_{CO_2})$, and the particular $Pout_{O_2}$ and $Pout_{CO_2}$ at given P in values depend on the (transfer rate)/(conductance) ratio (equations 14 and 19). Analytical expressions and typical values of conductance ratios G_{O_2}/G_{CO_2} for diffusive and convective transfers in air, unbuffered water, blood and tissue are given in Table 4 and plotted on an $O_2 \cdot CO_2$ diagram in Fig. 1. It can be seen that in water and tissues, O_2 transfer by convection as well as by diffusion is considerably more limited (i.e. requires a greater partial pressure difference) than CO_2 transfer, whereas in air, transfer of the two gases results in similar partial pressure changes, whatever the process involved. Interesting consequences can be deduced from these basic differences concerning O_2 and CO_2 exchanges and acid–base balance in water-breathers and in air-breathers (Rahn, 1966; Dejours, 1981) as well as in bimodal breathers (see for example Rahn & Howell, 1976).

In many gas exchangers, e.g. lung, gill or tissue, both convective and diffusive processes take place simultaneously. The question thus often arises as to know which of these processes opposes the greater resistance, i.e. is the more limiting, to gas transfer. Most respiratory organs can be reduced to a model in which gases are transferred by diffusion across a tissue barrier separating two flowing streams of fluid, medium and blood. Thus, three different conductances are involved, two convective, Gvent and Gperf, and one diffusive, Gdiff. Generally, the anatomical structure

Table 4. *Analytical expressions and typical values of conductance ratios G_{O_2}/G_{CO_2} in various media for convective and diffusive gas transfer*

Medium	Mode of transport	Conductance ratio G_{O_2}/G_{CO_2}	Typical values of conductance ratios
Air	Convection	$(\dot{V}air \cdot \beta g)/(\dot{V}air \cdot \beta g) = 1$	independent of temperature = 1
	Diffusion	$[D_{O_2} \cdot \beta g \cdot (A/E)]/[D_{CO_2} \cdot \beta g \cdot (A/E)] = D_{O_2}/D_{CO_2}$	at 37°C: 1.28[a]
Water	Convection	$(\dot{V}w \cdot \beta w_{O_2})/(\dot{V}w \cdot \beta w_{CO_2}) = \beta w_{O_2}/\beta w_{CO_2}$	at 20°C, pure water: 0.035[a,b]
	Diffusion	$[Dw_{O_2} \cdot \beta w_{O_2} \cdot (A/E)]/[Dw_{CO_2} \cdot \beta w_{CO_2} \cdot (A/E)]$ $= (Dw_{O_2} \cdot \beta w_{O_2})/(Dw_{CO_2} \cdot \beta w_{CO_2}) = Kw_{O_2}/Kw_{CO_2}$	at 20°C, pure water: 0.049[a,b]
Blood	Convection	$(\dot{V}b \cdot \beta b_{O_2})/(\dot{V}b \cdot \beta b_{CO_2}) = \beta b_{O_2}/\beta b_{CO_2}$	resting dogfish in normoxia at 17°C: 0.020[c] resting man in normoxia, 37°C: 0.136
Tissue	Diffusion	Kt_{O_2}/Kt_{CO_2}	at 20–22°C, frog muscle: 0.027[a]

[a] Dejours (1981).
[b] lower values prevail in carbonated waters at low Pw_{CO_2} due to carbonate buffering increasing βw_{CO_2}.
[c] Piiper and Scheid (1977).

determines the specific arrangement of the blood and medium flows, i.e. counter-current, cross-current, etc., which results in particular properties of the system. Numerous theoretical and experimental studies of these systems have been performed, with the aim of comparing their efficiencies (Piiper & Scheid, 1975, 1977, 1982). These studies have clearly shown that the limitations imposed on the whole transfer of each gas by each particular step can be assessed in terms of conductance ratios. The relative role of diffusion as compared to ventilation and perfusion in limiting gas transfer depends for example on the ratios Gdiff/Gvent and Gdiff/Gperf.

Normalization of respiratory data

Since Comparative Respiratory Physiology aims at comparing gas transfer in animals, it should take into account the very large range of body sizes and metabolic rates existing among animal species. Many quantities used in gas transfer equations depend on body size and some, but not necessarily all, may change adaptively as a function of metabolic rate.

Fig. 1. P_{CO_2} *vs* P_{O_2} diagram of O_2 and CO_2 exchange by diffusion and convection in gas phase and water phase (redrawn from Dejours, 1981). The dashed lines for diffusion and full lines for convection are described by equation 28 in the following conditions: respiratory exchange ratio RE = 0.8; values of conductance ratios G_{O_2}/G_{CO_2} in air and water given in Table 4; external P_{O_2} = 150 Torr, external P_{CO_2} negligible. The stippled areas represent typical O_2 and CO_2 tensions in the alveolar gas of a mammal, A; in the air chamber of hen eggs at the end of incubation, a.c.; and in the expired water of a fish, E.

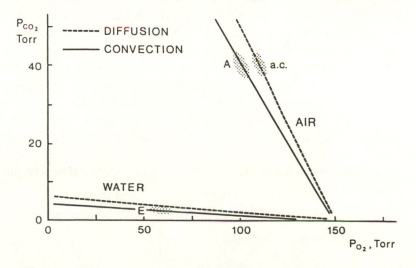

What, therefore, is the best reference that can be used to express respiratory data for comparative purposes?

The dependence of respiratory characteristics on body mass has been discussed extensively by Dejours (1981) and only some simple guidelines will be recalled here. They are valid for mammals, but can probably not be extended to all other animal groups. In resting conditions, gas transfer rates ($\dot{M}_{O_2}$, $\dot{M}_{CO_2}$) as well as volume flow rates ($\dot{V}m$, $\dot{V}b$), and thus the corresponding convective conductances, scale proportionately to a power function of body mass B (usually $B^{0.75}$ in mammals). As a consequence of equations for convective gas transfer (equations 8 and 14), it follows, also in resting, standard conditions, that gas concentration differences ($C_{I_{O_2}} - C_{E_{O_2}}$; $Ca_{O_2} - Cv_{O_2}$) should be relatively independent of body mass. The same holds true for gas partial pressures in respired air (Pa_{O_2}; Pa_{CO_2}), at least at sea level. Conversely, blood volume, lung volume, tidal volume or stroke volume of the heart are directly proportional to body mass, implying that respiratory and heart frequencies scale proportionately to $B^{-0.25}$.

Although these fundamental relationships clearly indicate how to take into account variable body masses, it may sometimes be more illuminating to refer size-dependent respiratory data to metabolic rate instead of to body mass, or a power of body mass. This is particularly so when functional characteristics are to be compared between species in different environments, or for a given species in different physiological conditions. The gas exchange system must adapt continuously in order to satisfy metabolic demand or to cope with changing environmental conditions. Basic requirements such as oxygen supply to the tissues, carbon dioxide elimination and maintenance of acid–base balance can be fulfilled in adjusting the various conductances in the gas exchange system. To assess and compare these responses, it is often useful to refer the conductances, or at least those of their components that are size-dependent, to the O_2 and CO_2 transfer rates, $\dot{M}_{O_2}$ and/or $\dot{M}_{CO_2}$. An illustration may be found in the concepts of ventilatory and circulatory convection requirements defined as the ratios $\dot{V}m/\dot{M}_{O_2}$ (or $\dot{V}m/\dot{M}_{CO_2}$) and $\dot{V}b/\dot{M}_{O_2}$ (or $\dot{V}b/\dot{M}_{CO_2}$), respectively (Dejours, Garey & Rahn, 1970; Dejours, 1972), which may also be called $\dot{M}_{O_2}$ (or $\dot{M}_{CO_2}$)-specific ventilation and $\dot{M}_{O_2}$ (or $\dot{M}_{CO_2}$)-specific blood flow.

From equations 8 and 14, convection requirements can be written:

$$\dot{V}/\dot{M}_x = 1/(Cin_x - Cout_x) = 1/[\beta_x \cdot (Pin_x - Pout_x)] \tag{29}$$

or, after introducing an adimensional *extraction coefficient* E_x defined as:

$$E_x = (Cin_x - Cout_x)/Cin_x \tag{30}$$

$$\dot{V}/\dot{M}_x = 1/(E_x \cdot Cin_x) = 1/(E_x \cdot \beta_x \cdot Pin_x) \tag{31}$$

The ratio $\dot{V}/\dot{M}_x$ has the dimension of *volume·as^{-1}* (unit L·mol^{-1}) and represents the volume of medium (in BTPS volume units for a gas phase) that must be circulated in order to effect delivery of one unit amount of x.

Fig. 2. Oxygen extraction coefficients and convection requirements of water and air as functions of the oxygen concentration of these media. Points 1 and 1' correspond to a human in hypoxia at 5800 m ($P_B = 380$ Torr) in acute conditions (1) and after altitude acclimatization (1'). Points 2 and 2' are for dogs without and with thermal polypnea. Points 3, 4, and 5 are for turtles at 10, 20 and 30 °C, respectively.

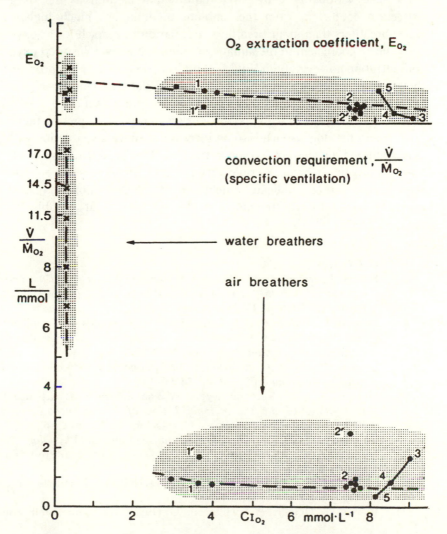

The $\dot{M}_{O_2}$-specific ventilatory requirement of external medium is for example written:

$$\dot{V}m/\dot{M}_{O_2} = 1/(Em_{O_2} \cdot Cl_{O_2}) \tag{32}$$

meaning that either $\dot{V}m/\dot{M}_{O_2}$, or Em_{O_2}, or both, must change according to inspired oxygen concentration. The ratio $\dot{V}/\dot{M}_{O_2}$ is the variable to be used if one wants to compare the regulation of breathing (or of cardiac output) in animals of various sizes, at variable temperatures, belonging to different groups, or living in different environments in which the oxygen concentration can be extremely variable.

Figure 2, redrawn from Dejours (1981), illustrates some of these relationships. The ventilatory convection requirement as well as the oxygen extraction coefficient from the ambient medium are much higher in water-breathers than in air-breathers, in relation to a much lower oxygen capacitance coefficient in the aquatic medium. At a given inspired oxygen concentration however, both $\dot{V}m/\dot{M}_{O_2}$ and Em_{O_2} may change reciprocally in the same organism, according to particular conditions. During acclimatization of man to high altitude, the ventilatory requirement progressively increases as the O_2 extraction coefficient decreases. Heat loss by thermal polypnea in the dog also leads to an increase of specific ventilation with a corresponding decrease of oxygen extraction. Finally, a decrease of body temperature in an heterothermic turtle entails an increased ventilatory requirement with an obligatory decrease of O_2 extraction, a response probably linked to adjustments of acid–base balance at variable body temperature.

References

Bureau International des Poids et Mesures (1985). *Le Système International d'Unités (SI) (The International System of Units)*, 5th edition. Pavillon de Breteuil, F-92310 Sèvres, France.

Dejours, P., Garey, W. F. & Rahn, H. (1970). Comparison of ventilatory and circulatory flow rates between animals in various physiological conditions. *Respiration Physiology*, **9**, 108–17.

Dejours, P. (1972). Comparison of gas transport by convection among animals. *Respiration Physiology*, **14**, 96–104.

Dejours, P. (1974). Recent concepts in the physiology of gas exchange. In *Proceedings of the International Union of Physiological Sciences*, XXVIth International Congress, New Delhi, vol. X, pp. 25–6.

Dejours, P. (1981). *Principles of Comparative Respiratory Physiology*. 2nd edition. Amsterdam, Elsevier/North Holland Biomedical Press (First edition 1975).

Fletcher, J. E. (1980). On facilitated oxygen diffusion in muscle tissues. *Biophysical Journal*, **29**, 437–58.

Imbert, G., Dejours, P. & Hildwein, G. (1982). The compressibility and the capacitance coefficient of helium-oxygen atmospheres. *Undersea Biomedical Research*, **9**, 305–14.

Otis, A. B. (1964). Quantitative relationships in steady state gas exchange. In *Handbook of Physiology. Respiration*, vol. I, eds W. O. Fenn & H. Rahn, pp. 681–98. Washington DC: American Physiological Society.

Pappenheimer, J. R. (1950). Standardization of definitions and symbols in respiratory physiology. *Federation Proceedings*, **9**, 602–5.

Piiper, J., Dejours, P., Haab, P. & Rahn, H. (1971). Concepts and basic quantities in gas exchange physiology. *Respiration Physiology*, **13**, 292–304.

Piiper, J. & Scheid, P. (1975). Gas transport efficacy of gills, lungs and skin: theory and experimental data. *Respiration Physiology*, **23**, 209–21.

Piiper, J. & Scheid, P. (1977). Comparative Physiology of Respiration: Functional analysis of gas exchange organs in Vertebrates. In *International Review of Physiology, Respiration Physiology II*, vol. 14, ed. J. G. Widdicombe, pp. 219–53. Baltimore: University Park Press.

Piiper, J. (1982). Respiratory gas exchange at lungs, gills and tissues: mechanisms and adjustments. *Journal of Experimental Biology*, **100**, 5–22.

Piiper, J. & Scheid, P. (1982). Models for a comparative functional analysis of gas exchange organs in vertebrates. *Journal of Applied Physiology*, **53**, 1321–9.

Piiper, J., Scheid, P., Perry, S. F. & Hughes, G. M. (1986). Effective and morphometric oxygen diffusing capacity of the gills of the elasmobranch *Scyliorhinus stellaris*. *Journal of Experimental Biology*, **123**, 27–41.

Rahn, H. (1966). Aquatic gas exchange: theory. *Respiration Physiology*, 1, 1–12.

Rahn, H. & Fenn, W. O. (1955). *A Graphical Analysis of the Respiratory Gas Exchange. The O_2–CO_2 Diagram.* Washington DC: The American Physiological Society.

Rahn, H., Wangensteen, O. D. & Fahri, L. E. (1971). Convection and diffusion gas exchange in air or water. *Respiration Physiology*, **12**, 1–6.

Rahn, H. & Howell, B. J. (1976). Bimodal gas exchange. In *Respiration of Amphibious Vertebrates*, ed. G. M. Hughes, pp. 271–85. New York: Academic Press.

G. M. HUGHES

Morphometry of respiratory systems

Introduction

Measurement of dimensions has always formed an important study for respiratory biologists and has received greater emphasis with the expansion of the comparative approach. Whether consideration is being given to the high surface/volume ratio and short diffusion distances in a protozoan or the large diameter of the whale aorta or great length of giraffe trachea, many theoretical studies have required quantitative data of this kind which has been fairly easy to obtain.

The basic physical processes involved in gas exchange are convection and diffusion, for which a knowledge of dimensions is essential for any quantitative understanding of the ventilatory and cardiovascular mechanisms that effect the transfer of oxygen, from the external medium to oxidative sites at the sub-cellular level. From an initial interest in the dimensions of species at the extreme ends of the size range, together with a need to have information for detailed studies of given species, the comparative approach has also given rise to a more systematic interest of differences that are inevitably due to animal size and the allometric characteristics of their growth.

The importance of allometry during evolution has also been emphasized by systematists and palaeontologists who have sought to establish the viability of the supposed physiological features of extinct groups.

This very wide interest in the study of the dimensions of respiratory systems has been given great impetus during the past 25 years by the availability of instruments capable of finer measurements, but in particular by the astonishing expansion of computers and development of automatic sampling and measuring systems. These have been particularly important, as a major problem has been the sheer number of measurements that must be made for any given specimen and the need for representative sampling. One reason why sufficient data is still lacking is the enormous effort required by scientists involved in these studies, and the way in which many hours of concentrated work may be condensed into one or two numbers which arouse little excitement in the uninitiated and certainly lack the admiration often given to the display of a single oscilloscope trace that might be obtained in a period of milliseconds!

There are now available several detailed texts on morphometry and stereology (e.g. Underwood, 1970; Weibel, 1979) which together with reviews give many details of the theoretical background. This article gives little attention to these aspects but tries to provide a more practical approach for those who wish to make some morphometric studies without too much attention to theory. As always there are dangers in such an approach and at least some attention to the theoretical background is essential in order to ascertain the limitations of any method especially when basic methods are to be modified. It must be remembered that morphometrically-determined values (e.g. diffusing capacity) usually represent the maximal capacity of a system (Hughes, 1977, 1980).

Rather than discuss different gas exchangers separately, some methods for making the main types of measurement – mass, volume, length and area – are given with examples from lung, gill, skin and muscle.

Measurement of mass

Many types of balance have been employed in respiratory studies over a wide range – from elephants to minute tissue samples. Preparation of the material before weighing influences the final measurement and is one of the problems using *wet weight*. Usually a standard procedure for removal of excess water is adopted but there are difficulties in relation to removal of body fluids. These include the blood from hearts, mantle cavity water of molluscs and the extensive haemocoelic fluid of crustaceans and molluscs. In relation to weighing whole organisms the method is usually satisfactory, as measuring errors are small relative to other causes of variations at a particular size range, such as feeding regime etc. Such variations are relatively small for larger animals but the increased surface/volume ratio at smaller sizes inevitably means greater errors of wet weight measurement. It is partly for this reason that many studies of smaller organisms – less than 5 g – are often expressed as *dry weight* and such measurements are also made for larger animals. Wet weight is also influenced by the osmotic conditions of the environment, for example the same salmon increases in weight shortly after transfer from sea- to freshwater. Measurement of dry weight requires facilities for incineration and drying to a constant value that becomes more difficult with larger structures. But there are also differences in the degree of 'dryness' adopted in a particular laboratory (e.g. 70 or 105 °C). A third method sometimes employed is to express results in terms of *total body nitrogen* or some comparable unit.

Each of these methods can give self consistent results in studies relating specific morphometric measurement to body mass, and others such as oxygen consumption. Few studies, however, have compared results obtained using two or more methods to express the body mass. A more detailed problem concerns whether skeletal structures (external or internal) should be included. Some studies suggest that all skeletal structures should be weighed, as this is the total mass that the animal must transport and for which respiration provides energy (Hughes, 1977). Other studies emphasize that skeletal systems are metabolically less active and their exclusion leads to a better understanding of metabolism/body mass relationships. Workers using molluscs may remove the shell and drain the haemocoel before measuring wet weight whereas others take the whole animal and weigh it after cursory drying on blotting paper! Use of measurements such as total nitrogen exclude most skeletal material and are based on a relatively simple chemical test (biuret).

As in all aspects of morphometry it is important that detailed methodology should be reported and if possible the authors indicate some factor for conversion to another well known system of measurement. In this context it is surprising how few authors have given any information on the relation-

Fig. 1. Ratio of dry weight to wet weight for a wide range of body sizes (shell length) in *Mytilus edulis*. (J. Widdows and P. N. Salkeld, unpublished data.)

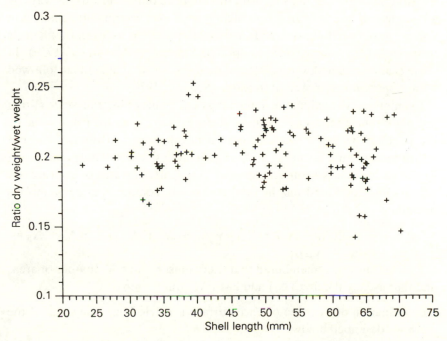

ship between wet and dry weight. The assumption is frequently made that the latter is a constant fraction of the former as in Fig. 1, but it seems unlikely that the assumption is valid in all cases. Seasonal variations also occur as indicated for example by variations in the inorganic content of *Mytilus* (Hawkins *et al.*, 1985).

Measurement of volume

Volume is measured in morphometric studies of respiratory structures, especially where data is required on the relative volume of different parts of an animal or organ system (e.g. red and white muscle). In many cases a portion of the structure is studied in detail to provide values for the relative volume of different components. These values can then be converted into absolute volumes when multiplied by the total volume of a given structure such as a lung. In practice the volume of the exchange tissue is often taken as a reference space.

A simple method of volume measurement is by the displacement of water or saline. With highly-branched structures problems arise because of the adherence of air, during the first immersion, and the adherence of liquid before the measurement is repeated, but for relatively simple shapes the method has few problems. Attention must be paid to the condition of fixation, as difficulties arise for very small objects and especially air-containing structures such as lungs. The method may be refined in various ways by attachment of fine capillaries for making the final measurement. Changes in volume have been followed in physiological studies, as for example changes in swimbladder volume following pressure changes (Alexander, 1959). In some cases Archimedes principle can be used, i.e. weighing in air followed by its apparent loss in weight in water (Scherle, 1970).

Estimates of the volume of some lungs can be made by transverse slicing and determining the area of each cut surface, multiplying the mean value by the length of the section. Such an approach is applicable to tubular lungs but where the structure is smaller and less regular in shape, the lungs can be cut into a series of slices a constant distance (D) apart. The area of each section (A_1-A_n) is determined and the total volume (V) computed using Simpson's formula:

$$V = \tfrac{1}{3} D[(A_1 + A_n) + 4(A_2 + A_4 + \ldots + A_{n-1}) + 2(A_3 + A_5 + \ldots + A_{n-2})]$$

It must be remembered that there must be an odd number of areas and that usually the first (A_1) and last (A_n) will be zero.

Determination of the area of each surface is carried out using one of the methods described below.

The most direct is to use graph paper and count squares; planimetry and digitisers are discussed later but the use of point-counting introduces a stereological technique that in some cases can also provide information on the volumes of individual structures within the lung as well as its total volume.

Point counting (Fig. 2a, b)

This method is similar to that using graph paper but instead of counting to the millimetre, for example, only the centimetre square is utilized. Thus, counts of points in the centre of each centimetre falling on a given section give an estimate of total area. Different types of grid are available – that using graph paper is a square grid. From theoretical considerations a triangular grid is slightly more accurate for a given spacing of points, but the square grid is better than a hexagonal one (Frolov & Maling, 1969). In calculating area, the total number of points (Tp) on the whole grid is equivalent to the total area (Ag). Hence the ratio of the points counted (P) to the total number of points, gives the ratio of the area of the section to the total grid area:

Fig. 2(*a*). Diagrammatic section through the lung of *Lepidosiren* with superimposed rectilinear grid showing the regions used for point and intersection counting. The coarser grid is used for the grosser regions e.g., duct air (Pd) and alveolar air (Pa), and the finer grid for capillaries (Pc) etc.
(*b*) Distribution of volumes of air and tissues in a lung of *Lepidosiren* (500 g) based upon morphometric study. (From Hughes & Weibel, 1976.)

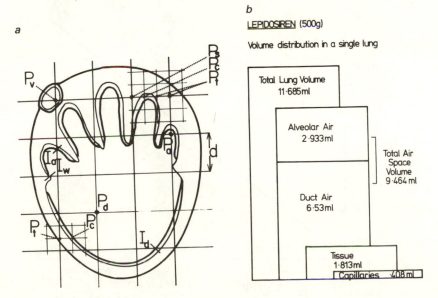

$$\frac{P}{T_P} = \frac{A}{A_G} \qquad A = \frac{P}{T_P} \cdot A_G$$

Clearly the accuracy (resolution) of this method depends upon the fineness of the grid relative to the size of the area to be measured and is an important consideration to be decided at the beginning of any study.

The ratio P/T_P is called by stereologists the area fraction (A_A). Apart from its use in area measurement, it is especially useful for volume measurement where the analogous parameter is the volume fraction (V_V).

Volume measurement

One of the most important principles in stereology derives from a geologist Delesse (1847) who noted that the 'area occupied by any given mineral on the surface of a section of the rock is proportional to the volume of the mineral in the rock'. Thus $V_V = A_A$. This parameter has many uses e.g. determination of the volume fraction, i.e. volume density of mitochondria in muscle which when multiplied by the total volume of the muscle gives the volume of mitochondria. If this is computed for all body muscles, an estimate of the total quantity of mitochondria can be derived. Such figures for red muscle of a fish for example give an estimate of its aerobic capacity.

Fig. 3. Plot showing relationship between capillary density and mitochondrial density in muscles of fish and mammals. (Data from Egginton & Johnston, 1983 and Hoppeler *et al.*, 1981.) For details see text.

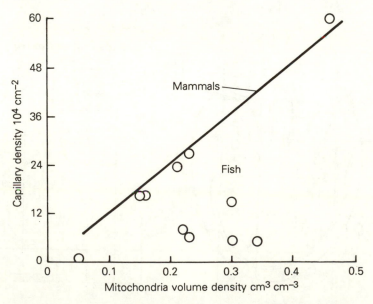

Data plotted in Fig. 3 illustrate the usefulness of measuring volume density of mitochondria and numbers of capillaries relative to the same reference space of muscle-fibre cross-sections. Data for a variety of fish slow muscles (listed in Egginton & Johnston, 1983) are shown together with a regression line (capillary density = $-0.11 \times 10^4 + 122.9 \times 10^4 \times$ volume density of mitochondria in fibres) for mammalian muscles (Hoppeler *et al.*, 1981). It is apparent that some fish muscles are close to the mammalian relationship but in others such a clear dependence on oxygen supply is not present.

Length measurement

Depending on the distances involved the use of a ruler, graph paper, micrometre, dividers, eyepiece micrometre, EM graticule, digitizer etc. are all standardized procedures available in most laboratories. The main complications concern the nature and condition of the biological material. Even well fixed material may not be flat and this problem may be exaggerated in some respiratory systems, as for example gill filaments. Sometimes allowance for such complications can only be approximate, but in others pieces of thread can be aligned with the object and then measured when stretched out.

Barrier thickness

In many calculations the diffusion distances for oxygen and/or carbon dioxide transfer need to be known. Where these distances are long they are usually less critical and estimates are easily achieved by direct measurements of photomicrographs using light microscopy. Where greater gas exchange occurs the distances are usually shorter and the use of electronmicroscopy is more accurate. In general there are two ways to express the results of such measurements – *arithmetic mean* and *harmonic mean*.

Arithmetic mean

Many measurements given in the literature have been made by taking distances across a barrier, perhaps at regular intervals and the mean value is used as the diffusion distance in any calculations. Some of these measurements ignore the fact that gas can also diffuse along longer pathways as well as directly at right angles to the external surface. Another way of establishing the arithmetic mean is to estimate the volume of the tissue through which the gas diffuses, which is divided by the mean area of the inner and outer surfaces.

This has often been used in studies on vertebrate lungs (Weibel, 1979)

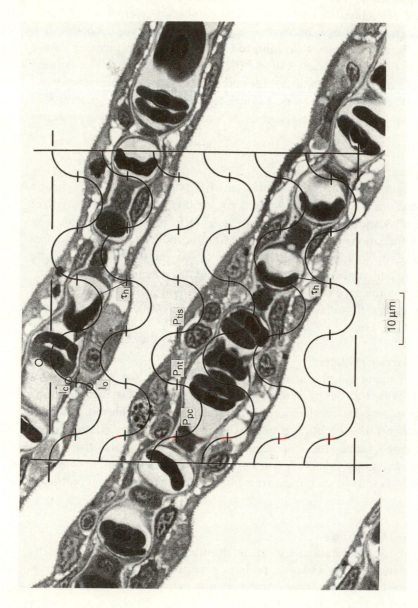

Fig. 4. Portion of Merz grid superimposed on a section through secondary lamellae illustrating points and intersections counted. Distances (τ_h) from intersections with outer surface to nearest red blood cell were used to compute harmonic barrier thickness. (After Hughes & Perry, 1976.)

and is appropriate in situations where the cell bodies occupy a relatively small portion of the total barrier. In other situations, as for example in fish gills, where the pillar cells occupy a much larger proportion of the total gas exchange region, such measurements would give erroneously high values. In gill lamellae, however, the same basic calculation can be applied provided that volume measurements of the barrier in the strict sense are divided by the mean area on the two sides. From a stereological point of view the volume may be estimated by counting the number of points (P_b) falling on the tissue barrier and the area of the two surfaces is proportional to the number of their intersections (I_0 and I_i) with a superimposed grid.

Hence the arithmetic mean $= P_b/(I_i + I_0)/2$ multiplied by a factor according to the grid (d for a square grid as in Fig. 2a).

Harmonic mean thickness (τ_h)

In this case, the fact that gas diffusion to a given point (e.g. an erythrocyte) occurs along a wide range of distances is taken into account, Weibel & Knight (1964) deduced that the harmonic mean was the appropriate measurement. In practice, the lengths of a range of randomly placed lines across the barrier are measured and the harmonic mean is calculated.

Fig. 5. Measuring devices used to obtain groups of lengths for determining harmonic mean distances. (*a*) is based on a logarithmic scale (Weibel, 1979) but (*b*) is only partly so (Perry, 1983.)

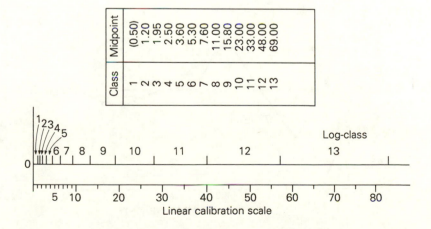

The exact way in which randomness is achieved depends upon the nature of the tissue and the sectioning procedure used. If the plane of the sections to the gas exchange surface are made at random, then a rectilinear grid may be used and the lengths of lines that pass across from the outer surface to the inner surface can be measured (Hughes, 1972; Weibel, 1971). In cases where the sectioning procedure is perpendicular to the outer surface (Hughes & Perry, 1976) then a more randomizing grid (e.g. Merz) is more appropriate. Clearly the latter type of grid can also be used with the less randomly sectioned material. Where the curved grid lines intersect the outer surface, measurements are made of the shortest distance between the intersection point and the nearest red blood cell (Fig. 4). If necessary the subdivisions of these lines are measured separately e.g. tissue and plasma distances. From the measurements of distances the harmonic mean is calculated.

It is necessary to make a large number of such measurements, and in order to speed up the procedure, complete accuracy of each measurement is not essential, especially for the longer distances, as in calculating the harmonic mean the contribution of each measurement is related to the reciprocal of the distances ($1/L$). Clearly, it is more important that the shorter distances should be measured more accurately and several devices have been developed to achieve this in practice (Fig. 5).

A simple expedient is to use a ruler based on a logarithmic scale in which the shorter intervals are more closely spaced (Fig. 5). However, the scale

Fig. 6. Plots showing change in estimate of harmonic mean thickness (τ_h) of the tissue and plasma layers of the water/blood barrier of a fish secondary lamella. Values are plotted as a percentage of the final result and show how 10–20 fewer measurements would have made only a small difference to the final result.

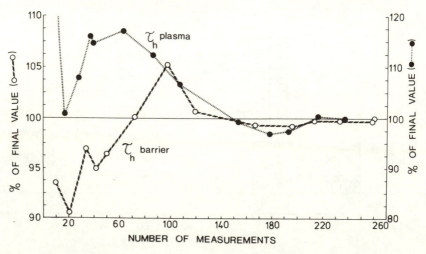

need not necessarily be logarithmic as in the device used by Perry (1976). Both of these and other types of rulers enable measurements to be grouped within given size ranges. The grouped data in each range are assumed to have a mean value of the distances (L_1–L_{10}) represented by the mid point. Using a scale divided into ten sections data are obtained in the form of numbers of measurements in each section ($n_1 - n_{10}$) hence the harmonic mean thickness is derived from the following relationship where N is the total number of measurements:

$$N/(n_1/L_1 + n_2/L_2 + n_3/L_3 + n_4/L_4 + \ldots + n_{10}/L_{10}) * \tfrac{2}{3} = \tau h$$

Where thickness of the plasma layer is measured, the factor $\tfrac{3}{4}$ instead of $\tfrac{2}{3}$ is used to compensate for the curvature of the erythrocyte surface (Weibel, 1971).

As is important in all morphometry, controls can show the number of measurements required to give reasonable accuracy (Fig. 6) and so reduce the time spent.

Measurement of area

The surface area through which materials, especially respiratory gases, are transferred, is of great importance in respiratory studies. Measurements most frequently required are the surface area of the skin, lungs or gills, with each of these having its own special problems. The easiest method would appear to be a direct measurement of the surface, or more usually some material that has been attached to the surface. Such methods have been used in determining the surface area of humans and other large animals. One method is to apply masking tape to completely cover the surface, which can then be removed and divided into small, more or less flat sections, and the individual areas summed. Smaller areas can be measured by a variety of means, e.g. weighing the individual pieces, where they are of a standard density, or by photometry. For example, Graan (1969) stained pieces of masking tape black and placed them in a simple device that measured the amount of absorbed light, which was directly proportional to the surface area. Other workers have used waxed or silver paper for making their moulds; polyethylene sheeting (Niimi, 1975) or some silicone material could also be used. This method is especially suited to large animals with fairly extensive surface areas that are not too folded. The surface of the individual pieces could also be determined using a planimeter and nowadays by a digitizer connected to a microcomputer. The latter are used extensively in studies that require the analysis of surface areas of micrographs or tracings of structures such as gill lamellae.

Digitizers transmit a series of X and Y co-ordinates depending upon where the cursor is placed on a special plate. From these co-ordinates the area below the line is computed as the cursor is moved from P_{i-1} to P_1 (Fig. 7b).

$$\text{Area} = \sum_{i=2}^{n} A_i$$

where $A_i = \frac{1}{2} X(Y_{i-1} + Y_i)$
and $\Delta X = X_i - X_{i-1}$

The values are positive or negative depending on the direction of movement, consequently when the cursor has returned to its original point the difference between the areas shown in Fig. 7a gives the area of the object that has been traced.

As with a number of other methods for computing area, accuracy is lost if the area traced is too small. We recently showed this by tracing shapes of a known area and calculating the error, which is mainly caused by deviations

Fig. 7. Diagrams to illustrate the principle of area measurement using a digitizing tablet. For details see text.

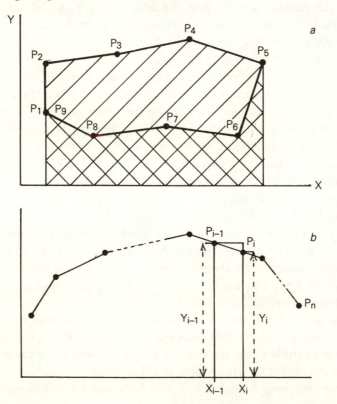

in tracing and the frequency with which the co-ordinates are being read. The common method of cutting out a shape and then weighing it also has errors due to the edge effect, i.e. errors in cutting, which are proportionately greater for smaller areas. The same problem is found with planimetry. For areas less than 400 mm^2 errors of about 1% increased to 5%, with the weighing method the errors are slightly less than those for the digitizer. The reverse was true for areas above 10 000 mm^2 when the errors were sometimes less than 0.5%.

As with all computer-aided equipment it is important to test the accuracy of the particular device employed, as a number of errors are known to be associated with some digitizers (Cornelisse & van den Berg, 1984).

Where the surface to be measured is highly folded then such procedures are inadequate and this is the situation with respect to the gills and lungs of animals. A distinguishing feature is that gills represent external branchings of the body surface whereas lungs are intuckings and far less accessible. However, some extremely simple lungs or lung-like structures, such as the suprabranchial chambers of air-breathing fish or the mantle cavity of molluscs can be dissected and treated in a manner similar to that described above, and is clearly applicable to the skin.

Intercept measurement

In about 1945 several scientists, mainly metallurgists, deduced that the surface area of a body was directly proportional to the number of intersections it made randomly with a lattice placed on the surface of a section through the whole structure. The basic relationship is $S = 2V/L_m$ where S is the total surface area enclosed within a given volume (V) and L_m is the mean linear intercept. The latter being obtained by dividing the total length of the test line of the grid by the number of intercepts (I). In the case of lungs, the equation is usually given as $S = 4V/L_m$ because each time a test line crosses an alveolus, two counts should be made and not one as where the test line crosses a single interface. This modified formula is used for gross and light microscopy but not when high resolution EM is employed because both sides of the interalveolar septum are distinguishable. For lungs the surface density of alveoli (Sv_a) is often estimated from electron-micrographs by counting the number of intersections with the alveolar surface of test lines of total length T$_L$ which are randomly superimposed on the sections.

$$Sv_a = 2I_a/T_L$$

It is sometimes useful to compare the surface areas of different structures within an organ. Assuming the same test line system is used, the ratio of the intercept counts can give the ratio of surface areas. Where absolute values for one of these is available using some other method, absolute values for the other structure measured relative to it can be calculated. An application of this approach enabled Hughes and Perry (1976) to use the term 'relative diffusing capacity' meaning the ratio between the diffusing capacity of experimental and control fish. In that case the assumption was also made that the permeation coefficients (K) were constant in the two groups.

Corrugation factors

Where the surface is slightly folded the area of the gas exchange surface is greater than that of the projected area, i.e. the area obtained by tracing its outline shape (Fig. 8b). The factor by which the projected area should be increased (I_o/I_{pr}) is readily calculated from sections across the barrier and by counting the intersections of the surface with a morphometric grid relative to the number that intersect the projection plane (Fig. 8a) In some respiratory structures the situation is more complicated because the areas involving gas exchange are separated from one another by non-respiratory areas. A good example being the so-called 'lanes' of the air sacs in *Heteropneustes*. In such instances the number of intersections of the respiratory and non-respiratory regions can be expressed relative to intersections with the projected area (Fig. 8c). Finally, using electron micrographs the outer gas exchange surface is often curved because the capillaries project as in the lungs of some lizards and fish air-breathing organs (Hughes, 1978). Again the ratio of intersections with the outer surface and projected area indicate a further factor by which the measured gross projected area (A_{pr}) should be increased (Fig. 8d). Hence area of respiratory surface,

$$A = A_{pr} \cdot \frac{I_r}{I_{pr}} \cdot \frac{I_a}{I_{r'}}$$

Awareness of such problems has developed from SEM studies of respiratory surfaces. In the case of *Heteropneustes*, attention was drawn to the fact that values previously published for total area should be reduced because of the non-respiratory lanes, whereas in fact it should be increased because of the corrugations of the respiratory zones (Hughes & Munshi, 1978). A more recent study of this problem indicates quantitatively the extent to which this is true. (Munshi, Hughes, Gehr & Weibel, 1989). Similarly underestimates of human lung area (70–100 m^2) were increased to 143 m^2 when electronmicroscopy replaced light microscopy (Gehr *et al.*, 1978).

Fig. 8. Diagrams to illustrate the use of intercept measurements to estimate the gas exchange surface in an accessory organ of air-breathing fish.
(*a*) Diagram showing general principle of corrugation factor estimation.
(*b*) Internal view of air sac; respiratory region is shaded and a portion of this enlarged to show gas exchange islets separated by surfaces not supplied with blood.
(*c*) Enlarged view of part of gas exchange region and use of superimposed grid to estimate these surfaces (I_r) relative to the projected area (I_{pr}) using light microscopy.
(*d*) Part of *c* under electronmicroscopy showing the intersections of the grid with the air (alveolar) surface (I_a) and the capillary surface (I_c). In this case the projection area (I_r) corresponds to part (I_r) of (*c*). Points falling on capillaries (P_c) are also counted. Distances measured for the estimations of harmonic mean thickness of the tissue (τ_t) and plasma layers ($\tau_p 1$) are indicated.

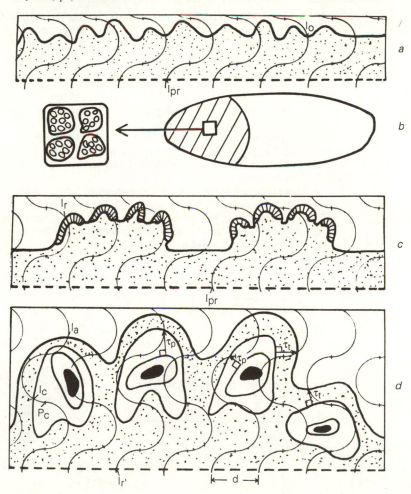

36   G. M. Hughes

Capillary loading

From intercept measurements of the gas exchange surface and point counts of capillaries, the ratio Capillary volume/Respiratory area can be established (Table 1). A low capillary loading indicates a large surface of contact between blood and the respiratory medium.

For most lungs and gills the surface has become so folded that it is impossible to attempt to measure the whole of the gas exchange surface. The situation with respect to gills is simpler than for lungs, in that they are

Table 1. *Values for capillary loading (ml/m^2) obtained morphometrically by dividing capillary volume ($\propto P_c$) by respiratory area ($\propto I_a$) for air-breathing vertebrates*

Lepidosiren	12.37–18.87	Hughes & Weibel, 1976
Turtle	13.3	Perry, 1978
Bats	0.32–0.8	Maina, King & King, 1982
Monopterus cuchia	2.72	Munshi *et al.*, 1989

Fig. 9(*a*). Diagram of a single hemibranch (72 filaments) of a teleost fish showing the position of filaments selected for measurement at regular intervals around the arch. Lengths of the first and last filaments are also determined. The positions of the secondary lamellae selected for measurement from the tip (t), middle (m) and base (b) of each selected filament are indicated.
(*b*) Diagram showing method for measurement of length of filaments 40 (l_{40}) and 50 (l_{50}) and the distance between filaments 40 and 50.
(*c*) Diagram illustrating the method for measurement of secondary lamellar frequency (1/d′) by measuring the distance for ten secondary lamellae (from Hughes, 1984*b*).

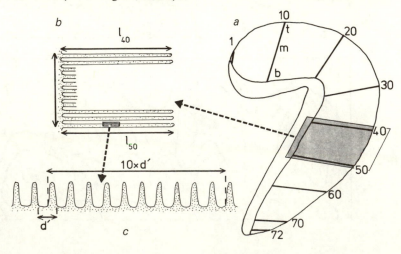

usually borne by a series of repeating structures so that the total computation can be broken into component parts.

Most gill systems are divided into a number of arches that bear primary sub-divisions, the length and number of which can be determined (Fig. 9). In most cases the assumption is made that the left and right sides are equal.

Having determined the total length of these primary elements (gill filaments of fish) an estimate must be made of the frequency of the secondary surfaces (lamellae of fish or platelets of crustaceans) and finally an estimate of the area of an average secondary lamella representative of the whole system. Multiplication of these three measurements gives the total surface area on one side of the animal. If the system were completely homogeneous this procedure would be relatively easy. However, a characteristic feature of respiratory systems is that they are heterogeneous (Hughes, 1973), at least to some extent. In order to cope with this heterogeneity, measurement of filament length is usually made over the whole system. This measurement is usually very accurate as it is not too difficult to measure the length of every second, fifth or tenth filament. The frequency of such measurements being determined by the total number of filaments and the rate at which the length changes along a given arch. Plots

Fig. 10. Plots of filament length of filaments at different positions along a branchial arch. Filament 1 is the short one at the dorsal end. Note the varying pattern of length distribution for different species: (*a*) skipjack, (*b*) a grouper, (*c*) barracuda, (*d*) mackerel. (From Hughes, 1984*b*).

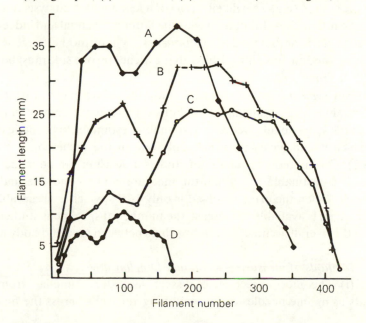

Table 2. *Percentage errors that would arise if secondary lamellar area &
frequency had only been sampled from a single arch of* Anabas
testudineus *(A.t.) or* Catla catla *(C.c.). Percentages are calculated relative
to the measurements that were actually made on all arches*

		Arch 1	Arch 2	Arch 3	Arch 4	Total area
1st arch	A.t.	0	16	57.6	—	9.83
sampled	C.c.	0	1.3	6.2	5	3.2
2nd arch	A.t.	13.9	0	35.7	—	5.4
sampled	C.c.	1.3	0	5.04	3.8	1.96
3rd arch	A.t.	36.5	26	0	—	30.2
sampled	C.c.	6.7	5.3	0	1.3	3.3
4th arch	A.t.	—	—	—	—	—
sampled	C.c.	5.3	3.9	1.3	0	2

After Hughes & Ojha, 1985.

of such measurements can give a useful basis for comparing gill systems
(Fig. 10).

Measurement of the frequency and area of individual lamellae is a more
difficult problem and several methods have been adopted each of which
makes some selection from the total population of secondary lamellae.
Broadly speaking the main choice is whether to make a detailed analysis of a
part of the system (e.g. a single gill arch) or a less detailed analysis covering
each of the four arches. In some species the latter is essential and indeed it is
essential to do it on both sides for flatfishes. Where the heterogeneity is
great, as in some air-breathing fishes, all the arches on one side must be used
(Hughes & Ojha, 1985).

Whatever system is adopted a trial study should be made in which the
whole system is investigated in detail. For example, measurement of all the
arches in the same detail gives data that enables estimates to be made of the
error which would have arisen had only one of the arches been chosen
(Table 2). On the basis of such pilot studies a decision can be made as to
whether it is justifiable to confine the measurement to one side and/or a
single arch. Often the error involved is only 1–3% which is acceptable, as
the time saved is available to increase the number of specimens studied and
improve the overall accuracy of relationships between area and body mass.

Methods of measuring secondary lamellar area

(i) The most direct is to dissect secondary lamellae from the
filaments using fine needles or by sectioning vertically across the filament

using a fine knife, e.g. a piece of razor blade. Sampling is usually carried out from the tip, middle and base of filaments selected from around the whole gill arch. Outlines of the sections are traced using a projection microscope and the area of the tracings determined, as described above.

(ii) Measurement of the maximum height and length of secondary lamellae was used by Price (1931) in his classical study. He estimated the area assuming that the shape of the filament was an isosceles triangle with surfaces folded into secondary lamellae.

Height and length can be measured quite accurately without dissecting the lamellae and so can be accomplished more rapidly with a larger selection being taken. If a relationship can be demonstrated between these dimensions and lamellar area, measured directly as above in a given species, then height and length measurements could be converted to area. The possible validity of such a relationship is illustrated in Fig. 11. Results of such a study have been published by Gehrke (1987).

(iii) Many measurements of gill area have been made at Krakow (e.g. Byczkowska-Smyk, 1957; Jakubowski, 1982) using specimens in which the gills had been injected with Indian ink. The secondary lamellae were more

Fig. 11. Plot of relationship between area of individual secondary lamellae (A) determined from tracings and the product (L × H) of their maximal height and length. Data based on measurements on a single hemibranch of the first arch from a rainbow trout of 8.0 cm by Morgan (1971)

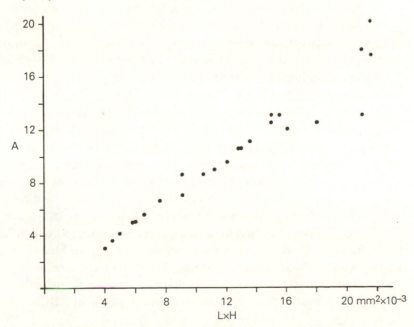

visible and their outline drawn following dissection. This method also allows an estimate of the proportion of the lamella directly overlying blood channels.

More recent studies have used secondary lamellae with resin-injected specimens (Farrell, 1980). Following removal of the tissue, the secondary lamellar castes can then be used to measure gill area. These casts have the advantage that they can be stored and used systematically in computer devices that determine the area by the reduction in transmitted light (Pohla *et al.*, 1987). The latter method will tend to overestimate the surface as quite a significant part of the pillar cell system lies below the level of the filament surface. Indeed this was regarded as a major factor resulting in low estimates of secondary lamellar area in *Scyliorhinus stellaris* (Hughes, Perry & Piiper, 1986) and has been shown to be an important variable in the regulation of gas exchange surface in some teleosts (Tuurala, 1983).

Diffusing capacity

Many morphometric studies of respiratory systems have been directed to providing values for diffusing capacity. This is because the basic Fick equation can be modified as follows:

$$\frac{\dot{V}_{O_2}}{\Delta P_{O_2}} = \frac{A}{t} \cdot K$$

The physiological values are equated to morphological measurements together with an appropriate co-efficient. Unfortunately few measurements of the permeation co-efficient (K) are available and for that reason Perry (1978) suggested that the ratio area: thickness be used for comparative purposes. However, this does not assist if the aim is to demonstrate the equivalence of the morphometric and physiological diffusing capacities. Nearly all morphometric determinations exceed physiological measurements, sometimes by a factor of more than 20-fold.

Diffusing capacity of mammalian lungs or fish gills has been subdivided into conductances across the tissue (D_t) plasma (D_p) and within the erythrocytes (D_e). The sum of these three parts being the pulmonary diffusing capacity (D_L). Sometimes the tissue and plasma values are combined together to form the membrane diffusing capacity (Dm) one of the reasons for this latter procedure is because the contribution of the erythrocyte is largely non-morphometric and available values are very restricted. This is particularly important, as it constitutes more than 70% of the overall calculation of D_L in many examples (Table 3). Measurement of D_p raises the problem of the relationship between electronmicrographs and the position of erythrocytes in the capillary circulation. Some authors (e.g. Moll, 1985) prefer to think in terms of a central pencil of RBCs.

Table 3. *Summary of values obtained for pulmonary diffusing capacity (DL) attributable to tissue (D_t), plasma (D_p), erythrocyte (D_e), and membrane ($D_m = D_t + D_p$) in various vertebrates. The percentage contributions of different parts of the overall resistance have been calculated.*

	D_t/W	D_p/W	D_e/W	D_m/W	D_L/W	
Mouse	0.028	0.094	0.0077	0.021	0.0064	Geelhaar & Weibel,
	20.3%	6.0%	73.3%			1971
Domestic fowl	0.011	0.015	0.003	0.006	0.001	Abdalla *et al.*,
	18.5%	13.6%	69.9%			1982
Fruit bat	0.188	0.429	0.024	0.131	0.02	Maina, King & King,
	10.6%	4.7%	83.3%			1982
Monopterus	0.0092	0.035	0.0028	0.0073	0.002	Munshi *et al.*, 1989
cuchia	21.7%	5.8%	72.5%			

In many comparative studies values are only given for Dt calculated from surface area and barrier thickness and is certainly the most reliable of the three components. In aquatic breathers another important component is the resistance to gas transfer within the water, which may be very large. A recent study (Piiper *et al.*, 1986) that takes this into account together with morphometric measurements of *S. stellaris*, has shown that the values for morphometric diffusing capacity are quite close to those determined physiologically in exercising fish. The ratio of morphometric to effective diffusing capacity being 1.64 for quiescent resting fish but 0.92 during swimming. It seems clear that morphometric values indicate the maximum capacity of the system that may only be approached during maximum exercise when the whole respiratory surface is fully perfused and ventilated (Hughes, 1980, 1984a).

Scaling

A major use of morphometric data has been to establish relationships with body mass that are valuable not only for their predictive value in other research but also in the theoretical assessment of animal functions. Thus, discussions about factors limiting maximal oxygen consumption, changes in aerobic scope, or ventilation requirement have arisen because of differences in the slope of regression relationships (Cragg, 1975; Hughes, 1977, 1984a). Scaling factors, frequently employed by engineers, have also been used in for example, the importance of resistance to gas transfer in the interlamellar water of fish gills (Hills & Hughes, 1970).

Microcomputers and calculators have greatly aided such studies but there is a danger of feeding these machines with insufficiently filtered data. The number of animals investigated needs to be as large as possible, as does the weight range. However, care must be exercised, as single specimens at the extremes of the size range can have a greatly exaggerated effect on the least square relationship usually employed in regression analyses. This may be illustrated by reference to recent data (Maina & Maloiy, 1986) on the gills of a catfish. Analysis of all 17 values in the weight range 105–3975 g gave a slope of 0.629, but the exclusion of one very large specimen resulted in a slope of 1.18 for fish of 105–680 g. Exceptional data points are not infrequent and it seems legitimate to exclude them, although the actual data should be reported as it may become of interest at a later stage. The more classical analyses based upon interspecific comparisons can give rise to distortions (Hughes, 1980, 1984b), and nowadays greater emphasis is being placed on intraspecific studies. Some of these are already indicating differences within given species and for that reason all measurements should be reported.

Some problems

Many problems arise from the heterogeneity of respiratory systems not only in their living condition but also differences in their response to anaesthesia, fixation and other preparative procedures. Before any extensive study is started some preliminary work is advisable, but once a method has been adopted it must be continued for all specimens if any statistical analysis is to be followed.

Care must be taken to prepare the material, maintaining a condition close to that of the normal animal. For example, lungs must be fixed in their resting volume and if by instillation, the pressure must be within the physiological range. Perfusion fixation must also be 'physiological' and where resins are used some estimate made of the extent to which this leads to abnormalities.

Shrinkage

Shrinkage is always a problem and controls must be made wherever possible. The assumption is often made that changes in a given linear dimension may be extrapolated to other dimensions such as area. In some cases, e.g. gill secondary lamellae, the mechanical situation is quite different from the more readily-determined shrinkage of gill filaments which are supported by gill rays (Hughes, 1984a). However, in the absence of a good measure for the structure under investigation, the shrinkage of the whole organ or some part of it at least gives a guide that can be useful in any future comparison.

Post-fixation procedures can produce unnecessary shrinkage, as for example the use of paraffin wax rather than embedding in resins, such as Epon or Araldite (Hughes *et al.*, 1986).

Sampling

Sampling is another major problem: material is usually fixed from many parts of a structure and a random selection made. An important consideration is the treatment of measurements from different regions and the extent to which each sample is representative of the whole organ. Weighted mean values are always preferable as they ensure that samples representing larger portions of the system have a greater effect on the final values than samples that, although important to represent extreme parts of the system, must not be allowed to over-influence final mean values.

Specialized examples of some of these problems can be illustrated by reference to the exchange system in skeletal muscle. Here models based upon the classical Krogh cylinder are well known and values are readily obtained for the number of capillaries per cross-sectional fibre area. However, for a true analysis of the capillary supply it is essential to take into account the difference in capillary density along the length of muscle fibres and especially their orientation with respect to the fibres. The latter is a type of heterogeneity which is usually termed 'anisotropy' (Hoppeler *et al.*, 1981; Egginton & Johnston, 1983) and several models have been used to apply a correction factor for it. Another aspect that does not seem to have been taken so much into consideration is the distribution of mitochondria within muscle fibres. For example in the red fibres of some fish the concentration of mitochondria is much greater around the periphery as is seen in cross-sections hence the diffusion distances for oxygen to most of the mito-chondria is far less than for that given by the diameter of the Krogh cylinder. Clearly a valid method would be to measure distances of randomly-selected mitochondria from the capillary cross-sections and to determine their harmonic mean.

Weighting

Because sampling is inevitable it is important to consider the way in which results are to be calculated from respiratory organs that may be very heterogeneous. If all parts of the system are the same then clearly means of values from each of the samples can be taken, i.e. given equal weighting. If however different samples represent portions of the total system that are significantly different from one another, then more weight should be given to those samples that represent a larger part of the whole organ. This general principle should be applied to both lungs and gills. For example in a

Fig. 12. Bilogarithmic plot of gill area and body mass early in development of rainbow trout. Regression lines are drawn for unpublished measurements (Morgan, 1971) for weights up to 0.1 g (slope = 3.443) and for specimens above this size (slope = 0.932) including measurements over 50 g by G. M. Hughes. 95% confidence limits are given for a fish of 0.1 g based on the lower weight group. A regression line (slope = 0.914) for data above 3.0 g is shown by a dashed line. Similar data for fish above 10 g also giving a slope of 0.932 has already been published (Hughes, 1980).

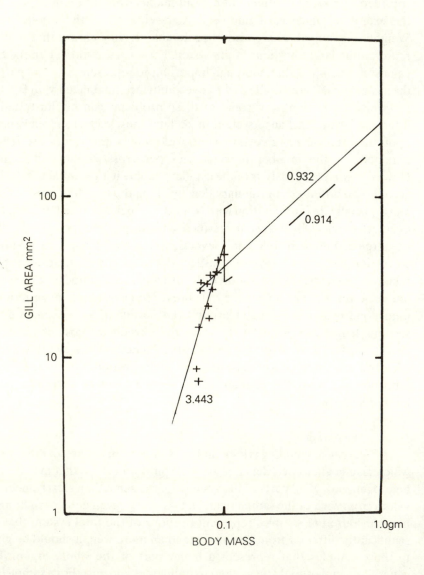

tubular lung (as in *Lepidosiren*) where the anterior portions contain far more respiratory surface area than the posterior portions, measurements of thickness etc., should be weighted accordingly. In gills some filaments have many more secondary lamellae and their dimensions should be given greater weighting in the overall calculation. It is sometimes assumed that weighting is not necessary if more samples are taken, but this is not true unless all or nearly all filaments are sampled. Thus, measurements on the short filaments near the ends of each arch should count far less than those in the centre that may carry 20 or more secondary lamellae.

Conclusions

Morphometry is time-consuming and its importance is not to be measured in the short term. For example, results obtained by Dr Miriam Morgan on the rainbow trout of very small size have been largely neglected but when plotted show interesting changes in slope at 0.1 g (Fig. 12). This data has become of significance in view of more recent data for other fish species (Hughes & Al-Kadhomiy, 1988). Finally I pay tribute to the measurements of Price (1931) that formed the basis upon which I was first able to analyse fish gill dimensions (Hughes, 1966) more than 25 years ago and which also provided the basis for the dimensional analysis alluded to earlier.

References

Abdalla, M. A., Maina, J. N., King, A. S., King, D. Z., & Henry, J. (1982). Morphometrics of the avian lung. 1. The domestic fowl (*Gallus gallus* variant *domesticus*). *Respiration Physiology*, **47**, 267–78.

Alexander, R. McN. (1959). The physical properties of the swimbladder in intact Cypriniformes. *Journal of experimental Biology*, **36**, 315–32.

Byczkowska-Smyk, W. (1957). The respiratory surface of the gills in teleosts. 1. The respiratory surface of the gills in the flounder (*Pleuronectes platessa*) and perch (*Perca fluviatilis*). *Zoologica Poloniae*, **8**, 91–111.

Cornelisse, J. T. W. A. & van den Berg, T. J. T. P. (1984). Profile boundary length can be overestimated by as much as 41% when using a digitizer tablet. *Journal of Microscopy*, **136**, 341–4.

Cragg, P. A. (1975). Respiration and body weight in the reptilian genus *Lacerta*: a physiological, anatomical and morphometric study. Ph.D. Thesis, Bristol University.

Delesse, M. A. (1847). Procede mecanique pour determiner la composition des roches. *Compte rendu hebdomadaire des seances de l'Academie des sciences (Paris)*, **25**, 544.

Egginton, S. & Johnston, I. A. (1983). An estimate of capillary anisotropy and determination of surface and volume densities of capillaries in skeletal muscles of the conger eel (*Conger conger* L.). *Quarterly Journal of Experimental Physiology*, **68**, 603–17.

Farrell, A. P. (1980). Gill morphometrics, vessel dimensions, and vascular resistance in ling cod, *Ophiodon elongatus*. *Canadian Journal of Zoology*, **58**, 807–18.

Frolov, Y. S. & Maling, D. H. (1969). The accuracy of area measurement by point counting techniques. *Cartographical Journal*, **6**, 21.

Geelhaar, A. & Weibel, E. R. (1971). Morphometric estimation of pulmonary diffusion capacity. III. The effect of increased oxygen consumption in Japanese Waltzing Mice. *Respiration Physiology*, **11**, 354–66.

Gehr, P., Bachofen, M. & Weibel, E. R. (1978). The normal human lung: ultrastructure and morphometric estimation of diffusion capacity. *Respiration Physiology*, **32**, 121–40.

Gehrke, P. C. (1987). Cardio-respiratory morphometrics of spangled perch, *Leiopotherapon unicolor* (Günther, 1859) (Percoidei, Teraponidar). *Journal of Fish Biology* **31**, 617–23.

Graan, C. H. van (1969). The determination of body surface area. *Supplement – South African Journal of Laboratory and Clinical Medicine*, **62**, 952–9.

Hawkins, A. J. S., Salkeld, P. N., Bayne, B. L., Gnaiger, I. & Lowe, D. M. (1985). Feeding and resource allocation in the mussel *Myutilus edulis*: evidence for time-averaged optimization. *Marine Ecology – Progress Series*, **20**, 273–87.

Hills, B. A. & Hughes, G. M. (1970). A dimensional analysis of oxygen transfer in the fish gill. *Respiration Physiology*, **9**, 126–40.

Hoppeler, H., Mathieu, O., Weibel, E. R., Krauer, R., Linstedt, S. L. & Taylor, C. R. (1981). Design of the mammalian respiratory system. VII. Capillaries in skeletal muscles. *Respiration Physiology*, **41**, 129–50.

Hughes, G. M. (1966). The dimensions of fish gills in relation to their function. *Journal of experimental Biology*, **45**, 177–95.

Hughes, G. M. (1972). Morphometrics of fish gills. *Respiration Physiology*, **14**, 1–25.

Hughes, G. M. (1973). Comparative vertebrate ventilation and heterogeneity. In *Comparative Physiology*, ed. L. Bolis, K. Schmidt-Nielsen & S. H. P. Maddrell, pp. 187–220. Amsterdam: North-Holland.

Hughes, G. M. (1977). Dimensions and the respiration of lower vertebrates. In *Scale Effects in Animal Locomotion*, ed. T. J. Pedley, pp. 57–81. New York: Academic Press.

Hughes, G. M. (1978). A morphological and ultrastructural comparison of some vertebrate lungs. In *XXI Colloquium Scientificum Facultatis Medicae Universitatis Carolinae: 21st Congressus Morphologicus Symposium*, ed. E. Klika, pp. 393–405. Praha: Universita Karlova.

Hughes, G. M. (1980). Morphometry of fish gas exchange organs in relation to their respiratory function. In *Environmental Physiology of Fishes*, ed. M. A. Ali, pp. 33–56. New York: Plenum.

Hughes, G. M. (1984*a*). Scaling of respiratory areas in relation to oxygen consumption of vertebrates. *Experientia*, **40**, 519–24.

Hughes, G. M. (1984*b*). Measurement of gill area in fishes: practices and problems. *Journal of the Marine Biological Association of the United Kingdom*, **64**, 637–55.

Hughes, G. M. & Al-Kadhomiy, N. K. (1988). Changes in scaling of respiratory systems during the development of fishes. *Journal of the Marine Biological Association of the United Kingdom*, **68**, 489–98.

Hughes, G. M. & Munshi, J. S. D. (1978). Scanning electron microscopy of the respiratory surfaces of *Saccobranchus* (= *Heteropneustes*) *fossilis* (Bloch). *Cell and Tissue Research*, **195**, 99–109.

Hughes, G. M. & Perry, S. F. (1976). Morphometric study of trout gills: A light microscopic method suitable for the evaluation of pollutant action. *Journal of experimental Biology*, **64**, 447–60.

Hughes, G. M. & Ojha, J. (1985). Critical study of the gill area of a free-swimming river carp, *Catla catla* and an air-breathing perch, *Anabas testudineus*. *Proceedings of the Indian National Science Academy*, **51**B, 391–404.

Hughes, G. M. & Weibel, E. R. (1976). Morphometry of fish lungs. In *Respiration of Amphibious Vertebrates*, ed. G. M. Hughes, pp. 213–32. New York: Academic Press.

Hughes, G. M., Perry, S. F. & Piiper, J. (1986). Morphometry of the gills of the elasmobranch *Scyliorhinus stellaris* in relation to body size. Jakubowski, M. (1982). Size and vascularization of the gill and skin respiratory surfaces in the white amur, *Ctenopharynodon idella* (Vol. I (Pisces, Cyprinidae). *Acta Biologica Cracoviensia (Zoologia)*, **24**, 93–106.

Johnston, I. A. & Bernard, L. M. (1982). Quantitative study of capillary supply to the skeletal muscles of crucian carp *Carassius carassius* L.: effects of hypoxic acclimation. *Respiration Physiology*, **57**, 9–18.

Maina, J. N., King, A. S. & King, D. Z. (1982). A morphometric analysis of the lung of a species of bat. *Respiration Physiology*, **50**, 1–11.

Maina, J. N. & Maloiy, G. M. O. (1986). The morphology of the respiratory organs of the African air-breathing catfish (*Clarias mossambicus*): A light, electron and scanning microscopic study, with morphometric observations. *Journal of Zoology, London (A)*, **209**, 421–45.

Moll, W. (1986). The various components of diffusional resistance to alveolar oxygen uptake. *Progress in Respiration Research*, **21**, 47–51.

Morgan, M. (1971). Gill development, growth and respiration in the trout *Salmo gairdneri*. Ph.D. Thesis, Bristol University.

Munshi, J. S. D., Hughes, G. M., Gehr, P. & Weibel, E. R. (1989). Structure of the air-breathing organs of a swamp mud eel, *Monopterns cuchia* (Ham.) (Synbrnachidae, Synbranchiforms). *Japanese Journal of Ichthyology* **35**.

Niimi, A. J. (1975). Relationship of body surface area to weight in fishes. *Canadian Journal of Zoology*, **53**, 1192–4.

Perry, S. F. (1978). Quantitative anatomy of the lungs of the Red-eared Turtle, *Pseudemys scripta elegans*. *Respiration Physiology*, **35**, 245–62.

Perry, S. F. (1983). Reptilian Lungs. Functional Anatomy and Evolution. *Advances in Anatomy Embryology and Cell Biology*, **79**, 1–81. Berlin: Springer-Verlag.

Piiper, J., Scheid, P., Perry, S. F. & Hughes, G. M. (1986). Effective and morphometric oxygen-diffusing capacity of the gills of the elasmobranch *Scyliorhinus stellaris*. *Journal of experimental Biology*, **123**, 27–41.

Pohla, H., Bernroider, G., Lametschwandtner, A. & Goldschmid, A. (1987). Computerised measurement of gill respiratory area with corrosion casts of gill vasculature. *Proceedings of the 5th Congress of European Ichthyologists, Stockholm* (in press).

Price, J. W. (1931). Growth and gill development in the small-mouthed black bass, *Micropterus dolomieu* Lacépède. *Ohio State University Studies*, **41**, 1–46.

Scherle, W. (1970). A simple method for volumetry of organs in quantitative stereology. *Mikroskopie*, **26**, 57–60.

Tuurala, H. (1983). Structure and blood circulation of the secondary lamellae of *Salmo gairdneri* (Richardson) gills in relation to oxygen transfer. Academic Dissertation, University of Helsinki.

Underwood, E. E. (1970). *Quantitative Stereology*. Massachusetts: Addison-Wesley.

Weibel, E. R. (1971). Morphometric estimation of pulmonary diffusion capacity. I. Model and method. *Respiration Physiology*, **11**, 54–75.

Weibel, E. R. (1979). *Stereological Methods*, Vol. 1. New York: Academic Press.

Weibel, E. R. & Knight, B. W. (1964). A morphometric study on the thickness of the pulmonary air–blood barrier. *Journal of Cell Biology*, **21**, 367–84.

Non-invasive measurements

R. KAUFMANN, H. FORSTNER and
W. WIESER

Respirometry – methods and approaches

Introduction
The principles of respirometry have been known for so long and
have been discussed so often that nothing new need be said about the basic
methodology in this field. In recent years, progress in respirometric
research has not so much been connected with the central problems of
measuring aerobic metabolism per se, but rather with what may be called
peripheral problems – increasing the stability of long-term measurements,
or the resolving power of short-term measurements, studying the multipli-
cative effects of environmental factors on the rates of gas exchange, or
linking the measurement of oxygen with that of other biological variables,
such as activity or heat production.

In the following we shall, firstly, discuss briefly some general trends in this
area of animal physiology, secondly, give a condensed account of some of
the technical features of the most frequently used measuring systems, and
finally, survey some of the problems connected with the use of input and
output variables in recent respirometric research.

General trends in respirometric research
Flexibility
Most investigators construct their own respirometers. Whereas
electrodes, pumps, thermostats and data processing instruments are standard equipment, experimental chambers and valves have to be tailored to
each specific problem. This means, above all, taking into consideration not
only the size and the metabolic rate, but also the behaviour and general
biology of the organisms to be studied. It is not always realized to what
extent animals react to the geometry of the respirometer vessel, to small
variations in the flow regime, and to seemingly unimportant features of the
experimental set-up. This can result in widely differing rates of oxygen
consumption under otherwise identical conditions (Hughes, Albers, Muster

& Götz, 1983; Forstner, 1983*a*). Only by trial and error can the most suitable experimental conditions for each object of study be arrived at.

Data acquisition and processing
The computerization of physiology has made possible the handling of a vast amount of data. This trend is bound to continue. Data processing is particularly required in investigations in which either very long-lasting processes or very short, dynamic responses of animals are to be studied (see below).

Long-term measurements
Data processing in combination with the development of stable oxygen sensors and pumps has made it possible to record the oxygen consumption of animals over fairly long time periods (Soofiani & Hawkins, 1982: up to 16 days; Forstner 1983*a*: 16 days; Sondermann, Becker & Guenther, 1984: up to 42 days; Du Preez, Strydom & Winter, 1986; Du Preez, McLachlan & Marais, 1986: up to 8 days). This approach will make continued progress, but specific solutions are essential for the problem of starvation on the one hand and bacterial contamination on the other.

Short-term measurements
At the other end of the time scale, the problem of monitoring fast responses and transient metabolic states of animals is encountered. In aquatic organisms this is still a wide open field, requiring high resolution of the measuring equipment as well as sophisticated methods for signal reconstruction (see below).

Link-up with other measuring systems
The scientific value of measurements of oxygen consumption can be increased considerably by linking respirometers with systems measuring other output variables – the most important of these systems are calorimeters (see Gnaiger, 1983 and this volume) and activity monitors (Koch & Wieser, 1983; Kaufmann, 1983 and below). In aquatic organisms there is still great scope for the synchronized application of NH_3-, CO_2- and pH-sensitive electrodes together with a polarographic oxygen sensor (POS) in studying metabolic responses to food and other input variables.

Measuring systems
Oxygen consumption of individual organisms in the laboratory
Closed systems. Because of the high O_2 capacity of air and the ease with which the metabolic end product, CO_2, can be removed, closed

respirometers are still very popular in studying air-breathing animals. The classical manometric and volumetric systems, such as Warburg, Scholander and Gilson, have been automated by, e.g. monitoring the movements of the meniscus of the manometric fluid by means of photo diodes or by using sensitive electronic manometers (Plischke & Buhr KG, Bonn, West Germany).

A more elegant method along these lines is *capacity coulometry*, based on replacing the O_2 consumed with O_2 produced electrolytically by discharging a capacitor through a $CuSO_4$ solution. In the hands of a small number of investigators, notably Heusner (Heusner, Hurley & Arbogast, 1982), this method has been developed into a versatile respirometric system. The smallest measuring chamber used by Heusner and co-workers has a volume of 0.25 ml; the smallest animals measured (ticks) weigh only a few mg, and volume changes of 0.02 nl are readily detectable in a 2 ml chamber. In a coulometric respirometer O_2 is released at the anode which dips into a solution of $CuSO_4$, Cu being deposited at the cathode. An additional advantage of this system is that it can be autoclaved and that the O_2 produced is sterile, thus making it suitable for the measurement of oxygen consumption under sterile conditions.

The most sensitive closed respirometer is the Cartesian Diver that permits the measurement of oxygen consumption of organisms in the μg range (Klekowski, 1971). A comparable level of sensitivity cannot be achieved in open-flow respirometry. Closed systems are much less appropriate for the measurement of oxygen consumption in water. Although the disadvantages of closed systems in aquatic respirometry have been discussed on many occasions (Kamler, 1969; Oertzen, 1984; Lampert, 1984), Oertzen claims that even until 1984 about 90% of all measurements on aquatic animals were based on such systems. Of the disadvantages most often cited, the absence of water movements is the least important as water can (and should) be circulated even in a closed system. However, the disappearance of O_2 and the accumulation of metabolic products, like CO_2, NH_3, H^+ etc., limit the use of closed systems in aquatic respirometry, particularly for long-term measurements. This disadvantage may be turned into an advantage where the reduction of P_{O_2} and the accumulation of metabolic products in a finite volume of water are part of the problem to be studied, as, for example, in investigations on the switch from aerobic to anaerobic metabolism in benthic invertebrates (Pörtner, Heisler & Grieshaber, 1985). The major advantage of closed systems in respirometric studies is simplicity of construction and operation. Moreover, it has to be borne in mind that in closed systems the decrease in P_{O_2} represents the actual amount of oxygen consumed. As a consequence, the longer the measuring period, the less is the contribution of any base line error to the final change of P_{O_2} recorded, in contrast to the situation in open-flow systems where base-line errors accumulate with time. Thus, with low rates of oxygen consumption it is sometimes easier to achieve high accuracy of measurement with a closed than with an open-flow system.

Open-flow systems. Obviously, the major advantage of open-flow systems is that they allow the maintenance of stable dynamic and chemical conditions. Only in this way is it possible to conduct long-term measurements. Furthermore, in flow-through systems it is generally easier to detect, and monitor, phasic responses of the organisms to short-term manipulation of input variables (Lomholt & Johansen, 1983). Aquatic flow-through systems can be operated with one or with two POS. High precision under non-stable conditions requires frequent calibrations. This has been accomplished best in so-called 'Twin Flow Respirometers' (Gnaiger, 1983) in which two POS switch position at a predetermined frequency, each serving alternately as the inflow or the outflow sensor (see also Lampert, 1984). By means of a suitable valve mechanism one POS may be connected to the outflows of many experimental chambers (Scharf, Oertzen, Scharf & Stave, 1981; Hughes, Albers, Muster & Götz, 1983; Du Preez, McLachlan & Marais, 1986). Constant flow rates through open systems can be maintained by high precision pumps (expensive) or by a constant head of water (cheap). The actual flow-rates can be measured by various devices, the best is by electromagnetic sensors (Statham SP 2202 and others; see Lomholt & Johansen, 1983). In studying air-breathing animals in flow-through systems oxygen is monitored either with a paramagnetic sensor, or by gas chromatography, or by mass spectrometry, in order of increasing expense (for a recent survey of these methods see Cameron, 1986). Recently, however, instruments have become available for measuring oxygen consumption by exploiting the electrolytic properties of zirconium dioxide (ZrO_2) at high temperatures. Since the zirconium sensor is both sensitive and very stable, a new range of possibilities for animal experimentation is thus opened up. The ultimate open-flow apparatus eliminates the respirometer chamber altogether, connecting the animal's exchange system directly with oxygen sensors and flow probes. This has been accomplished both with large aquatic (Lomholt & Johansen, 1983) and – about a century ago for the first time – with terrestrial animals (Zuntz & Hagemann, 1898; Seeherman, Taylor, Maloiy & Armstrong, 1981; see Wieser, 1986).

Intermittent systems. By separating a measurement phase from a flushing phase some of the advantages of closed systems (simplicity) and open systems (stable conditions) may be combined in aquatic respirometry (Jobling, 1982; Forstner, 1983a). The periodicity of the whole cycle can be so chosen that during the closed measuring phase P_{O_2} does not fall below a given level. When this level is reached a valve opens, allowing water from an air-saturated reservoir to flow through the experimental chamber, thus maintaining P_{O_2} – and other variables – within a set range (Quetin, 1983). The system described by Forstner (1983a) allows the sequential measurement of several chambers by means of a single POS, which during the closed phase monitors the drop of P_{O_2} in one of the chambers, and during the flushing phase is calibrated against the oxygen partial pressure of the flow-through water. Even during the closed phase, water circulates through a loop that connects an experimental chamber with the oxygen sensor. In this way the animals are subjected to identical flow conditions during both phases. However, continuous records cannot be obtained with this system. This may sometimes be a disadvantage, particularly with greatly fluctuating rates of oxygen consumption. On the other hand, intermittent systems offer an enormous advantage whenever the precise regulation of the flow rate of the medium (on which all open-flow systems depend) proves difficult. This problem is increasingly encountered in measurements of oxygen consumption in air since the introduction of the zirconium sensor (see below) has popularized the use of open-flow systems for monitoring gas exchange even of very small poikilothermic animals. At the low flow rates required for this purpose the pumps of the commercial instruments available may not be sufficiently stable. To circumvent this difficulty several authors have begun to use the zirconium sensor purely as a gasometric device by letting the animals respire for a given time in a closed vessel (a syringe, for example) and injecting an aliquot of the spent air into the sensor. The signal produced is directly proportional to the oxygen consumed (Lighton, Bartholomew & Feener, 1987). By means of a switching device an intermittent system can be constructed, separating a measuring phase protected from all variations of flow rate, from a flushing phase that returns the experimental chamber to the starting conditions.

Oxygen consumption of organisms and ecosystems in the field

There are three major areas in which *in situ* measurement of oxygen consumption is desirable:

(i) The study of individual organisms that live under such extreme conditions, or are so delicate, that they would not survive if transported to the laboratory. This holds particularly for deep-sea animals, for which rather elaborate equipment has been constructed by Smith and collaborators (see review by Smith & Baldwin, 1983). Most of the sessile Cnidaria are also nearly impossible to transfer in a viable state into a lab-based respirometer (Svoboda 1978).

(ii) In community respiration *in situ*. This application has a long history in aquatic ecological research (Odum & Odum, 1955) and many 'bell jar' or similar systems have been designed during the last decades.

(iii) In measuring the biological oxygen demand (BOD) of sediments, sludges etc. This application is of interest in connection with problems of pollution, waste water treatment, fermentation etc. Typically, the BOD of a sample is deduced from the dissolved oxygen (OD) consumed during a five-day incubation. The BOD is considered to be a function of the organic matter available to organisms. Many commercial instruments are available, equipped, for example, with POS (WTW, Leeds & Northrup) or with transducers sensitive to volume changes of a gas phase from which oxygen is consumed (Techline).

Technical features
Electrochemical oxygen sensors

Most oxygen sensors used in aquatic research still embody the basic principle of the 'Clark-type' electrode (Clark, L. C., U.S. Patent no. 2913 386, 1959; see Hitchman, 1978; Gnaiger & Forstner, 1983). Four new design features have been reported in the technical literature and may lead to improvements in sensor function (Fig. 1): A sensor introduced by WTW has been provided with a second sensing anode the purpose of which is to stabilize the polarographic potential despite changes in resistance of the working anode due to accumulation of AgCl (Rommel, 1984; Fig. 1*a*). For measurements at low P_{O_2} a guard cathode may be interposed between working cathode and electrolyte reservoir that scavenges oxygen diffusing out from the electrolyte (Hitchman, 1983; Fig. 1*b*). A new anode configuration designed by Leeds & Northrup (Phelan, Taylor & Fricke, 1982) incorporates an oxygen generating reaction that compensates the consumption of molecular oxygen at the cathode (Fig. 1*c*). In this way a steady state is achieved with regard to the sample. Stirring sensitivity and influence of fouling on the membrane are almost completely eliminated. Moreover, the electrolyte is not consumed, which additionally prolongs the working life of the sensor. A fourth innovative feature is aimed at eliminating the thin electrolyte layer between membrane and cathode that acts both as a diffusion barrier for O_2 and as an electrical series resistance for OH^- ions between cathode and anode (Fig. 1*d*). This is one of the reasons for the slow response and the pressure sensitivity of conventional Clark-type sensors. In the new design (Anonymous, 1985) a thin gas-permeable layer of platinum is deposited directly on the electrolyte facing side of the membrane. In our

Fig. 1. Schematic representation of five innovative design features for polarographic oxygen sensors (POS).

(*a*) POS with working and sensing anode. A_w: working anode; A_m: sensing anode; C: cathode; E_p: polarizing potential; O_a: operational amplifier.

(*b*) POS with guard cathode. A: anode; C_g: guard cathode; C_w: working cathode.

(*c*) Electrode layout and the diffusion path of O_2 in a Leeds & Northrup POS. A: multiple anodes; C: multiple cathodes; M: membrane; E1: electrolyte.

(*d*) POS with thin film cathode deposited on membrane. M: membrane, C: cathode; E1: electrolyte; A: anode.

(*e*) Zirconium sensor. Pt: platinum electrodes; E_o: output voltage; ZrO_2: zirconium oxide.

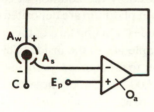

a

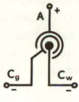

b

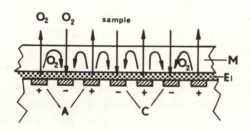

c

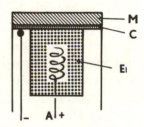

d

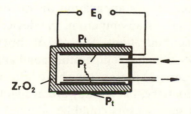

e

opinion this design could be further improved by incorporating the temperature sensing element into the metallic layer. At present, in all commercially available POS the temperature sensing element is situated some distance from the cathode, thus introducing errors of temperature measurement and unpredictable time lags into the process of temperature compensation.

For use in air, as mentioned earlier, the 'zirconium sensor' has been incorporated into a commercially available oxygen analyzer (Ametek – Electrochemistry Instruments) that is finding increasing use in the respirometry of air-breathing animals. The basic sensor consists of a closed cylinder of zirconium-oxide (ZrO_2) stabilized by the addition of calcium-oxide (Fig. 1*e*). Thin gas-permeable layers of platinum are deposited on the inside and outside of the cylinder as electrodes. On one face of the cylinder two tubes are provided for passing the sample gas to the inside. The outer electrode is exposed to ambient air. In order to function properly the sensor must be heated to a temperature of 360–800 °C. Under these conditions ZrO_2 becomes electrically conductive for oxide ions and functions as an electrolyte. Furthermore, at these temperatures the reaction $O_2 + 4e \rightleftarrows 2 O^{2-}$ becomes fully reversible, the sensor turning into a true potentiometric device. The voltage produced follows the Nernst equation and is thus proportional to the difference in oxygen concentration across the walls of the cylinder. In the Ametek instrument two sensors are used in a differential configuration which is especially suitable for measuring small differences in oxygen concentration. In other sensors (Fujikura Ltd.) a somewhat modified principle is employed. The sensor cyclinder is provided with just a small aperture that restricts gas diffusion into the inside. A polarizing voltage is then applied across the electrodes, the cathode being situated inside the cylinder. Oxygen is reduced at the cathode and O^{2-} migrates in the electrical field through the layer of ZrO_2 to the outer electrode. There it takes up two electrons and is released as gaseous oxygen. In this way the oxygen concentration on the inside decreases, which causes oxygen from the outside to diffuse through the aperture until an equilibrium is attained. This equilibrium and the resulting diffusion current is proportional to the outside oxygen concentration. In this application the zirconium sensor functions like a Clark-sensor, i.e. as an amperometric device, and can be used for measuring oxygen concentrations. Sensors of this type offer advantages over Clark-type POS at higher temperatures and are also used in catalytic converters of motor cars. At present they have little merit by themselves for application in respirometry, but as they are available at reasonable prices we can foresee that such sensors could be adapted for biological applications, allowing the construction of instruments at less exotic cost than the now commercially available equipment.

Chambers and flow units

Respirometer chambers should be constructed for easy access so as to facilitate the quick removal of test animals, in cases where, for example, the measurement of oxygen consumption is to be followed by biochemical studies. In order to keep the time constant of the system as short as possible, the volume of the chamber should be small, but not so small as to restrict the animals (Hughes, Albers, Muster & Götz, 1983). It should also be borne in mind that measurements of oxygen consumption alone are of limited value unless combined with other – ecological or biological – variables. Thus it may be desirable to construct chambers equipped with devices for influencing the animals' activity, like paddles, wheels or electrical stimulators (Wieser, 1985), or for adding food and other material (Lampert, 1986; Puckett & Dill, 1984). A large number of chambers may be measured in sequence by means of switching devices or electrical controllers (Forstner, 1983b). The development of electronic hardware has made it possible to construct control devices that are accurate, inexpensive, flexible and virtually non-mechanical (Emerson & Strydom, 1984).

Two older types of measuring systems should be mentioned specifically as they still offer a large potential field for contributions to respirometry, i.e. tunnel respirometers and large-volume respirometers. The former are based on the principles embodied in the respirometers of Blazka, Volf & Cepela (1960) and Brett (1965), which allow calculation of the aerobic cost of swimming. Three basic types of such respirometers have been distinguished by Smith & Newcomb (1970). A refinement has been the addition of tilting facilities, enabling the observer to distinguish between horizontal and upward swimming (Priede & Holliday, 1980). The gap in our knowledge concerning the energetics of swimming at low Reynolds numbers ought to stimulate the construction of respirometers in which small aquatic organisms are subjected to clearly defined hydrodynamic forces (Vlymen, 1974; see also below, Fig.3). Large-volume respirometers allow the measurement of routine oxygen consumption of groups of animals under more or less natural conditions (Durbin, Durbin, Verity & Smayda, 1981; Koch & Wieser, 1983). There is unlimited scope for the imaginative investigator in rendering the environment within such respirometers as natural as possible, and undoubtedly we still have much to learn about the energetics of spontaneous activity in groups of animals under the influence of combinations of environmental and biological factors.

Signal reconstruction

The slowed response characteristic, especially of aquatic respirometers, causes two major problems. Firstly, it takes some time until the respirometer has equilibrated after insertion of the animals or changes of operating conditions, and, secondly, it acts as a low-pass filter attenuating all rapid changes of the animals' respiration rates. Most of the correction techniques proposed in the literature (e.g. Niimi, 1978; Propp, Garber & Ryabuschko, 1982; Hughes, Albers, Muster & Götz, 1983) deal with quite specific situations, so a more general point of view will be outlined here. All examples given are taken from open-flow measurements, but the methods employed are applicable to closed or intermittent respirometers in an analogous form.

Specific delays and distortions of the recorded signals are due to the sensors' inertia, the tubings, and particularly to mixing effects and turbulences within the experimental chamber as is schematically shown in Fig. 2. Therefore it should be obvious that an optimized respirometer design (small

Fig. 2. Comparison of an ideal and a real open-flow respirometer. In the ideal apparatus the oxygen uptake ($\dot{N}_{O_2}$) simply causes a difference between inlet ($P_{O_2,i}$) and outlet ($P_{O_2,0}$) partial pressures. In reality, however, the recorded signals are distorted by the slowed response of the sensors (g_1), the mixing effects within the animal chamber (g_2, g_3) and delays caused by the tubing Δ_1, Δ_2. Moreover, there may be a blank oxygen uptake due to microorganisms ($\dot{N}_{O_2,b}$). (From Kaufmann 1986).

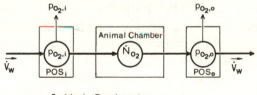

a Ideal Respirometer

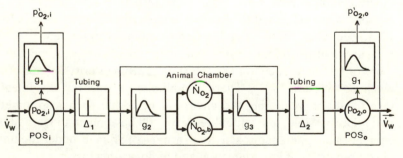

b Real Respirometer

chamber volume, high flow-through rate, etc.) is vital for achieving high temporal resolution of the measurements, but numerical corrections are useful for further improvement.

The mathematical background of the correction calculations is the theory of linear filters which states that the properties of such systems are fully described by their so-called impulse response or weighting function $g(t)$. This impulse response can be measured directly in a respirometer by injecting a certain amount of oxygen-free water into the experimental chamber, thus simulating a sudden burst of oxygen uptake. Modern computing facilities and algorithms have made it possible to use this basic description of a linear system that connects the original respiration rate, here denoted as the filter input $x(t)$, and the distorted measurement, the filter output $y(t)$, by the convolution with the impulse response:

$$y(t) = \int_0^\infty g(t') \cdot x(t - t') \cdot dt'.$$

Reconstruction of the original respiration rate, the deconvolution, requires

Fig. 3. Respiration of juvenile cyprinids (*Chalcalburnus chalcoides*) swimming in a tunnel respirometer at different speeds (15° C; 30 individuals; average fresh weight 25 mg; average body length 17 mm). Correction of the respirometer's response with a single time constant, experimentally determined as $\tau = 5.7$ min, was accomplished by using the simple differential equation given in the text.

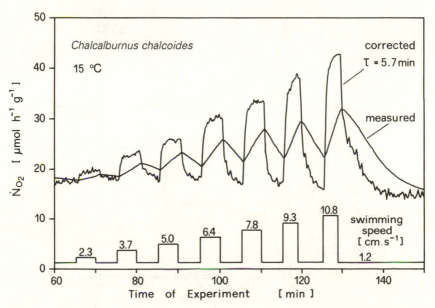

solving this integral equation for $x(t)$, which is usually done by employing Fourier transformations. An extensive literature is available on these methods of time series analysis (Jenkins & Watts, 1968; Brigham 1974).

Fortunately in some cases a much simpler approach is sufficient. If a perfectly mixed chamber is the only important reason for distortion, the impulse response will decay exponentially with a time constant τ equal to the chamber volume divided by the flow-through rate. In this case the formulation of the filter action by means of linear differential equations attains its most simple form and the original respiration rate can be recovered from the measurement by:

$$x(t) = t(t) + \tau.dy(t)/dt.$$

This is the most commonly used correction formula in respirometry (e.g. Evans, 1972) which in the field of calorimetry is also known as the Tian-equation (Calvet & Prat, 1963). Generalizations of this equation (Randzio & Suurkuusk, 1980) are for practical reasons restricted to a

Fig. 4. Respiratory and activity response of the oligochaete worm, *Lumbriculus variegatus*, to electrical stimulation (averaged from ten stimuli). To make the measured respiration rate comparable to the delay-free recording of the locomotor activity (obtained by the video method of Kaufmann, 1983) the latter was distorted numerically with the respirometer's response (convolution). Model functions (fit) were used to reconstruct the original respiratory response which declined more slowly than the activity, thus revealing an oxygen debt.

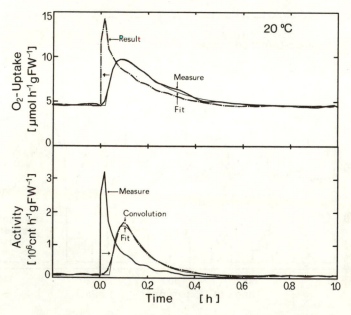

maximum of two different time constants (e.g. a second one for the sensor's response).

As an example of this most simple form of correction Fig. 3 shows an experiment with juvenile cyprinid fish (*Chalcalburnus chalcoides*) swimming in a tunnel respirometer at varying speeds. The water current assured perfect mixing within the experimental chamber, thus giving a response with only a single time constant. Note that the noise has increased drastically in the corrected signal. This effect limits the increase of temporal resolution by numerical methods, illustrating the fact that information actually lost in the noise of the measurement cannot be recovered.

The second example given in Fig. 4, the reaction of an oligochaete worm (*Lumbriculus variegatus*) to an activity burst induced by electrical stimulation, is much more complicated. In addition to a more complex respirometer response there was an effect very typical of open-flow systems. The influence of the animals' movements on the mixing pattern of the chamber caused artificial fluctuations in the recorded signals. These were

Fig. 5. Open flow measurement of the response of an oligochaete (*Lumbriculus variegatus*) to declining oxygen concentration. The outlet oxygen recording ($P_{O_{2,0}}$) is interrupted by short calibrations of the inlet oxygen tension (C), which facilitates the reconstruction of the total reference line ($P_{O_{2,cal}}$) that already includes the blank oxygen uptake ($\dot{N}_{O_{2,b}}$). α_{O_2} is the oxygen solubility and $\dot{V}_w$ the flow-through rate. (From Kaufmann, 1986.)

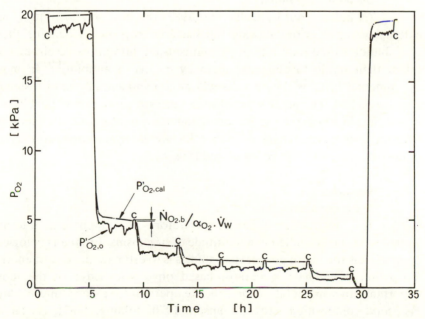

filtered out by fitting model functions to averages of ten identical stimu-
lations. To make this model of the respiration rate comparable to the
activity response obtained by a delay-free video method (Kaufmann, 1983)
the activity signal was convoluted with the respirometer's impulse response,
thus avoiding the numerical problems of deconvolution. In this way the fast
component of the respiratory response due to activity could be separated
from a slow response due to oxygen debt, and the model functions yielded
an estimate of the original respiration rate.

If respirometers are not operated under stable conditions additional
problems arise in conjunction with the reference line being subjected to
distortions different from the measurement itself. To illustrate this point
Fig. 5 shows an experiment in which oligochaetes were exposed to different
oxygen concentrations. The inlet P_{O_2} has to be reconstructed either by
separate calibration experiments in single sensor systems, as shown in
Fig. 5, or, more easily, by the simultaneous calibration recording of two
sensors.

It should be emphasized that these correction methods do not yield the
real respiration rate but give a better estimate than could be obtained
otherwise. Reliability tests on the results may become rather complicated,
but they are imperative, requiring attention to both systematic and statistical errors in the respirometer's responses.

Bacterial respiration

As pointed out by Dalla Via (1983) the best way to control this
external variable is by allowing the bacteria to grow, calculating their
contribution to the total rate of oxygen consumption in the respirometer,
rather than trying to eliminate them by means of antibiotics. In most
experimental systems the bacteria will grow exponentially until a steady
state is reached. The parameters of such a growth curve can be established
in test runs in which the experimental animals are intermittently removed
from the respirometer chamber so that the rate of oxygen consumption due
to the bacteria alone can be measured (Fig. 6).

Variables
Input variables

The relationship between respiratory rate, input and internal
variables has been studied in a multitude of organisms. From an ecological
point of view the major goal of such studies is to estimate the rate of energy
dissipation of animals in nature, based on a knowledge of the most
important environmental factors, like temperature, season, time of day,
P_{O_2}, food composition, etc. (Phillipson, 1970; Jobling, 1982). As far as

methodology is concerned the following aspects of the relationship between input variable and $\dot{N}_{O_2}$ (or other output variables, see below) are the most relevant:

(i) Response of $\dot{N}_{O_2}$ to short-term (acute) input changes. This requires sensitive detectors and experimental systems with short time constants. Thus, the use of open-flow respirometers is indicated.

(ii) Course of $\dot{N}_{O_2}$ during long-term (chronic) exposure, and the role of

Fig. 6. Bacterial growth curve in a respirometer intermittently inoculated with experimental animals (*Palaemonetes antennarius*) under different conditions of temperature and salinity. Circles and vertical lines denote means and standard deviations. (From Dalla Via, 1983).

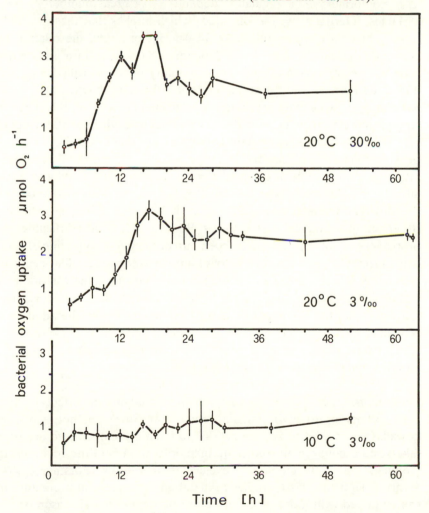

acclimation. This requires maintenance of stable conditions in laboratory experiments, provision for long-term respirometry, and the solution of problems connected with food supply and bacterial contamination. This suggests the use of intermittent respirometers.

(iii) Interactive effects of several variables on $\dot{N}_{O_2}$.

Some of the technical problems connected with important input variables will be discussed briefly.

P_{O_2}. Although determination of the relationship between P_{O_2} and $\dot{N}_{O_2}$ in aquatic animals has a long history, in most experiments the approach has remained simplistic. The answer to the question whether an animal behaves as a 'conformer' or a 'regulator' depends on a great number of other variables, internal as well as on input, which implies that these variables have to be controlled in some way. In fish, for example, the relationship between $\dot{N}_{O_2}$ and P_{O_2} is critically dependent on temperature and activity (Ott, Heisler & Ultsch, 1980). The variability of the so-called '$P_{O_2 crit}$' is illustrated by new data on an invertebrate, the oligochaete *Lumbriculus variegatus*, showing that the relationship between P_{O_2} and $\dot{N}_{O_2}$ is much more strongly influenced by the animals' activity than by environmental temperature (Fig. 7). If non-steady state conditions are employed the difference between inlet P_{O_2} and inspired P_{O_2} may become critical, requiring special attention to the problems of signal reconstruction, as discussed above (see also Hughes, Albers, Muster & Götz, 1983). A second problem of interest is the long-term maintenance of aquatic animals at reduced P_{O_2}'s in studying the effects of acclimation to this factor. Because of technical difficulties this problem does not appear to have been pursued actively. Bubbling of water with nitrogen or gas mixtures is a possibility, but requires feedback controls (Gruber & Wieser, 1983) and maintenance of normocapnic conditions by adding CO_2 and controlling the pH by means of a pH-stat. It has also been suggested that the P_{O_2} could be controlled chemically by the addition of Na-sulfite with cobalt chloride as a catalyst (Vanderhorst & Lewis, 1969; Kirk, 1974). In any case, a good long-term experiment involving P_{O_2} requires some sophistication and experience.

Temperature, season, photoperiod. Although studying the effects of long-term acclimation to environmental temperature on $\dot{N}_{O_2}$ is not as popular nowadays as it used to be, the role of acclimation temperature has to be taken into account in all studies on metabolism/temperature relationships (Jobling, 1982). By employing cyclical temperatures (Forstner, 1983a; Henry & Houston, 1984) studies of metabolism/temperature relationships can be linked with those of the effects of photoperiod and season on $\dot{N}_{O_2}$

(Newell & Roy, 1973; Navarro, Ortega & Iglesias, 1987). In fact, if photoperiod has not been taken into account, the investigation of long-term effects of temperature on animal metabolism is problematical. From a technical point of view the rates and shapes of temperature change in long-term experiments are crucially important. If slow temperature ramps are employed the question of acclimation during ramp time becomes critical (Jobling, 1982).

Fig. 7. Comparison of the P_{O_2} dependence of resting metabolism and activity-induced oxygen uptake in the aquatic oligochaete *L. variegatus* at three temperatures. Data were obtained from three experiments as shown in Fig. 4 by linear regression of oxygen consumption versus locomotor activity. Regression parameters represent resting rates (intercept) and oxygen uptake per unit activity (given in arbitrary counts = cnt). Activity outbursts comparable in magnitude throughout the whole P_{O_2} range were induced by electric stimulation and their effect measured by means of a video method (Kaufmann, 1983). Whereas the animals exhibit good oxyregulation in the resting state (shown are fitted curves of the Michaelis-Menten type), the oxygen uptake due to activity shows more of a conformer-like behaviour (half-maximum values are marked in the drawing). From Kaufmann, 1986.

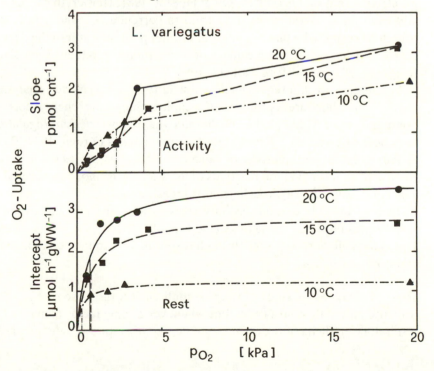

Food. As with the previous two variables the response of the rate of aerobic metabolism to food contains a dynamic element that has hardly been investigated in lower, particularly aquatic, animals. For example, how long does it take after ingestion of a meal before $\dot{N}$ increases? How is the shape of the response curve influenced by factors like temperature, age, starvation, etc.? Studying these aspects in aquatic animals calls for respirometers into which food can be infused at a known rate (Lampert, 1986; Du Preez, McLachlan & Marais, 1986; Soofiani & Hawkins, 1982). Long-term measurements are necessary if the duration of the calorigenic effect of food, or the shape of the response pattern of $\dot{N}_{O_2}$ to extended periods of starvation, is to be determined. These are important parameters for the definition of, e.g. post-absorptive $\dot{N}_{O_2}$ (Jobling, 1981), or of species specific starvation effects (Marais, 1978; Du Preez, McLachlan & Marais, 1986; Du Preez, Strydom & Winter, 1986).

pH, P_{O_2}, salinity. The effect of pH on physiological functions of aquatic animals has become a popular field of study in connection with acid rain (Carrick, 1981, Haines & Johnson, 1982). Such studies require long-term maintenance of animals under stable conditions of pH and P_{CO_2}, and this can only be achieved by employing feedback systems with the most stable pH electrodes available (Ultsch, Ott & Heisler, 1981; Heisler, this volume). Although many investigators have tried to correlate $\dot{N}_{O_2}$ and acidity the consensus seems to be that $\dot{N}_{O_2}$ *per se* is a rather poor indicator of the degree of acid toxicity, except when damage of the respiratory epithelia is involved (Hughes, 1981; Wood & McDonald, 1982).

Acclimation to different regimes of salinity is necessary for analysing another time-honoured problem of ecophysiology, the energy cost of osmoregulation. Although the low thermodynamic costs of osmoregulation have been stressed since Krogh's fundamental studies (Krogh, 1939; Potts & Parry, 1964), the question of 'ancillary costs', as for example those connected with behavioural reactions of the animals, cannot be answered in a general way. An approach to this question is to subject variously acclimated animals to acute salinity stresses and to study the dynamic responses of both $\dot{N}_{O_2}$ and activity (Oertzen, 1984; 1985; Dalla Via, 1986). This is possible only with flow-through respirometers connected to activity monitors (see below).

Stress, toxicity. Most kinds of stress stimuli elicit a response of $\dot{N}_{O_2}$. The response is usually non-specific, but whenever a toxic factor damages the respiratory epithelia of the animal, the degree of respiratory distress is an indicator of stress intensity (Hughes, 1981; Leivestad, 1982; Wood & McDonald, 1982).

General stress stimuli or disturbances (electrical, mechanical, chemical) may be used to induce flight or fright reactions in an animal, thus offering a means of evaluating the energy cost of maximum activity (Wieser, 1985; Wieser & Forstner, 1986, Du Preez, Strydom & Winter, 1986). Additionally, the reduction in the scope for activity has been used for estimating the harmful effects of toxic factors on the respiratory system of aquatic animals, mainly fish (Wedemeyer & McLeary, 1981).

Output variables

The importance of linking the measurement of oxygen consumption with that of other output variables cannot be over emphasized. Amongst these, $\dot{N}_{CO_2}$ and $\dot{N}_{NH_3}$ are the most obvious candidates, and the methods for measuring them are discussed in the chapter by Heisler (this volume). Other output variables of relevance are heat production (Gnaiger and Shick, this volume), end-products of anaerobic metabolism (Grieshaber and Gäde, this volume) and activity (see below).

Activity. In order to understand the levels and fluctuations of oxygen consumption during an investigation the activity of the animals has to be taken into account. In fact, as early as 1971, Fry came to the following conclusion: 'It seems unlikely that any useful determination of the metabolic rate of fish can now be made which is not accompanied by a measure of physical activity.' It thus seems fitting to conclude this review with a short discussion of the measurement of activity in conjunction with that of oxygen consumption.

In aquatic animals the first approach was to define activity levels on the basis of continuous records of oxygen consumption which led to terms such as standard (s), active (a), and routine (r), metabolism (Fry, 1971). From these levels various types of 'scopes for activity' can be calculated, such as absolute scope ($\dot{N}_{O_2}a–\dot{N}_{O_2}s$: Fry, 1947), relative scope (e.g. $\dot{N}_{O_2}a–\dot{N}_{O_2}r$: Wieser, 1985), or dimensionless quotients, like the normalized relative scope for activity, ($\dot{N}_{O_2}a–\dot{N}_{O_2}r)/(\dot{N}_{O_2}a–\dot{N}_{O_2}s$): Priede, 1985; Wieser & Forstner, 1986.

A 'standard' or 'resting' level of activity can also be determined by extrapolation from measurements of animals swimming at different speeds (Beamish, 1964; Brett, 1972), or by comparison with anaesthetized animals (Holliday, Blaxter & Lasker, 1964; De Silva, Premavansa & Keembiyahetty, 1986). However, both approaches have specific disadvantages.

Extrapolation to zero swimming speed usually underestimates true resting metabolism; and anaesthesia has quite different effects on different animals (Beis & Newsholme, 1975) so that metabolic levels defined in this way are difficult to interpret.

A much more satisfactory definition of activity is achieved by correlating $\dot{N}_{O_2}$ with an independent measurement of activity. Aquatic animals, particularly fish, can be forced (and trained) to swim against currents of defined velocity in tunnel respirometers (Blazka, Volf & Cepela, 1960; Brett 1965; Fry, 1971), thus allowing the calculation of the aerobic energy cost of swimming at different speeds. A critical swimming speed is defined as the velocity of a current against which a fish is capable of holding its position for a given time – say one hour. The energy cost of such an activity level may be compared with the maximum metabolic effort induced by other means, for example by electrical or mechanical stimulation (Wieser, 1985; Wieser & Forstner, 1986; Du Preez, McLachlan & Marais, 1986; Du Preez, Strydom & Winter, 1986), or by feeding maximum rations (Jobling, 1981; Soofiani & Priede, 1985). Another type of respirometer for the determination of activity metabolism is based on the principle of the rotating chamber (Fry, 1971). A body swimming in a rotating chamber is influenced by additional hydrodynamic forces ('Taylor effect'), thus complicating the interpretation of energy consumption. To overcome this problem Dabrowski (1986*a, b*) developed a non-rotating circular respirometer surrounded by a rotating cylinder with vertical stripes. Some fish can be induced to follow the rotating cylinder, their speed being determined by the rotation frequency of the stripes (optomotoric response). However, maximum swimming speeds cannot be induced by this method.

Another and more ecological approach to the study of active metabolism is the simultaneous measurement of $\dot{N}_{O_2}$ and spontaneous activity of the animals in large volume respirometers. In aquatic animals this has been achieved by means of paddle wheels (Spoor 1941; 1946), heat loss flow meters (Beamish, 1964; Kausch, 1968; Oertzen, 1985), photocells or diodes (Quetin, 1983), movie cameras (Durbin, Durbin, Verity & Smayda, 1981), and by more complex video-based activity monitors (Kaufmann, 1983; Koch & Wieser, 1983). In air-breathing animals the measurement of oxygen consumption has been combined with the measurement of activity by means of various types of actographs, like rotating wheels, acoustic detectors, capacitance detectors and others. The average metabolic costs of normal activity in the field can be estimated on the basis of measurements of $\dot{N}_{O_2}$ during spontaneous activity in the laboratory (Jobling, 1982; Soofiani & Priede, 1985).

Acknowledgment
The writing of this review, and the original work reported herein, have been supported by the 'Fonds zur Förderung der wissenschaftlichen Forschung in Österreich', project no. S–35/04.

References
Anonymous. (1985). Ist die Clark-Zelle jetzt passe? *Labor-Praxis*, März 1985, 159–60.

Beamish, F. W. H. (1964). Respiration of fishes with special emphasis on standard oxygen consumption. II. Influence of weight and temperature on respiration of several species. *Canadian Journal of Zoology*, **42**, 177–88.

Beis, I. & Newsholme, E. A. (1975). The contents of adenine nucleotides, phosphagens and some glycolytic intermediates in resting muscles from vertebrates and invertebrates. *Biochemical Journal*, **52**, 23–32.

Blazka, P., Volf, M. & Cepela, M. (1960). A new type of respirometer for the determination of the metabolism of fish in an active state. *Physiologia Bohemica*, **9**, 553–60.

Brett, J. R. (1965). The relation of size to rate of oxygen consumption and sustained swimming speed of sockeye salmon (Oncorhynchus nerka). *Journal of the Fisheries Research Board of Canada*, **22**, 1491–501.

Brett, J. R. (1972). The metabolic demand for oxygen in fish, particularly salmonids, and a comparison with other vertebrates. *Respiration Physiology*, **4**, 151–70.

Brigham, E. O. (1974). *The fast Fourier transform.* Englewood Cliffs: Prentice Hall.

Calvet, E. & Prat, H. (1963). *Recent Progress in Microcalorimetry.* Oxford: Pergamon Press.

Cameron, J. N. (1986). *Principles of Physiological Measurement.* New York: Academic Press.

Carrick, T. R. (1981). Oxygen consumption in the fry of brown trout (*Salmo trutta* L.) related to pH of the water. *Journal of Fish Biology*, **18**, 73–80.

Dabrowski, K. R. (1986*a*). Active metabolism in larval and juvenile fish: ontogenetic changes, effect of water temperature and fasting. *Fish Physiology and Biochemistry*, **1**, 125–44.

Dabrowski, K. R. (1986*b*). A new type of metabolic chamber for the determination of active and postprandial metabolism of fish, and consideration of results for coregonid and salmon juveniles. *Journal of Fish Biology*, **28**, 105–17.

Dalla Via, G. J. (1983). Bacterial growth and antibiotics in animal respirometry. In *Polarographic Oxygen Sensors*, ed. E. Gnaiger & H. Forstner, pp. 202–18. Heidelberg: Springer Verlag.

Dalla Via, G. J. (1986). Salinity responses of the juvenile penaeid shrimp *Penaeus japonicus*. I. Oxygen consumption and estimations of productivity. *Aquaculture*, **55**, 297–306.

De Silva, C. D., Premavansa, S. & Keembiyahetty, C. N. (1986). Oxygen consumption in *Oreochromis niloticus* (L.) in relation to development, salinity, temperature and time of day. *Journal of Fish Biology*, **29**, 267–77.

Du Preez, H. H., McLachlan, A. & Marais, J. F. K. (1986*a*). Oxygen consumption of a shallow water teleost, the spotted grunter, *Pomadasys commersonni* (Lacepede, 1802). *Comparative Biochemistry and Physiology*, **84**A, 61–70.

Du Preez, H. H., Strydom, W. & Winter, P. E. D. (1986*b*). Oxygen consumption of two marine teleosts, *Litognathus mormyrus* (Linnaeus, 1758) and *Lithognathus lithognathus* (Cuvier, 1830) (Teleosti: Sparidae). *Comparative Biochemistry and Physiology*, **85**A, 313–31.

Durbin, A. G., Durbin, E. G., Verity, P. G. & Smayda, T. J. (1981). Voluntary swimming speeds and respiration rates of a filter-feeding planktivore, the Atlantic menhaden, *Breevortia tyrannus* (Pisces: Clupeidae). *Fishery Bulletin*, **78**, 877–86.

Emmerson, W. D. & Strydom, W. (1984). An electrical controller for the automatic determination of oxygen consumption in aquatic animals. *Aquaculture*, **36**, 173–7.

Evans, D. O. (1972). Correction for lag in continuous-flow respirometry. *Journal of the Fisheries Research Board of Canada*, **29**, 1214–18.

Forstner, H. (1983*a*). An automated multiple-chamber intermittent-flow respirometer. In *Polarographic Oxygen Sensors*, ed. E. Gnaiger & H. Forstner, pp. 111–26. Heidelberg: Springer Verlag.

Forstner, H. (1983*b*). Electronic circuits for polarographic oxygen sensors. In *Polarographic Oxygen Sensors*, ed. E. Gnaiger & H. Forstner, pp. 90–101. Heidelberg: Springer Verlag.

Fry, F. E. J. (1947). Effects of the environment on animal activity. *University of Toronto Studies Biological Series*, **55**, 1–62.

Fry, F. E. J. (1971). The effect of environmental factors on the physiology of fish. In *Fish Physiology*, vol. 6, ed. W. S. Hoar & D. J. Randall, pp. 1–99. New York: Academic Press.

Gnaiger, E. (1983). The twin-flow microrespirometer and simultaneous calorimetry. In *Polarographic Oxygen Sensors*, ed. E. Gnaiger & H. Forstner, pp. 134–66. Heidelberg: Springer Verlag.

Gnaiger, E. & Forstner, H. (1983). *Polarographic Oxygen Sensors. Aquatic and physiological Applications*. Heidelberg: Springer Verlag.

Gruber, K. & Wieser, W. (1983). Energetics of development of the alpine charr, *Salvelinus alpinus*, in relation to temperature and oxygen. *Journal of Comparative Physiology*, **149**B, 485–93.

Haines, T. A. & Johnson, R. E., eds. (1982). *Acid Rain Fisheries*. Bethesda, MD: American Fishery Society.

Henry, J. A. C. & Houston, A. H. (1984). Absence of respiratory acclimation to diurnally cycling temperature conditions in rainbow trout. *Comparative Biochemistry and Physiology*, **77**A, 727–34.

Heusner, A. A., Hurley, J. P. & Arbogast, R. (1982). Coulometric microrespirometry. *American Journal of Physiology*, **243**, R185–92.

Hitchman, M. L. (1978). *Measurement of Dissolved Oxygen*. New York: John Wiley.

Hitchman, M. L. (1983). Calibration and accuracy of polarographic oxygen sensors. In *Polarographic Oxygen Sensors*, ed. E. Gnaiger & H. Forstner, pp. 18–30. Heidelberg: Springer Verlag.

Holliday, F. J. T., Blaxter, J. H. S. & Lasker, R. (1964). Oxygen uptake of developing eggs and larvae of the herring (Clupea harengus). *Journal of the Marine Biological Association of the United Kingdom*, **44**, 711–23.

Hughes, G. M. (1981). Effects of low oxygen and pollution on the respiratory systems of fish. In *Stress and Fish*, ed. A. D. Pickering, pp. 121–46. London: Academic Press.

Hughes, G. M., Albers, C., Muster, D. & Götz, K. H. (1983). Respiration of the carp, *Cyprinus carpio* L., at 10 and 20 °C and the effect of hypoxia. *Journal of Fish Biology*, **22**, 613–28.

Jenkins, G. M. & Watts, D. G. (1968). *Spectral Analysis and its Applications*. San Francisco: Holden-Day Inc.

Jobling, M. (1981). The influence of feeding on the metabolic rate of fishes: A short review. *Journal of Fish Biology*, **18**, 385–400.

Jobling, M. (1982). A study of some factors affecting rates of oxygen consumption of plaice, *Pleuronectes platessa* L. *Journal of Fish Biology*, **20**, 501–16.

Kamler, E. (1969). A comparison of the closed-bottle and flowing-water methods for measurement of respiration in aquatic invertebrates. *Polish Archives of Hydrobiology*, **29**, 31–49.

Kaufmann, R. (1983). VAMP: A video activity monitoring processor for the registration of animal locomotor activity. *Journal of Experimental Biology*, **104**, 295–98.

Kaufmann, R. (1986). Die Atmung des Glanzwurmes, *Lumbriculus variegatus* Mueller, bei Bewegungstätigkeit und Sauerstoffmangel. Thesis, University of Innsbruck.

Kausch, H. (1968). Der Einfluss der Spontanaktivität auf die Stoffwechselrate junger Karpen (*Cyprinus carpio* L.) im Hunger und bei der Fütterung. *Archiv für Hydrobiologie, Supplement*, **33**, 263–330.

Kirk, W. L. (1974). The effects of hypoxia on certain blood and tissue electrolytes of channel catfish, *Ictalurus punctatus* (Rafinesque). *Transactions of the American Fishery Society*, **103**, 593–600.

Klekowski, R. Z. (1971). Cartesian diver microrespirometry for aquatic animals. *Polish Archive of Hydrobiology*, **18**, 93–114.

Koch, F. & Wieser, W. (1983). Partitioning of energy in fish: can reduction of swimming activity compensate for the cost of production? *Journal of Experimental Biology*, **107**, 141–6.

Krogh, A. (1939). *Osmotic Regulation in Aquatic Animals*. Cambridge: University Press.

Lampert, W. (1984). The measurement of respiration. In *A Manual on Methods for the Assessment of Secondary Productivity in Fresh Waters*, 2nd edn., ed. J. A. Downing & F. H. Rigler, pp. 413–68. Oxford: Blackwell.

Lampert, W. (1986). Response of respiratory rate of *Daphnia magna* to changing food conditions. *Oecologia*, **70**, 495–501.

Leivestad, H. (1982). Physiological effects of acid stress on fish. In *Acid*

Rain Fisheries, ed. T. A. Haines & R. E. Johnson. pp. 157–64. Bethesda, American Fisheries Society.

Lighton, J. R. B., Bartholomew, G. A. & Feener, D. H. (1987). Energetics of locomotion and load carriage and a model of the energy cost of foraging in the leaf-cutting ant *Atta colombica* Guer. *Physiological Zoology*, **60**, 524–37.

Lomholt, J. P. & Johansen, K. (1983). The application of polarographic oxygen sensors for continuous assessment of gas exchange in aquatic animals. In *Polarographic Oxygen Sensors*, ed. E. Gnaiger & H. Forstner, pp. 127–33. Heidelberg: Springer Verlag.

Marais, J. F. K. (1978). Routine oxygen consumption of *Mugil cephalus*, *Liza dumerili* and *L. richardsoni* at different temperatures and salinities. *Marine Biology*, **50**, 9–16.

Navarro, E., Ortega, M. M. & Iglesias, J. I. P. (1987). An analysis of variables affecting oxygen consumption in *Actinia equina* L. (Anthozoa) from two shore positions. *Comparative Biochemistry and Physiology*, **86**A, 233–40.

Newell, R. C. & Roy, A. (1973). A statistical model relating the oxygen consumption of a mollusk (Littorina littorea) to activity, body size, and environmental conditions. *Physiological Zoology*, **46**, 253–75.

Niimi, A. J. (1978). Lag adjustment between estimated and actual physiological responses conducted in flow-through systems. *Journal of the Fisheries Research Board of Canada*, **35**, 1265–9.

Odum, H. T. & Odum, E. P. (1955). Trophic structure and productivity of a windward coral reef community on Eniwetok Atoll. *Ecological Monographs*, **25**, 291–320.

Oertzen, J. A. von (1984). Metabolic similarity of *Palaemon* populations from different brackish waters. *Internationale Revue der gesamten Hydrobiologie*, **69**, 735–55.

Oertzen, J. A. von (1985). Resistance and capacity adaptation of juvenile silver carp, *Hypophthalmichthys molitrix* (Val.) to temperature and salinity. *Aquaculture*, **44**, 321–32.

Ott, M. E., Heisler, N. & Ultsch, G. R. (1980). A re-evaluation of the relationship between temperature and the critical oxygen tension in freshwater fishes. *Comparative Biochemistry and Physiology*, **67**A, 337–40.

Phelan, D. M. Taylor, R. M. & Fricke, S. (1982). A maintenance-free dissolved oxygen monitor. *International Laboratory*, **IX**, 1982, 60–75.

Phillipson, J. (1970). The 'best estimate' of respiratory metabolism: its applicability to field situations. *Polish Archive of Hydrobiology*, **17**, 31–41.

Pörtner, H. O., Heisler, N. & Grieshaber, M. K. (1985). Oxygen consumption and mode of energy production in the intertidal worm *Sipunculus nudus* L.: Definition and characterization of the critical P_{O_2} for an oxyconformer. *Respiration Physiology*, **59**, 361–77.

Potts, W. T. W. & Parry, G. (1964). *Osmotic and Ionic Regulation in Animals*. London: Pergamon Press.

Priede, I. G. (1985). Metabolic scope in fishes. In *Fish Energetics*, ed. P. Tytler & P. Calow, pp. 33–64. London & Sidney: Croom Helm.

Priede, I. G. & Holliday, F. G. T. (1980). The use of a new tilting tunnel respirometer to investigate some aspects of metabolism and swimming activity of the plaice (*Pleuronectes platessa* L.). *Journal of Experimental Biology*, **85**, 295–309.

Propp, M. V., Garber, M. R. & Ryabuschko, V. I. (1982). Unstable processes in the metabolic rate measurements in flow-through systems. *Marine Biology*, **67**, 47–9.

Puckett, K. J. & Dill, L. M. (1984). Cost of sustained and burst swimming to juvenile coho salmon (*Oncorhynchus kisutch*). *Canadian Journal of Fisheries and Aquatic Sciences*, **41**, 1546–51.

Quetin, L. B. (1983). An automated, intermittent flow respirometer for monitoring oxygen consumption and long-term activity of pelagic crustaceans. In *Polarographic Oxygen Sensors*, ed. E. Gnaiger & H. Forstner, pp. 176–83. Berlin: Springer Verlag.

Randzio, S. L. & Suurkuusk, J. (1980). Interpretation of calorimetric thermograms and their dynamic corrections. In *Biological Microcalorimetry*, ed. A. E. Beezer, pp. 311–41. London: Academic Press.

Rommel, K. (1984). Ein neues Sauerstoff-Messsystem. *Labor-Praxis*, **VII/VIII**, 1984, 736–9.

Scharf, E. M., Oertzen, L. A. von, Scharf, W. & Stave, A. (1981). A microflow respirometer for measuring the oxygen consumption of small aquatic organisms. *Internationale Revue der gesamten Hydrobiologie*, **66**, 895–901.

Seeherman, H. J., Taylor, C. R., Maloiy, G. M. O. & Armstrong, R. B. (1981). Design of mammalian respiratory system: Measuring maximum aerobic capacity. *Respiration Physiology*, **44**, 11–24.

Smith, K. L. & Baldwin, R. J. (1983). Deep-sea respirometry: *in situ* techniques. In *Polarographic Oxygen Sensors*, ed. E. Gnaiger & H. Forstner, pp. 298–319. Berlin: Springer Verlag.

Smith, L. S. & Newcomb, T. W. (1970). A modified version of the Blazka respirometer and exercise chamber for large fish. *Journal of the Fisheries Research Board of Canada*, **27**, 1321–24.

Sondermann, U., Becker, K. & Günther, K. D. (1985). Untersuchungen zum oxienergetischen Äquivalent beim Spiegelkarpfen (*Cyprinus carpio* L.) im Hunger. *Zeitschrift für Tierphysiologie, Tierernährung und Futtermittelkunde*, **54**, 161–75.

Soofiani, N. M. & Hawkins, A. D. (1982). Energetic cost at different levels of feeding in juvenile cod, *Gadus morhua* L. *Journal of Fish Biology*, **21**, 577–92.

Soofiani, N. M. & Priede, I. G. (1985). Aerobic metabolic scope and swimming performance in juvenile cod, *Gadus morhua* L. *Journal of Fish Biology*, **26**, 127–38.

Spoor, W. A. (1941). A method for measuring the activity of fishes. *Ecology*, **22**, 329–31.

Spoor, W. A. (1946). A quantitative study of the relationship between the activity and oxygen consumption of the goldfish, and its application to the measurement of respiratory metabolism in fishes. *Biological Bulletin*, **91**, 312–25.

Svoboda, A. (1978). *In situ* monitoring of oxygen production in cnidaria with and without zooxanthellae. In *Physiology and Behaviour of Marine Organisms*, ed. D. S. McLusky & A. J. Berry, pp. 75–82. New York: Pergamon Press.

Ultsch, G. R., Ott, M. E. & Heisler, N. (1981). Acid–base and electrolyte status in carp (*Cyprinus carpi*) exposed to low environmental pH. *Journal of Experimental Biology*, **93**, 65–80.

Vanderhorst, R. & Lewis, S. D. (1969). Potential of sodium sulfite catalyzed with cobalt chloride in harvesting fish. *The Progressive Fish-Culturist*, **31**, 149–54.

Vlymen, W. J. (1974). Swimming energetics of the larval anchovy *Engraulis mordax*. *Fishery Bulletin*, **72**, 885–99.

Wedemeyer, G. A. & McLeary, D. J. (1981). Methods for determining the tolerance of fishes to environmental stressors. In *Stress and Fish*, ed. A. D. Pickering, pp. 247–75. London: Academic Press.

Wieser, W. (1985). Developmental and metabolic constraints of the scope for activity in young rainbow trout (*Salmo gairdneri*). *Journal of experimental Biology*, **118**, 133–42.

Wieser, W. (1986). *Bioenergetik*. Stuttgart: Georg Thieme.

Wieser, W. & Forstner, H. (1986). Effects of temperature and size on the routine rate of oxygen consumption and on the relative scope for activity in larval cyprinids. *Journal of Comparative Physiology* B, **156**, 791–6.

Wood, C. M. & McDonald, D. G. (1982). Physiological mechanisms of acid toxicity to fish. In *Acid Rain Fisheries*, ed. T. A. Haines & R. E. Johnson, pp. 197–226. Bethesda: American Fisheries Society.

Zuntz, N. & Hagemann, O. (1898). Untersuchungen über den Stoffwechsel des Pferdes. *Landwirtschaftliches Jahrbuch*, **27**, Supplement III, 1–438.

P. TATNER and D. M. BRYANT

Doubly-labelled water technique for measuring energy expenditure

Introduction
Applications of the doubly-labelled water technique
The largely non-invasive nature of this technique is one of its most appealing features. It can be used to measure free-living metabolism of animals in their natural environment. Previously, it was difficult or impossible to obtain information of this type.

The technique can be used to address fundamental ecological questions regarding the energetic cost of free existence. These in turn provide an indication of the energetic impact of an animal on its environment (e.g. seabirds on Antarctic plankton resources, Costa, Dunn & Disher, 1986). The technique can also be used in conjunction with activity costs determined in the laboratory, time-activity budgets established from field observations, and meteorological information, to identify the components of daily energy expenditure. This can then be applied to the investigation of feeding, territorial, and life history strategies. Novel physiological applications include determination of flight cost, utilization of energy reserves, acclimatization strategies, and water economy. In clinical research the non-invasive and non-toxic nature of the method permits accurate measures of average daily energy expenditure over protracted periods. These can be obtained outside the confines of a metabolic chamber, and over the range from healthy to chronically ill subjects. There are many unique opportunities afforded by the technique to study nutrition, from the obese condition in developed society to starvation in the Third World (Prentice, Coward & Whitehead 1984).

Fundamental aspects of the technique
The rationale for obtaining a measurement of the rate of carbon dioxide production using doubly-labelled water is based on the observation that the oxygen of respiratory carbon dioxide mixes freely with the oxygen of body water (Lifson *et al.*, 1949). An exchange equilibrium involving

oxygen is facilitated by the carbonic anhydrase catalysed hydration of carbon dioxide (Maren, 1967).

Body water is labelled by injection of deuterium (D, $= {}^2H$) and oxygen-18 (^{18}O) as heavy water. After an equilibrium is reached, the loss of oxygen-18 isotope occurs either as respiratory carbon dioxide or as water. Loss of the isotope as water is determined independently by measuring the decline in hydrogen label (D). In essence, calculation of carbon dioxide production is based on the difference in turnover rates of the oxygen and hydrogen labels. The theoretical basis for the use of doubly-labelled water to measure energy and material balance is described by Lifson & McClintock (1966), who derive the following expression to determine the rate of carbon dioxide production:

$$r_{CO_2} = (N/2.08) \times (K_o - K_d) - 0.015 K_d N \tag{1}$$

Where r_{CO_2} is the rate of carbon dioxide production ($mMol \cdot h^{-1}$), which is usually expressed as the Average Daily Metabolic Rate

$$ADMR = (rCO_2 \times 22.4)/\text{average body mass}, cm^3 CO_2\ g^{-1}h^{-1}$$

The body water pool (N, mMol) may be calculated by the isotope dilution method (Schoeller *et al.*, 1980, Nagy & Costa, 1980). This method relies upon accurate identification of the plateau isotope concentration at equilibrium, which is more difficult to establish for small (<100 g) than for larger animals because of the rapid turnover rate for the former. We use an average figure for the body water fraction of the total mass, established by freeze-drying or desiccation. The body water pool (N) is then estimated as the average body mass (g) $\times$ body water fraction $\times$ 1000/18 (g to mMol). Equations that take account of variation in lipid content with body mass yield more accurate estimates of body water (Bryant & Westerterp 1982, Westerterp & Bryant 1984).

Fractional turnover rates of the hydrogen (K_d) and oxygen (K_o) labels are obtained from measurements of the initial and final concentrations of isotope in the body water:

$$\frac{\text{isotope}}{\text{turnover rate}} = \frac{\left[\begin{array}{c}\ln(\text{excess initial}\\ \text{isotope conc.})\end{array}\right] - \left[\begin{array}{c}\ln(\text{excess final}\\ \text{isotope conc.})\end{array}\right]}{\text{time period}} \tag{2}$$
(K_d or K_o)

Deuterium and oxygen-18 are present in all natural waters at concentrations of about 150 ppm D and 2000 ppm ^{18}O (= background). Hence it is necessary to establish initial and final enrichments above the background level, that is to obtain the excess figures. The other variable required to calculate the turnover rate is the time period (hours) between collection of the initial and final samples of body water.

The denominator (2.08) in equation (1) is the product of a fractionation factor ($H_2^{18}O$ liquid $\rightarrow C^{16}O^{18}O$ gas $= 1.04$) and a stoichiometric factor (i.e. carbon dioxide has an oxygen equivalent of two molecules of water). The remaining term in the expression ($-0.015K_dN$) accounts for the fractionation effects of evaporative water loss, on the assumption that this avenue involves 50% of the total water loss (Lifson & McClintock, 1966).

The figure for carbon dioxide production is converted into energy units by using a constant for the heat equivalent of $1 cm^3 CO_2$, which is dependent on the *RQ* value (Brody 1945). The *RQ* is seldom measured under typical field conditions, so a mean figure for the resting condition on natural food is usually employed (King & Farner, 1961). This aspect warrants further investigation (Speakman & Racey, 1987, Tatner, 1988), although the generally long term nature of DLW measurements means that an average value will probably not introduce great error (e.g. approx 3% for man, Schoeller 1983).

Assumptions of the technique

The major assumptions associated with the efficacy of equation (1) are detailed by Lifson & McClintock (1966) and more recently have been considered in relation to field data by Nagy (1980). We present a short summary of the salient points.

Assumption 1: body water volume is constant

Changes in the calculated carbon dioxide output are relatively insensitive to small changes in the body water content, as they exceed 5% only when the body water more than halves or doubles in volume (Nagy, 1980).

Assumption 2: rates of water flux and carbon dioxide production are constant

Water and carbon dioxide flux rates are unlikely to remain constant, but since equation (1) is based on the difference between the turnover rates, it will not vary as water flux changes because the hydrogen and oxygen labels are affected to the same extent (Lifson & McClintock, 1966). The figure for carbon dioxide production represents the average metabolic rate over the period (Lifson & McClintock 1966). Resting metabolic rate exhibits a diurnal cycle (Aschoff & Pohl, 1970), so to obtain an appropriate value for daily energy expenditure (DEE) the isotope flux should be measured over a period as close to a multiple of 24 h as possible, to ensure that the result is not biased by inclusion of a disproportionate amount of the active or resting phase.

Assumption 3: isotopes label only water and carbon dioxide in the body

Comparisons between the total body water volume based on oxygen-18 dilution and carcass desiccation suggests that this label is not rapidly exchanged with non-aqueous compounds in the body (Nagy, 1980), although its overall incorporation rate has yet to be measured in a whole animal. Some of the deuterium label rapidly leaves the aqueous pool as indicated by the overestimation (up to 13%) of body water using the isotope dilution method with hydrogen labels (Nagy & Costa, 1980). Hydrogen is incorporated into fats during their synthesis (Jungas, 1968), but the extent and effect of this on the calculated rate of carbon dioxide production is unknown. There may be additional sources of error caused by isotope incorporation when applying the technique to growing (subadult) or reproducing (laying birds, pregnant/lactating mammals) individuals. However, isotope ratios in the body tissues are generally similar to those of the dry components in the diet (Estep & Dabrowski, 1980, Schoeller *et al.*, 1986*c*) and different (lighter) from either the local water or body fluids, which themselves are similar. This suggests that isotope exchange between body tissues and the aqueous pool is small and probably very slow. Culebras *et al.* (1977a, 1977b) indicate that a maximum of only 5.22% of the hydrogen in body tissues is free to exchange with hydrogen in the aqueous pool, and they provide experimental data to prove the point.

Assumption 4: isotopes leave the body only as water and carbon dioxide

Loss of isotope labels in forms other than water and carbon dioxide, will only affect the calculated rate of carbon dioxide production if their ratio in this other form departs from two D to one ^{18}O. When they are lost in this ratio the difference between the turnover rates remains unaffected, although the water turnover will be overestimated (Lifson & McClintock, 1966). Other avenues of isotope loss are likely to involve faeces and urine, but where it has been investigated these are minimal (Nagy, 1980, Schoeller *et al.*, 1986*b*).

Assumption 5: the isotopic enrichments in water and carbon dioxide leaving the body are the same as in the body water

Similarity between the isotopic enrichments (specific activity in the case of tritium) of water and carbon dioxide leaving the body and those of the body water, depends upon the consequences of isotopic fractionation and water flux rates. Even if the animal has a high fluid intake, the latter is unlikely to cause significant error except in the estimate of water turnover (McClintock & Lifson, 1958*a*, Weathers and Nagy, 1980). Isotope fractionation describes the preferential selection of one isotope of an atom over another (e.g. when passing from a liquid to a gaseous phase, lighter molecules escape more frequently so the liquid becomes enriched relative to the vapour above it). Lifson & McClintock (1966) discuss this aspect and provide an expression (equation 1) that corrects for the effects of fractionation when using deuterium and oxygen-18 for *in vitro* measurements. More recently fractionation effects have been investigated in dynamic biological systems (Luz, Kolodny & Horowitz, 1984, Schoeller *et al.*, 1986*a*). Evaporative water loss via the breath and transcutaneous vapour (23% of water turnover, SE = 10%) is thought to be the main cause of error due to isotope fractionation in man as the isotopic abundance ratios for hydrogen and oxygen in both sweat and urine are not significantly different from those in the plasma (Schoeller *et al.*, 1986*a*). Ignoring fractionation effects produces an overestimate in the rate of carbon dioxide production (e.g. 7% in a laboratory mouse, Lifson & McClintock 1966).

Assumption 6: water or carbon dioxide do not enter the animal across the skin or lung surfaces

Entry of water or carbon dioxide can occur across the skin and lung surfaces. Overestimates in the rate of carbon dioxide would be dependent on the rate of turnover of inspired carbon dioxide, which in most cases is inconsequential (Lifson & McClintock, 1966). In an analogous manner, the rate of water turnover will be increased by the entry of unenriched inspired water vapour, which could be quite significant under humid conditions. But as this would affect the turnover of Oxygen-18 and Deuterium equally it would not cause an error in the figure for carbon dioxide production (Lifson & McClintock 1966, Nagy 1980). Certain unusual, but not improbable conditions, such as high ambient carbon dioxide concentrations (3.4%) combined with low humidity (3.8 mg H_2O/L air) can counterintuitively, produce huge errors in the carbon dioxide production rate (+81%, Nagy, 1980). Evaluation of the doubly-labelled water (DLW) technique with a mammal under humid conditions, contrary to expectation, produced estimates of both water turnover and r_{CO_2} that were in close agreement with figures obtained using an energy balance method (Gettinger 1983).

In summary, these points show the DLW method to be a robust technique, which in most situations, is affected to only a minor extent by the invalidity of any one of the six inherent assumptions.

Field protocol

Success in obtaining a DLW measurement with free-living animals relies on the ability to recapture an individual after the experimental period (24, 48, or more hours). Knowledge of the species habits are essential in establishing an effective trapping technique. Permits for catching and handling wild birds must be obtained from the British Trust for Ornithology (Beech Grove, Tring, Herts., HP23 5NR) and a licence (Animals (Scientific Procedures) Act 1986) from the Home Office must be obtained to use the DLW water technique. A checklist of the equipment required to collect the samples for a DLW measurement is provided in Appendix 2.

Having captured a subject it is weighed and marked. A suitable dose of $D_2^{18}O$ (see later) is loaded into a 1 ml disposable syringe. The bird is injected intraperitoneally and placed in a cloth bag, having made a note of the injection time. Care is taken when withdrawing the needle to ensure that the injectate is not lost via the puncture. Measurements, including notes on the age and sex of the individual are usually obtained during the equilibration period (1 hour for birds up to 70g, Williams 1985). The use of prepared field sheets is recommended as these ensure that uniformly complete data sets are obtained, especially when several subjects are processed simultaneously.

Eight to ten blood samples of 5–10 µl each are obtained using graduated micro-pipettes (Vitrex). Blood is collected from a puncture of the femoral vein just above the 'ankle' joint (tibiotarsus/tarsometatarsus), effected using a sterile hypodermic needle. A less preferred site, as the feathers are easily soiled, is the ventral side of the carpal wing joint. The small droplet of blood that emerges should be collected quickly by capillary action. Blood must immediately be shaken to the centre of the capillary. The time is noted when the samples have been collected, and the blood flow is staunched by gentle pressure on the wound after cleaning with sterile cotton wool moistened in alcohol. Having replaced the subject in a cloth bag, the capillaries are sealed. A butane gas torch fitted with a fine flame nozzle is used to heat the ends of the capillary, which are then drawn-off and sealed (Fig. 1). The importance of this aspect is often overlooked by those new to the technique, but it is critical because an imperfectly sealed capillary allows the contents to fractionate and the sample is then useless. The capillaries should be sealed as quickly as possible after taking the blood, by an accomplice if this is feasible. A note indicating details of the individual, the date and time of the last blood sample are placed together with the sealed capillaries in an unbreakable tube. The subject is then removed from the cloth bag and its wound examined and treated with antiseptic. The bird is released and note is made of the time. In most cases the subject returns to its normal activities soon after release, although some individuals may abandon their normal routine for up to a day. We have had no known cases of complete desertion, or of individual mortality as a result of using this technique.

After 24 or 48 hours the individual is recaptured and the second set of blood samples is obtained. If the individual is recaptured at 24 hours, an additional sample at 48 hours will provide a second estimate of daily energy expenditure. During the intervening period (isotope turnover time), weather variables, and if possible the time invested in various activities by the individual are recorded. This information is used to interpret variability in the ADMR figures.

The blood samples may be kept for an indefinite period before analysis, although it is advisable to ensure that they are kept cool to forestall rotting of the blood. The absolute sample size (5–10µl) is not critical because isotope ratios and not absolute isotope abundance are determined during analysis. However, the precision of the analyses is improved when the size of the samples is consistent.

Fig. 1. Procedure for sealing a blood sample in a glass capillary. Special care is necessary to ensure a good seal at the end of the capillary that is contaminated with blood. The flame generates a large amount of water that should not be allowed to contaminate the sample. It is possible to achieve complete success at sealing with practice, which is essential to avoid fractionation during storage. Samples should be kept in a cool place (4 °C) after checking the seals under low power binocular microscope.

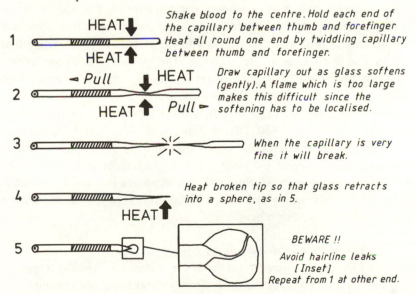

It is most important to ensure that a non-experimental individual is trapped and a blood sample obtained to establish the background levels of oxygen-18 and deuterium. Urinary or faecal water from the subject, obtained prior to the injection of isotopes can sometimes be collected as a preferable alternative.

Processing of body water samples

Having obtained background, initial, and final enriched body water samples, the next stage involves determination of the minor isotope (label) concentrations. This is achieved by preparing gases from the body water samples, which are then analysed using isotope ratio mass spectrometry (IRMS). The ratio of the minor isotope (e.g. deuterium) to the major isotope (e.g. protium) is measured, so it is critical that this ratio does not change between the time the sample is obtained and when it is analysed on the mass spectrometer (i.e. no fractionation). To facilitate this, and prevent contamination, the preparation of gases from body water samples is performed under vacuum (10^{-2} mbar). Exact quantification of the body water sample size (approx 5μl of blood) is unnecessary because the isotope concentration is established from the ratio of heavy to light isotope.

As an alternative to the use of deuterium and oxygen-18 as the isotopes in the doubly-labelled water method, it is possible to use tritium with oxygen-18 (Nagy, 1983). The label concentrations can then be established using liquid scintillation counting. A prerequisite in the case of oxygen-18 involves proton activation to produce gamma emitting fluorine-18 (Wood *et al.*, 1975, Nagy, 1983). In some countries, including Britain, the use of tritium creates problems associated with the release of radio-active materials in the environment. In addition, the proton activation technique demands higher doses of the oxygen-18 isotope than are required when IRMS is used to measure the isotope concentration, and the cost of oxygen-18 enriched water, especially with large animals can limit the scope of an investigation (i.e. £30–140 for 5 mls of 20 atom% $H_2^{18}O$). Sources of error during the preparation of blood samples for liquid scintillation counting are discussed by Nagy (1983). The most important being a pipetting error when measuring tritium abundance and uneven irradiation of $H_2^{18}O$ due to traces of organic material present in the sample during its exposure to the cyclotron beam.

Preparation of deuterium/hydrogen gas

The most common method for obtaining hydrogen gas from aqueous samples involves the reduction of water using hot uranium (Wong

& Klein, 1986). This is undertaken in three stages on the high vacuum sample preparation line illustrated in Fig. 2.

Stage 1. A sample in a sealed capillary is placed through the eye in the handle of a breaker tube, which is then attached to the line. Section A is evacuated (10^{-2} mbar) and isolated. A dewar of liquid nitrogen (-196 °C) is placed on trap 1. The breaker assembly is then heated and the capillary broken. Water in the blood vapourises and freezes down into trap 1. This process is monitored by the Pirani gauge (vacuum sensor), which indicates maximum recovery of the vacuum after about five minutes. The sample is left for a further ten minutes to avoid fractionation errors and then the non-condensible gases are pumped away.

Stage 2. Trap 1 is now isolated under vacuum with the pure ice sample. It is opened to the evacuated section B, which comprises the uranium furnace at 800 °C, a cold trap (No. 2), and gas collection vessels. The liquid nitrogen dewar is removed from trap 1 and placed on trap 2 to retain any water that passes through the furnace. An additional dewar of liquid nitrogen is used to cool activated charcoal in an open gas collecting vessel. The water sample sublimates as trap 1 warms to room temperature and it is routed through the

Fig. 2. Vacuum line for the preparation of hydrogen from a body water sample. The line is made of pyrex glass with ground glass stopcocks greased with Apiezon T. Vacuum is achieved by the combination of a two stage rotary pump (atmosphere to low vacuum) and a mercury diffusion pump (low vacuum to high vacuum, 10^{-2} mbar) fitted with a liquid nitrogen fore-line trap.

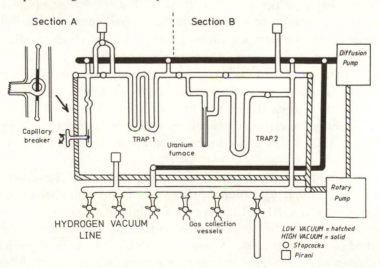

furnace where hydrogen is liberated ($2H_2O + U \rightarrow UO_2 + 2H_2$). The gas is adsorbed onto the cooled activated charcoal (Sackett, 1978) over a ten minute period.

Stage 3. Section B is then isolated, the liquid nitrogen is removed from trap 2 and any water present is directed back through the furnace. After a further ten minutes for this second passage, the gas collection vessel is sealed. During this period Stage 1 is undertaken with the next sample. This method of preparation allows the production of 12 samples per working day.

Samples from a single individual are processed in duplicate to ensure that analysis errors are rapidly identified and additional samples processed. Initial (high enrichment), final (low enrichment), and background (natural abundance) samples are processed in separate batches to avoid problems associated with furnace memory (Friedman & Woodcock, 1957). At the start of a new batch several samples are flushed through the furnace to 'condition' it to the appropriate enrichment. When large second passes are observed, after approximately 150 samples, it is necessary to recharge the uranium in the furnace (see Wong, Lee & Klein, 1987).

Preparation of carbon dioxide from body water sample

Although the ^{18}O:^{16}O ratio can be measured in both oxygen and water, it is more usual to determine this ratio from a sample of carbon dioxide. Oxygen is a difficult gas to handle due to its corrosive nature. The mass spectrometry of water vapour is fraught with problems of sample memory that give rise to contamination of successive samples. Carbon dioxide is a useful gaseous form for the measurement of ^{18}O:^{16}O ratios because it can be pumped away quickly with no memory effect, it is non-corrosive, and it is simple to purify using cryogenic methods.

Measurement of the ^{18}O:^{16}O ratio is commonly achieved by equilibrating the water sample of unknown enrichment with a volume of known enrichment carbon dioxide (Epstein & Mayeda, 1953, Taylor, 1973). The final enrichment of the carbon dioxide is measured and corrected (Craig, 1957, Taylor & Hulston, 1972) to derive the unknown enrichment of the water sample.

An alternative to the equilibrium technique involves conversion of the body water sample to carbon dioxide via reaction with guanadine hydrochloride (($NH_2)_2C$:$NH.HCl$, Boyer et al., 1961). An updated version of this technique has recently been described and compared with the equilibration method (Dugan et al., 1985, Wong et al., 1987). The method given here is fundamentally similar, but the gas preparation procedures are different.

The carbon dioxide preparation line is illustrated by Fig. 3. As with the hydrogen line, a vacuum is achieved by means of a mercury diffusion pump backed by a double stage rotary pump. The upper and lower halves of the line are separate and simply duplicate the preparation facility. To understand the operations it is best to consider either half as being composed of three sections (A→C). The gases are purified cryogenically. Vacuum changes that occur as a result of gases vapourising and condensing are monitored via the Pirani gauges. Three stages are involved in the preparation of carbon dioxide from body water samples, the latter two are undertaken on the same day.

Stage 1. Preparation of guandine/water reaction tubes. A blood sample is loaded into the capillary breaker and attached to a port (e.g. port 1, Fig. 3) while 50 mg of guanidine hydrochloride in a borosilicate tube (18 cm long, id = 4 mm) is attached to the adjacent port. The section is then evacuated and the tip of the borosilicate tube immersed in a dewar of liquid nitrogen. When the capillary is broken, water is vacuum distilled out of the blood

Fig. 3. Vacuum line for the preparation of carbon dioxide from a body water sample. The details given in caption for Fig. 2 also apply here.

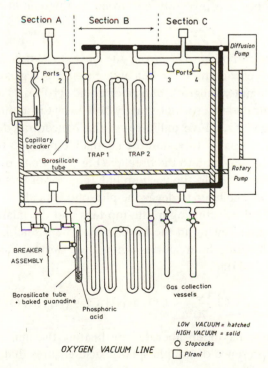

sample and frozen down onto the guanidine hydrochloride. Other samples can be prepared simultaneously on the remaining three double ports (Fig. 3).

Fifteen minutes after breaking a capillary the non-condensible gases are pumped away. The top of the borosilicate tube is then heated with a flame torch and it is drawn off to produce a vacuum-sealed tube containing the guanidine hydrochloride and water from the blood sample. It is labelled with a diamond-tipped pen and put to one side. Between 20 and 30 of these tubes may be prepared in a working day. At the end of the day the tubes are placed in a muffle furnace set at 250 °C for 14 hours. The water reacts with the guanadine to produce ammonia and carbon dioxide (Dugan *et al.*, 1985):

$$NH_2C{:}(NH)NH_2.HCl + 2H_2O \rightarrow 2NH_3 + CO_2 + NH_4Cl$$

However, when this cools below 70 °C the carbon dioxide combines with the ammonia (Dugan *et al.*, 1985):

$$2NH_3 + CO_2 + NH_4Cl \underset{< 70\,°C}{\overset{> 70\,°C}{\rightleftharpoons}} NH_4NH_2CO_2 + NH_4Cl$$

The sealed reaction tubes can probably be stored in this condition indefinitely.

Stage 2. Production of carbon dioxide from the baked guanidine/water reaction tubes. A baked reaction tube is taken and scored with a glass knife to facilitate cracking. Commercial grade phosphoric acid (H_3PO_4 85%) has previously been concentrated to a specific gravity of 1.91 by the addition of phosphorous pentoxide (P_2O_5). Five millilitres of this concentrated phosphoric acid is added to a breaker tube and the guanadine reaction tube is placed inside before it is connected to the breaker (Fig. 3). A pair of these breaker assemblies are attached to the vacuum line (e.g. lower section A, Fig. 3) and evacuated for one hour to degas the phosphoric acid.

The breakers are sealed by screwing in the top taps and the guanadine tubes are then broken by screwing in the lower taps. The assemblies are placed in an oven (80 °C) for 1 hour, during which time carbon dioxide is liberated (Dugan *et al.*, 1985):

$$NH_4NH_2CO_2 + NH_4Cl \xrightarrow[70\,°C]{H_3PO_4} CO_2 + NH_4Cl + (NH_4)_3PO_4$$

Once the carbon dioxide has been liberated by heating the ammonium complex above 70 °C, presence of the phosphoric acid ensures that it will not be recombined due to the temperature drop when the breakers are removed from the oven. On a normal working day 12 breakers prepared in this manner will then be processed according to stage 3.

Stage 3. Purification of the carbon dioxide. A pair of breakers and two gas collection vessels are attached to the vacuum line (Sections A & C respectively, Fig. 3) which is then evacuated. Liquid nitrogen (-196 °C) is placed on trap 1 and a dry-ice/acetone mixture (-80 °C) on trap 2. Section A is isolated except for access to the liquid nitrogen trap. One of the breakers is then opened allowing the CO_2 to freeze down in trap 1, section B. Non-condensible gases are pumped away after ten minutes.

Trap 1 is isolated from the breakers (Section A), connected to trap 2, and opened to a gas collection vessel (Section C). The liquid nitrogen is removed from trap 1 and placed on the empty gas collection vessel. Immediately, a second dry-ice/acetone mixture is placed on trap 1 so that the carbon dioxide frozen here vapourises and passes through trap 2 before freezing down in the gas collection vessel. Replacing the liquid nitrogen on trap 1 with a dry-ice mixture retains any contaminants, while trap 2 acts as a fail-safe purifier to ensure that only the carbon dioxide passes through to the gas collection vessel.

Measurement of isotope concentrations using isotope ratio mass spectrometry

In our laboratory, isotope ratios of prepared gases have been measured using either a Micromass 602 (D/H ratios) or a Micromass 903 ($C^{18}O^{16}O/C^{16}O^{16}O$ ratios). Recently these machines have been superceded by a Sira 9 (D/H) and a Sira 10 ($C^{18}O^{16}O/C^{16}O^{16}O$) with autorun facilities. It has been the practice to prepare gases from blood samples of 5µl, which yield 1cm^3 of hydrogen gas, or 3 cm^3 of carbon dioxide. These are sufficient to achieve the recommended major beam currents of 4×10^{-9} A for D/H analysis ($\sim 6 \times 10^{-8}$ mbar), and 6×10^{-9} A for $C^{18}O^{16}O/C^{16}O^{16}O$ analysis ($\sim 4 \times 10^{-8}$ mbar).

Isotope ratio mass spectrometers are designed for high precision measurements of abundance ratios of light elements (H, C, N, O, & S). It is therefore necessary to recognise a number of machine correction factors, as outlined in a recent review by Fallick (1984). Some aspects, such as background peaks, and the pressure-dependent interference of triatomic hydrogen (H_3^+) are corrected on modern instruments by the machine software.

Isotope ratio measurements are expressed as delta (δ) values in parts per thousand (per mil, ‰) relative to the working standard:

$$\text{Delta raw value} = ((R_{sample}/Rws) - 1) \times 1000$$

where R_{sample} and Rws are the ratios of the heavy isotope to the light isotope (e.g. D/H) of the sample and the working standard respectively. This delta (raw) value is then corrected (= normalized) so that it is expressed in relation to an internationally accepted standard (e.g. V-SMOW). The normalised delta per mil values are converted into atom per cent or parts per million (ppm) as indicated by Wong et al. (1986), except that their equation should be:

$$^{17}R = (^{18}R/^{18}R_{V-SMOW})^{1/2} \times {}^{17}R_{V-SMOW}$$

The international standards used for normalising the raw delta values are restricted to the range of natural abundance (e.g. SLAP = -55.5‰SMOW, and -428‰.SMOW Gonfiantini, 1978; SMOW = 2005.2 ppm ^{18}O, 155.9 ppm D). To characterise machine responses outside this range enriched standards were prepared gravimetrically using concentrates of D_2O and $H_2^{18}O$ diluted with double-distilled water of known isotopic composition. These enriched working standards were run in conjunction with a batch of samples to correct for inconsistencies of machine response at high enrichments. The standardization of enriched IRMS is an area of current research, most notably with the development of international enriched standards, and the determination of machine correction factors (Fallick, 1984). These aspects are especially important to improve the precision when measuring highly enriched gases. The latter are common for DLW studies with small homeotherms, because their rapid turnover rate of body water necessitates the use of high initial isotope concentrations.

Sensitivity analyses

A sensitivity analysis on the estimated rate of carbon dioxide production (r_{CO_2}, mMol h^{-1}) was undertaken for equation (1), the parameters of which were mean values obtained from 26 DLW studies using birds of less than 100 g body mass. Most of these involved individuals of approximately 20 g, so this is used as the standard body mass. Using the parameter values given in Table 1, standard values for r_{CO_2} are either 5.44 mMol CO_2 h^{-1} for the 24 hour period, or 4.51 mMol CO_2 h^{-1} for 48 hours. Sensitivity was investigated by changing a single parameter by the amount indicated in Table 1, and observing the magnitude and direction of the change in the standard r_{CO_2} value. Variations in the isotope concentrations given in Table 1 are all increases, but decreases of the same magnitude produce identical changes in r_{CO_2}, but with the opposite sign.

Table 1. *Sensitivity analysis using typical values for parameters in the equation one to calculate the rate of carbon dioxide production (rCO_2, mMol h^{-1}).*
Standard rCO_2 values were calculated for both 24 and 48 hours using the corresponding parameter values given in the table. Over 24 hours for a 20 g bird this was 5.44 mMol h^{-1}, and for 48 hours it was 4.51 Mol h^{-1}. The scale of variation in the isotope concentrations (ppm) correspond to a maximum acceptable 2% discrepancy between the two delta per mil (‰) values obtained for each sample (see text). The effect of this variation on the standard rCO_2 value is indicated in the last column.

Variable	Time period	Standard parameter value	Scale of variation in parameter	Effect on standard rCO_2
Initial oxygen-18	24	4573 ppm	+1.6%	+6.8%
concentration	48	4441 ppm	+1.6%	+4.2%
Final oxygen-18	24	2568 ppm	+0.3%	−3.5%
concentration	48	2155 ppm	+0.1%	−2.5%
Initial deuterium	24	988 ppm	+1.7%	−5.3%
concentration	48	965 ppm	+1.7%	−3.2%
Final deuterium	24	429 ppm	+1.4%	+5.7%
concentration	48	250 ppm	+0.8%	+3.2%
Natural abundance	24	2008 ppm	Min: 1992.9	−5.3%
of oxygen-18	24	2008 ppm	Max: 2021.1	+4.8%
	48	2008 ppm	Min: 1992.9	−14.3%
	48	2008 ppm	Max: 2021.1	+15.9%
Natural abundance	24	150 ppm	Min: 145.2	+3.1%
of deuterium	24	150 ppm	Max: 154.6	−3.2%
	48	150 ppm	Min: 145.2	+7.0%
	48	150 ppm	Max: 154.6	−7.5%
Body water	24/48	0.63	18 g=0.66	−5.4%
fraction	24/48	0.63	22 g=0.61	+6.1%

Variability in isotope concentration

It will be recalled that the sampling procedure involves encapsulation of eight to ten, 5μl aliquots of blood in separate capillaries. A pair of these capillaries are processed and the mean delta value used to determine the isotope concentration. There are many reasons why the two capillaries may not yield identical delta values, including, fractionation during the sealing of the capillaries, an imperfect seal in one of the capillaries, and analytical errors during the preparation of the enriched gas. It is our practice to repeat the sample analysis if the delta values for a pair of capillaries exhibit a difference of more than 2% of their mean value. Experience has shown that the mean figure obtained by increasing the number of samples processed is not altered significantly unless the first two samples differed by more than the 2% criterion. Most samples agree to within 1% or less.

The scale of variation in isotope concentrations (ppm) given in Table 1 correspond to our criterion of a maximal 2% discrepancy in delta values. This variability is due to two aspects: (1) variation between two aliquots of the same blood sample, and (2) reduced precision of mass spectrometry at high enrichment levels (e.g. 1.6% variability in initial ^{18}O concentration compared to only 0.1–0.3% variability in final ^{18}O concentration). The latter is currently receiving corrective attention after acquisition of machines dedicated to enriched work that will operate with enriched reference gases and achieve the precision (0.1–0.01%) normally associated with isotope ratio mass spectrometry (Craig 1957, Terwilliger 1977).

The first point to note from Table 1 is that an overestimate in initial oxygen-18 concentration produces an overestimate in r_{CO_2}, but an overestimate in the final concentration produces an underestimate in r_{CO_2}. This situation is reversed when the deuterium concentrations are considered. Random errors arising in the determination of the isotope concentration will therefore tend to cancel. Errors in the determination of the isotope concentrations for an experimental period of 48 hours have a smaller effect on calculated r_{CO_2} than when the experimental period is shorter. In our laboratory, the maximum extent of variability on calculated rate of carbon dioxide production (±6%, Table 1) due to an error in the determination of one isotope concentration is of the same order as that obtained in validation experiments that employ comparisons between DLW measurements and respiratory gas analysis (Nagy 1980, see later). However, greater variability in the rate of carbon dioxide production, than indicated in Table 1 can occur if errors in isotope determination are compounded. For example, the worst error would be produced by an overestimate of the initial oxygen-18 and an underestimate of the final oxygen-18 concentration, together with an underestimate of the initial deuterium concentration and an overestimate of the final deuterium concentration. In this situation the 24 hour rate of carbon dioxide production would be in error by +22% and the 48 hour rate by +13%.

Variation in background isotope abundance

Unlike radio isotopes, stable isotopes of the non-typical species are relatively common in the environment, which means that tracer studies with the latter must take into account background levels. The fractional turnover rate of an isotope is the difference between the logarithms of initial and final excess abundance divided by the turnover time (equation 2, Fig. 4). Subtraction of a constant background level from the initial and final isotope concentrations has a much greater effect on the logarithm of the final excess abundance than on the logarithm of the initial excess abundance (Fig. 4). In an analogous manner, small differences in the background level due to natural fluctuations will have a disproportionate effect on the final excess concentration causing significant changes in the turnover rate (i.e. the slope).

The range of natural abundance values given in Table 1 has been obtained from a variety of temperate bird species. Variability in background abundance caused by interspecific, habitat, and temporal differences are currently under consideration (P. Tatner – in preparation). For the present purpose it is sufficient to note that the error which arises in r_{CO_2} by failing to take into account the correct background level may be as high as 16%. The extent of this error is more evident as final isotope concentrations approach background. However, a positive correlation between natural abundance levels of oxygen-18 and deuterium (P. Tatner – in preparation) may reduce this source of error.

Fig. 4. The effect of correcting isotope concentration for background abundance on calculated fractional turnover rate. The example shows an initial deuterium concentration of 2490 ppm (excess = 2335 ppm) which drops to 200 ppm (excess = 45 ppm) over 48 hours, hence Kd = 0.08227 h^{-1}. Background deuterium concentration is assumed to be 155 ppm. Solid line is the uncorrected situation, and the broken line represents correct fractional turnover rate based on excess figures.

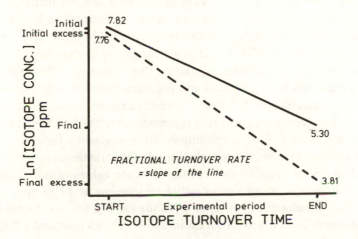

Body water fraction

Relationships between fresh mass and body water content determined by carcass desiccation have been established for birds between 10 and 200 g. The average proportion of fresh body mass accounted for by water is 0.63 (SD = 0.92, $n = 7$ species). This proportion may change with body mass, so that for an 18 g robin *Erithacus rubecula* it is 0.66, and 0.61 for a 22 g individual (Tatner, unpub.). The lower proportion for heavier birds is due largely to the presence of fatty material that contains no water. Incorrect estimation of the body water fraction can cause an error of up to 6% in r_{CO_2} (Table 1).

Fractional turnover rate

The expression for fractional turnover rate (equation 2) may be rewritten as:

$$K_x(h^{-1}) = \frac{\ln (\text{excess initial/excess final})}{t} \tag{3}$$

where the subscript x may be either 'o' (^{18}O) or 'd' (Deuterium). In terms of excess initial this is:

$$K_x(h^{-1}) = \frac{\ln (\text{excess initial}/(Z \times \text{excess initial}))}{t}$$

$$= \frac{\ln (1/Z)}{t} \tag{4}$$

where 'Z' is the proportion of the excess initial that is equal to the excess final enrichment. If $Z = 0.5$, the excess final enrichment has dropped to 50% of the initial excess enrichment and the time taken represents the biological half-life of the stable isotope, at the fractional turnover rate in question. The biological half-life represents the time taken for the excess concentration of the isotope in the body to be diluted to half its initial concentration, and NOT a period of isotope decay as suggested by the use of this term when considering radioactive isotopes. The relationship between turnover time (t, hours) and fractional turnover rate (K_x) is given in Table 2, together with some extreme examples from avian studies. Oxygen-18 fractional turnover rate is considered in the following discussion because it is greater, since it encompasses the deuterium turnover.

As equation (4) indicates, determination of the fractional turnover rate is dependent on the relative and not the absolute values of initial and final excess abundance. So an excess initial enrichment of 49.29 ppm with an excess final enrichment of 10 ppm gives the same fractional turnover rate (0.06646) over 24 hours as an excess initial of 9857.05 ppm and a 2000 ppm

Table 2. *Calculation of the oxygen-18 turnover time, and some examples illustrating the range in free-living birds. The value 'Z' is the proportion of the excess initial remaining after time 't'. The turnover time is calculated using the fractional turnover rate (K_x), with the subscript indicating either oxygen-18 (K_o) or deuterium (K_d). For example, 10% of the excess initial oxygen-18 isotope remains after 20.63 hours, given the maximum fractional turnover rate of oxygen-18. Maximum K_o is a mean for breeding Blue tits* Parus caeruleus *(body mass = 11.7 g), and the minimum K_o is that found for breeding Black guillemots* Cepphus grylle *(body mass = 410.4 g). The mean K_o is based on 26 studies including both breeding and wintering stages, for 11 species with a range of body mass from 12 to 75 g.*

Z	Turnover time (t)	Turnover time (hours)		
		Maximum K_o (0.111 64)	Minimum K_o (0.031 08)	Mean K_o (0.067 23)
0.9	0.105/K_x	0.94	3.38	1.56
0.8	0.223/K_x	2.00	7.18	3.32
0.7	0.357/K_x	3.20	11.49	5.31
0.6	0.511/K_x	4.58	16.44	7.60
0.5	0.693/K_x	6.21	22.30	10.31
0.4	0.916/K_x	8.20	29.47	13.62
0.3	1.204/K_x	10.78	38.74	17.91
0.2	1.609/K_x	14.41	51.77	23.93
0.1	2.303/K_x	20.63	74.10	34.26

excess final. For economic reasons (5 ml $H_2^{18}O$ costs £30–140) we aim to use the lowest possible initial enrichment of oxygen-18. This aspect is constrained by the possibility of losing a result for an individual with a higher than average fractional turnover rate in which the final enrichment becomes indistinguishable from the background. Our results from 25 avian studies indicate that when 95% confidence limits are applied to the mean oxygen-18 fractional turnover for a sample, this range extends from ±5% to ±50% of the mean value. We found that few results were lost when the average final excess oxygen-18 concentration was approximately 500 ppm for a 24 hour period, or about 150 ppm over 48 hours.

Nagy (1980) has drawn attention to the significance of turnover time in relation to variability in the r_{CO_2} estimate caused by errors in the measurement of isotope concentration. To investigate this aspect, equation (4) is modified to account for a +1% error in the determination of initial excess concentration:

$$K_x = \frac{\ln\left((\text{excess initial} + 0.01 \times \text{excess initial})/Z \times \text{excess initial}\right)}{t}$$

$$= \frac{\ln(1.01/Z)}{t} \tag{5}$$

or a −1% error in initial excess concentration:

$$K_x = \frac{\ln(0.99/Z)}{t} \tag{6}$$

For a −1% error in the final excess concentration:

$$K_x = \frac{\ln\left(\text{excess init}/((Z \times \text{excess init}) - (0.01 \times Z \times \text{excess init}))\right)}{t}$$

$$K_x = \frac{\ln(1/0.99 \times Z)}{t} \tag{7}$$

Thus an error of −1% in the final excess concentration has practically the same effect as an error of +1% in the initial excess concentration. This is also true when comparing a positive error of 1% in final excess concentration with a negative error of 1% in the initial excess concentration.

The effect of error in isotope measurement, for a given oxygen turnover rate (e.g. mean $K_o = 0.064\,307$) over different experimental periods was

Fig. 5. The effect on the rate of carbon dioxide production (r_{CO_2}) caused by a 1% error in the estimation oxygen-18 excess concentration, in relation to the extent of isotope turnover (Z). When the fractional turnover rate of oxygen-18 is 0.0643, the excess isotope concentration has dropped to 30% of its initial value after 19 hours.

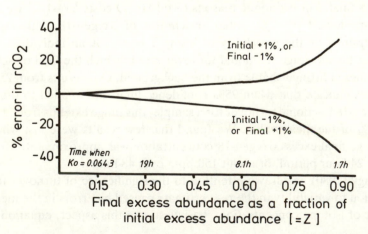

investigated. The turnover times corresponding to values of Z (0.1–0.9) were calculated for a given oxygen-18 fractional turnover rate using equation (4). Equations (5) & (6) were then solved to derive oxygen-18 fractional turnover rates that result from either a $+1\%$ (equation 5) or a -1% (equation 6) error in initial excess concentration. Deuterium fractional turnover rate was held constant, and the percentage error in r_{CO_2} at times from 10 to 90% of the initial enrichment were calculated (Fig. 5).

Using the mean oxygen-18 fractional turnover rate (0.06431) it can be seen that when the experimental period is as short as 1.7 hours ($Z = 0.9$) a 1% error in any of the excess isotope enrichments will produce a 37% error in r_{CO_2} (Fig. 5). The extent of this error decreases rapidly with a protraction of the turnover period, so that by 25.4 hours ($Z = 0.2$, or about two biological half-lives) the 1% error in isotope enrichment causes only a 2.4% error in r_{CO_2}. In general, errors in r_{CO_2} caused by imprecise excess isotope determination can be minimised by extending the experimental period (= turnover time, Fig. 5). Using a range of fractional turnover rates with this sensitivity analysis it can be demonstrated that the precision of isotope measurement is more critical for lower turnover rates.

The average fractional turnover rates of breeding birds (9 species) are

Fig. 6. Average oxygen-18 fractional turnover rate (Ko, hr^{-1}) in relation to body mass (g) for breeding birds. The symbols indicate species: bt = blue tit *Parus caeruleus*, sf = spotted flycatcher *Muscicapa striata*, r = robin *Erithacus rubecula*, gt = great tit *Parus major*, wh = wheatear *Oenanthe oenanthe*, cs = common sandpiper *Tringa hypoleucos*, dp = dipper *Cinclus cinclus*, rp = ringed plover *Charadrius hiaticula*, bg = black guillemot *Cepphus grylle*. Statistics for the regression: F = 56.6, $P < 0.001$, $r^2 = 85.4$.

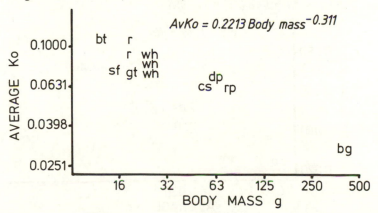

related to their body mass (g) in an exponential fashion (Fig. 6). The following expressions characterize the relationships:

$$\text{average } K_o = 0.2214 \text{ body mass}^{-0.311}, r^2 = 85.4\% \tag{8}$$

$$\text{average } K_d = 0.1740 \text{ body mass}^{-.317}, r^2 = 84.8\% \tag{9}$$

For breeding birds, the fractional turnover rates of both isotopes decline as an inverse function of approximately the third root of their body mass. Since turnover rate is the inverse of turnover time, the latter is approximately proportional to the third root of body mass. This is of interest because metabolic rate is proportional to body mass with an exponent of about 0.7, so mass specific metabolic rate is proportional to body mass with an exponent of ~0.3. Hence the turnover times of both oxygen-18 and deuterium isotopes in breeding birds are directly proportional to their mass specific metabolic rate. Birds with a low mass specific rate (larger species) have longer isotope turnover times. Using equations (2) and (8) it is possible to derive a relationship that predicts the oxygen-18 biological half-life for breeding birds, as a function of their body mass:

$$\text{biological half-life of }^{18}O \text{ (h)} = 3.13 \text{ body mass}^{0.311} \tag{10}$$

The data for wintering birds are more restricted in terms of body mass range, but comparing species of approximately 20 g body mass, the mean winter oxygen-18 fractional turnover rate (0.05279, SD = 0.00739, $n = 7$) was 40% lower than the mean value for breeding birds (0.08759, SD = 0.01072, $n = 6$).

Fig. 7. Relationship between the average fractional isotope turnover rates over a range of breeding birds. Symbols indicate species as in Fig. 6. Statistics for the geometric mean regression: $F = 577.8$, $P < 0.001$, $r^2 = 98.3$; the constant in the equation was not significantly different from zero ($t = 0.49$, NS).

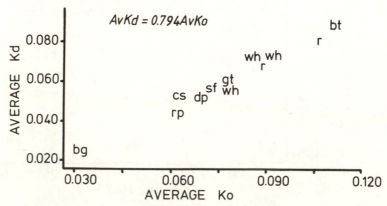

For breeding birds there is a strong correlation between the average oxygen-18 fractional turnover rate and the corresponding deuterium fractional turnover rate (Fig. 7):

$$AvK_d = -0.00183 + 0.794 \, AvK_o, r^2 = 98.3, n = 12, P < 0.001 \quad (11)$$

The constant in the geometric mean regression equation is not significantly different from zero (t = 0.49, NS), so the average K_d is 79.4% of K_o. This result could also have been predicted from the two expressions relating fractional turnover rate to body mass, since their exponents were not significantly different (t = 0.11, P > 0.25), and the ratio of the constants indicates that the average K_d is 78.6% of K_o. This point is remarkable for the fact that in breeding birds at least, *average* water turnover remains a constant fraction of the oxygen-18 turnover throughout the body mass range considered (blue tits = 11.7 g to black guillemots = 410.4 g).

Isotope dosages for the doubly-labelled water technique

The injectate is prepared by adding 0.34 mls of 99.7 atom% deuterated water to the commercially packaged 5 ml vials of 20 atom% $H_2^{18}O$ (Amersham Int.). This produces a mixture of approximately 18.7 atom% ^{18}O with 6.4 atom% deuterium.

To calculate a suitable isotope dosage for a DLW measurement three parameters must be considered: the turnover rate of the isotope, the experimental period, and the final excess isotope concentration. The intention is to use these parameters to predict a suitable excess initial isotope concentration from which the dose can be calculated. There are no hard and fast general rules as to the correct values for the parameters, only guidelines. The parameters are related according to the expression shown below (modified from equation 2):

$$\text{excess initial enrichment} = \text{excess final enrichment} \times e^{K_o \cdot t} \quad (12)$$

where 'e' is the base of natural logarithms.

As we have already noted, the oxygen-18 fractional turnover rate is a function of both the birds' body mass and their environmental/behavioural situation. Where a comparison is possible (birds of ~20 g), it has been shown that the average K_o during breeding is much higher than during the winter. The higher breeding value may be predicted from the bird's body mass (equation 8).

The most suitable time period is dictated by the fractional turnover rate, being longer for larger birds (Table 2). However, the maximum time period for smaller species (20 → 100 g) is in most cases 48 hours, as this roughly corresponds to 5 half-lives, and it allows recapture attempts at 24 h and 48 h.

The final excess isotope concentration must be significantly above background, and is therefore the most critical parameter. If an average final excess concentration of only 10 ppm is chosen, results for individuals with a high metabolic rate/water turnover will be lost, yielding a biased sample. The overall average final excess oxygen-18 concentration for a 48 hour period from our studies is circa 150 ppm. Using this value for smaller species (<20 g) will demand very high initial enrichments, in which case we suggest that the lowest acceptable final excess should be 40 ppm, and retraps should be attempted principally after 24 hours. Our initial oxygen-18 concentrations are of the order 1500–3000 ppm excess (Fig. 8). Since much of the oxygen-18 turnover (78%) is due to water turnover, it should be assumed that species with a high water content in the diet will require a higher initial isotope enrichment. We have previously noted that individual variability in K_o within a single study encompasses the range from $\pm 5\%$ to $\pm 50\%$ of the mean value. In cases where a high turnover rate of isotope is anticipated, a pilot study employing an initial value of 3000 ppm excess of oxygen-18 is advisable.

Having derived a suitable estimate of the initial oxygen-18 excess

Fig. 8. Average initial oxygen-18 concentration in relation to the injected dose of isotope (18.7 atom% ^{18}O with 6.4 atom% D). Symbols refer to species: bf = bullfinch *Pyrrhula pyrrhula*, ps = pacific swallow *Hirundo tahitica*, be = blue throated bee-eater *Merops viridis*, d = dunnock *Prunella modularis*, m = thick-billed murre *Uria lomvia*, bt = blue tit *Parus caeruleus*, sf = spotted flycatcher *Muscicapa striata*, r = robin *Erithacus rubecula*, gt = great tit *Parus major*, wh = wheatear *Oenanthe oenanthe*, cs = common sandpiper *Tringa hypoleucos*, dp = dipper *Cinclus cinclus*, rp = ringed plover *Charadrius hiaticula*, bg = black guillemot *Cepphus grylle*. Statistics for the regression: $F = 114$, $P<0.001$, $r^2 = 84.5$.

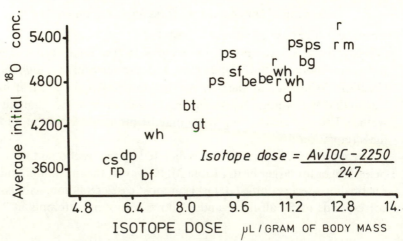

abundance, the value of 2005.1 (mean of 122 measurements of natural abundance over 15 species) is added to produce a figure for the initial oxygen-18 concentration. The dosage values we have used for 23 studies in relation to the initial abundance are shown in Fig. 8. This produces an expression for deriving dosage given the initial enrichment:

$$\frac{\text{isotope dosage}}{(\mu\text{lg}^{-1})} = \tfrac{1}{247} \text{ (average initial } {}^{18}\text{O conc.} - 2250) \qquad (13)$$

In the light of experience we would increase the dosage for small breeding birds (e.g. blue & great tits) to approximately 10 µl g^{-1}.

The discussion to this point has been concerned only with the oxygen-18 dosage, as this is more critical than that of deuterium. The oxygen-18 isotope dosage calculated using equation (13) will automatically include sufficient deuterium if the injectate mixture, given at the beginning of this section, is employed.

Validation of the doubly-labelled water technique

A representative selection of the DLW validation studies under-taken on a range of animals, from insects to man, have been summarized in Table 3. Validation involves comparisons of either the rate of carbon dioxide production with a gas analysis method, or the estimate of energy expenditure with a measure derived from an energy intake/output method. In the five cases where both validation methods have been undertaken simultaneously, similar discrepancies with the DLW method have been apparent (Table 3).

The mean absolute discrepancy for the DLW method in vertebrate studies is generally less than 10%, but it may be much greater for invertebrate studies. The problems with application of the technique to invertebrates seem to involve a tendency for the hydrogen label to exhibit compartmentation (King & Hadley, 1979). The possibility of high uptake of unlabelled water vapour (Cooper, 1983) may cause significant error if fractionation effects are ignored.

Recent interest in applying the technique to man has stimulated a reappraisal of the basic parameters in the Lifson & McClintock expression (equation 1). The longer experimental period (5–14 days) and relative ease of sample collection has encouraged the analysis of more frequent body water samples, which facilitates the application of curve fitting techniques (Klein *et al.*, 1984, Haggarty *et al.* in press). Differences in the size of exchangeable hydrogen and oxygen pools as compared to the body water pool, are taken into account and the effect of fractionated water loss has been reappraised (Coward *et al.*, 1985, Schoeller *et al.*, 1986a, 1986b).

Table 3. *Doubly-labelled water validation studies*

Species	n	Mean body mass (g)	Predicted metabolism (Method G, F)	Method	Mean % error in DLW estimate		Range	CIF	Source
					Algebraic	Absolute			
Man (adults)	5	71600	2830 kcal/day	G F	+5.9 +8.0	8.5 10.2	−6.5, +14.1 −5.3, +14.5	Y	Schoeller & Webb, 1984
Man (babies)	4	1635	17.8 LCO_2/day	G F	−1.4 −5.3	4.3 6.4	−4.8, +5.8 −25.7, +13.0	Y	Roberts *et al.*, 1986
Caribou *Rangifer t. granti*	3	93000	0.581 mlCO_2/g/h	G	+11.7	11.7	+4.9, +20.2	Y	Fancy *et al.*, 1986
Reindeer *Rangifer tarandus*	1	54000	0.398 mlCO_2/g/h	G	+1.1	1.1	—	—	
Rat *Rattus rattus*	8	245	1134.5 mMol CO_2	G	+1.8	3.2	−2.2, +10.1	Y	McClintock *et al.*, 1958b
Gopher[a] *Thomomys bottae*	6	127	1.22 mlCO_2/g/h	F	+3.7	6.9	−8.7, +14.5	N	Gettinger, 1983
Kestrel *Falco tinnunculus*	8	193	0.2301 MCO_2/day	G	+2.2	4.4	−5.1, +9.5	Y	Masman & Klaassen, 1987

Starling *Sturnus vulgaris*	4	69.8	3.08 mlCO_2/g/h 124.36 kJ/day	G F	−7.1 −5.1	9.3 8.1	−21.6, +7.0 N −14.4, +8.0	Williams, 1985
Sparrows Four species[b]	13	22.3	4.30 mlCO_2/g/h 53.54 kJ/day	G F	−3.5 −3.1	5.1 9.8	−17.0, +4.2 N −29.0, +16.1	
House martin *Delichon urbica*	4	16.6	2.24 mlCO_2/g/h	G	+3.4	6.6	−5.1, +14.0 Y	Hails & Bryant, 1979
Sand martin *Riparia riparia*	2	13.7	—	G	+4.4	4.4	+2.4, +6.4 Y	Westerterp *et al.*, 1984
Lizard *Sceloporous* sp.	4	7.1	0.167 mlCO_2/g/h	G	+3.2	8.0	−5.7, +18.6 N	Congdon *et al.*, 1978
Scorpion *Hadrurus arizonensis*	5	3.0	0.321 mlCO_2/g/h	G	+36.5	36.5	+11.1, +71.7 N	King & Hadley, 1979
Locust *Locusta migratoria*	9	1.2	41.70 mgCO_2/j 37.72 mg/j	G F	−4.1 +2.6	23.4 5.4	−60.3, +31.9 N −3.6, +13.0	Buscarlet *et al.*, 1978
Beetles *Eleodes* sp.	4	<1.0	0.435 mlCO_2/g/h	G	+14.5	14.5	+12.7, +74.2 N	Cooper, 1983
Cryptoglossa sp.	1	<1.0	0.224 mlCO_2/g/h	G	+25.9	25.9	—	

[a] Undertaken in humid conditions.

[b] song sparrow *Melospiza melodia*, white-throated sparrow *Zonotrichia albicolis*, house sparrow *Passer domesticus*, and savannah sparrow *Passerculus sandwichensis*.

Method: G = Gas analysis, F = Food balance.

CIF: Correction for isotope fractionation, Y = Yes, N = No

Accounting for these aspects offers a slight improvement in the agreement between simultaneous calorimetric and doubly-labelled water measurements of energy expenditure, but it can be shown that individual variation is sufficiently great to preclude identification of a superior mode of calculation (Speakman, 1986).

Application of the technique to ruminants involves consideration of methane production. Loss of the hydrogen label in CH_4 increases the r_{CO_2} estimate (7.6–7.7%), but this is counteracted by losses of CO_2 due to its reduction to methane (Fancy et al., 1986). Problems may also arise due to antler growth when there is a continuous exchange of HCO_3^- and CO_3^{-2} ions between the blood and bony tissue (Fancy et al., 1986).

Although the mean algebraic and absolute discrepancies between DLW estimates and the more traditional methods of measuring energy expenditure are generally acceptable (Table 3), the range of discrepancy for individual measurements indicates that closer investigation of the basic assumptions and further categorization of the sources of error should be undertaken before complete confidence in individual measurements is justified. For example, most studies employing tritium instead of deuterium do not include correction factors for isotopic fractionation (Nagy, 1980, Table 3). It should also be noted that validation studies generally involve animals during their resting phase, whereas the ADMR of most small homeotherms is between two and four times their resting level.

Application of the doubly-labelled water technique to free-living birds

The doubly-labelled water technique was first used, on a large scale, to explore the concept of reproductive effort in birds. It was assumed that estimates of energy expenditure would integrate the costs of a range of processes and behaviours that, individually or cumulatively increased the risk of dying. The chance of maintaining daily energy balance, for example, might be lessened if more effort was devoted to raising offspring. If these or analogous assumptions held, then the theoretical proposition that reproductive effort should be inversely related to residual reproductive value, could be examined (Pianka & Parker, 1975). A study of breeding house martins, *Delichon urbica*, revealed a close match with theory (Fig. 9a); rearing of a second brood involved greater effort (Hails & Bryant, 1979). This study, however, revealed little about the proximate factors that could influence energy expenditure by free-living birds. Subsequently, a study of house martin energy budgets (Bryant & Westerterp, 1980) showed the effects of food supply (Fig. 9b) and brood provisioning rates on energy expenditure (Bryant & Westerterp, 1983a, 1983b). A counter-intuitive

Fig. 9. Graphical representation of some of the main points from avian DLW investigations.

(*a*) Relationship between parental ADMR and metabolic brood mass for house martins. Square symbols are males, triangles are females; closed symbols indicate first broods and the open symbols are for second broods. The lines *a* to *d* represent overall trends for the four categories (Hails & Bryant, 1979).

(*b*) Average daily metabolic rate for breeding house martins in relation to food abundance; solid symbols = brood size of 4, open symbols = other brood sizes (Bryant & Westerterp, 1983*b*).

(*c*) Average daily metabolic rate in relation to the proportion of the day spent in flight for house martins (Bryant & Westerterp, 1980).

(*d*) Relationship between avian metabolism and body mass during sustained flights: open symbols indicate aerial foraging species (Hails, 1979).

(*e*) Avian metabolic rate as a function of body mass: M_{fly} indicates cost during short flights, M_{rest} indicates resting cost, and BMR_p indicates basal metabolic rate during the night resting phase. Symbols 'r' and 's' are DLW measurements ($\pm$ SD) for the robin and starling (Tatner & Bryant, 1986).

(*f*) Daily energy expenditure in relation to the keel length of house martins (data from Table 5, Bryant & Westerterp, 1983*a*).

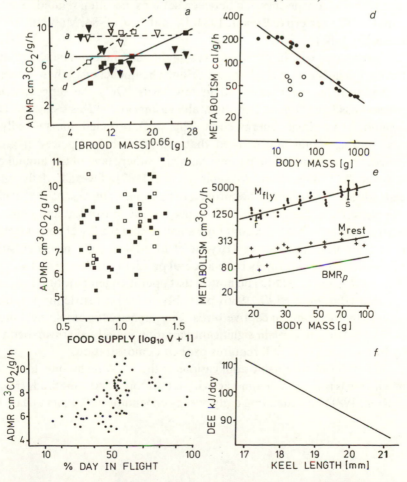

finding was that large individuals had lower total energy costs than small individuals (Fig. 9*f*, Bryant & Westerterp 1982, 1983*b*), although the mechanism underlying this trend is not yet established. Subsequent intra-specific studies have revealed both small and large individuals with higher costs (Bryant & Tatner, 1986, P. Tatner – in preparation). So any general view that free-living metabolism scales to around the 0.7 power of body mass within species, or is even necessarily a *positive* function of body mass, is unfounded (Calder & King, 1974; Walsberg, 1983; Heusner, 1982).

Energy expenditure is markedly affected by changes in activity, but less so by changes in climate that do not themselves influence activity. If the daily routine includes a high cost component, this is likely to be a principal cause of variation in daily energy costs (Bryant & Westerterp, 1980, 1983*a*, Bryant, Hails & Prys-Jones, 1985, Fig. 9*c*) and it may be exploited to establish the cost of these activities. In this way the costs of sustained flight (Hails, 1979, Fig. 9*d*; Bryant, Hails & Tatner, 1984; Masman & Klaassen, 1987), short flights (Tatner & Bryant, 1986, Fig. 9*e*), and diving (Bryant & Tatner, 1988) have been established. Also differences in energy expenditure between conspecifics and between seasons are often related to activity patterns (Westerterp & Bryant, 1984; Bryant *et al.*, 1984; Masman, Daan & Beldhuis, 1987).

The annual cycle in birds involves simple self-maintenance through to territory acquisition and raising offspring. The impact of all these on energy expenditure has been explored in few birds. Only the dipper, *Cinclus cinclus*, has been followed through all seasons and stages using the DLW technique to estimate energy costs. This work revealed unexpectedly low costs during moult and showed that rearing a brood placed a greater demand for energy on the parents than any other stage in the annual cycle (Bryant & Tatner, 1987, Masman *et al.*, 1987). Overall, daily energy expenditure by small birds has proved to be remarkably insensitive to direct climatic effects. Temperature differences between tropical and temperate environments probably account for a small part of the lower costs measured in two tropical insectivores (Bryant *et al.*, 1984). Yet in other studies temperature has only rarely been a useful predictor of energy cost variation (Bryant & Tatner, 1988). The notion that operative temperature is a crucial variable (Buttemer *et al.*, 1986) probably relies on restricting a subject's activity by focussing on captive birds. The capacity for biophysical models of heat balance to explain significant variation in daily energy expenditure by free-ranging birds still requires explicit demonstration.

One of the most valuable applications of the DLW technique in the study of wild birds is to validate time-activity-laboratory (TAL) methods (Mugaas & King, 1981) for estimating daily energy expenditure. Results to date have shown a reasonable correspondence (Williams, 1985, Bryant *et al.*, 1985, Bryant & Tatner, 1986) but further refinement of TAL models using DLW estimates as a standard, offers the chance of reducing the often excessive errors involved in the former (Travis 1983).

Acknowledgements
We should like to thank the following people for their valuable contributions to the development of the technique; Dr C. J. Hails, Dr K. Westerterp, and Dr R. Prys-Jones. The analytical work is undertaken in the Stable Isotope Laboratory, Scottish Universities Research & Reactor Centre and we are most grateful for the invaluable advice and services provided by the staff. Prof. M. Baxter, Dr. A. E. Fallick, Mr T. Donnelly, Mr J. Borthwick, and Mr F. Cornwallis are specifically acknowledged. Some of the data mentioned were collected during collaborative research projects with the following: Dr P. Evans & Dr A. Wood, Dr R. Prys-Jones, Dr P. Grieg-Smith, Dr A. Gaston. This manuscript has been improved as a result of comments from: Dr D. Masman and Dr A. E. Fallick. We are pleased to acknowledge their contributions. The work was supported by a NERC project grant GR3/4185 awarded to DMB.

References
Aschoff, J. & Pohl, H. (1970). Der Ruheumsatz von Vogeln als Funktion der Tageszeit und der Korpergrosse. *Journal für Ornithologie*, **111**, 38–47.

Boyer, P. D., Graves, D. J. Suelter, C. H. & Demsey, M. E. (1961). Simple procedure for the conversion of orthophosphate or water to carbon dioxide for oxygen-18 determination. *Analytical Chemistry*, **33**, 1906–9.

Brody, S. (1945). *Bioenergetics and growth*. New York: Hafner Publishing Company.

Bryant, D. M. & Westerterp, K. R. (1980). The energy budget of the House martin *Delichon urbica*. *Ardea*, **68**, 91–102.

Bryant, D. M. & Westerterp, K. R. (1982). Evidence for individual differences in foraging efficiency amongst breeding birds: a study of House martins *Delichon urbica* using the doubly-labelled water technique. *Ibis*, **124**, 187–92.

Bryant, D. M. & Westerterp, K. R. (1983a). Time and energy limits to brood size in House martins *Delichon urbica*. *Journal of Animal Ecology*, **52**, 905–25.

Bryant, D. M. & Westerterp, K. R. (1983b). Short term variability in energy turnover by breeding House martins *Delichon urbica*: A study using doubly-labelled water ($D_2{}^{18}O$). *Journal of Animal Ecology*, **52**, 525–43.

Bryant, D. M., Hails, C. J. & Tatner, P. (1984). Reproductive energetics of two tropical bird species. *Auk*, **101**, 25–37.

Bryant, D. M., Hails, C. J. & Prys-Jones, R. (1985). Energy expenditure by free-living Dippers (*Cinclus cinclus*) in winter. *Condor*, **87**, 177–86.

Bryant, D. M. & Tatner, P. (1986). The costs of brood provisioning: effects of brood size and food supply. In *Proceedings of the 19th. Congressus Internationalis Ornithologicus*, ed. O. Ouellet, Ottawa: Canada.

Bryant, D. M. & Tatner, P. (1988). Energetics of the annual cycle of Dippers *Cinclus cinclus*. *Ibis*, **130**, 27–38.

Buscarlet, L. A., Pronx, J. & Gerster, R. (1978). Utilisation du double marquage $HT^{18}O$ dans une etude de bilan metabolique chez *Locusta migratoria migratorioides*. *Journal of Insect Physiology*, **24**, 225–32.

Buttemer, W. A., Hayworth, A. H., Weathers, W. W. & Nagy, K. A. (1986). Time budget estimates of avian energy expenditure: physiological and meterological considerations. *Physiological Zoology*, **59**, 131–49.

Calder, W. A. & King, J. R. (1974). Thermal and caloric relations in birds. In *Avian biology*, Vol. **IV**, ed. Farner, D. S. & King, J. R., pp. 259–413. London: Academic Press.

Congdon, J. D., King, W. W. & Nagy, K. A. (1978). Validation of the HT0–18 method for determination of CO_2 production of lizards (Genus *Sceloporus*). *Copeia*, **1978**, 360–2.

Cooper, P. D. (1983). Validation of the doubly-labelled water $H^3H^{18}O$ method for measuring water flux and energy metabolism in tenebrionid beetles. *Physiological Zoology*, **55**, 35–44.

Costa, D. P., Dunn, P. & Disher, W. (1986). Energy requirements of free-ranging Little penguins *Eudyptula minor*. *Comparative Biochemistry & Physiology*, **85**A, 135–38.

Coward, W. A., Roberts, S. B., Prentice, A. M. & Lucas, A. (1985). The $^2H_2^{18}O$ method for energy expenditure measurements – clinical possibilities, necessary assumptions, and limitations. *Proceedings of the 7th. Congress of the European Society for Enteral and Parenteral Nutrition*, Munich.

Craig, H. (1957). Isotopic standards for carbon and oxygen and correction factors for mass spectrometric analysis of carbon dioxide. *Geochimica Cosmochimica Acta*, **12**, 133–49.

Culebras, J. M. & Moore, F. D. (1977*a*). Total body water and the exchangeable hydrogen. 1. Theoretical calculation of non-aqueous exchangeable hydrogen in man. *American Journal of Physiology*, **232**, R54–9.

Culebras, J. M., Fitzpatrick, G. F., Brennan, M. F., Boyden, C. M. & Moore, F. D. (1977*b*). Total body water and the exchangeable hydrogen. 2. A review of comparative data from animals based on isotope dilution and desiccation, with a report of new data from the rat. *American Journal of Physiology*, **232**, R60–5.

Dugan, J. P., Jr., Borthwick, J., Harmon, R. S., Gagnier, M. A., Glahn, J. E., Kinsel, E. P., MacLeod, S. & Viglino, J. A. (1985). Guanadine hydrochloride method for determination of water oxygen isotope ratios and the oxygen-18 fractionation between carbon dioxide and water at 25 °C. *Analytical Chemistry*, **57**, 1734–6.

Epstein, S. & Mayeda, T. (1953). Variation of ^{18}O content of waters from natural sources. *Geochimica Cosmochimica Acta*, **4**, 213–24.

Estep, M. F. & Dabrowski, H. (1980). Tracing food webs with stable hydrogen isotopes. *Science*, **209**, 1537–8.

Fallick, A. E. (1984). A review of instrument correction factors in light element stable isotope ratio mass spectrometry. *Proceedings of the British Mass Spectrometry Society*, **13**, 159–66.

Fancy, S. G., Blanchard, J. M., Holleman, D. F., Kokjer, K. J. & White, R. G. (1986). Validation of the doubly-labelled water method using a ruminant. *American Journal of Physiology*, **251**, R143–9.

Friedman, I. & Woodcock, A. H. (1957). Determination of deuterium/hydrogen ratios in Hawaiian waters. *Tellus*, **IX**, 553–6.

Gettinger, R. D. (1983). Use of doubly-labelled water ($^{3}HH^{18}O$) for determination of H_2O flux and CO_2 production by a mammal in a humid environment. *Oecologia*, **59**, 54–7.

Gonfiantini, R. (1978). Standards for stable isotope measurements in natural compounds. *Nature*, **271**, 534–6.

Haggarty, P., McGaw, G. A., James, W. P. T., Ferro-Luzzi, A., Scaccini, C., Virgili, F. & Franklin, M. (in press). Calculation of isotope flux rates in truly 'free-living' subjects. *Proceedings of the 441st Nutrition Society*, Cambridge.

Hails, C. J. (1979). A comparison of flight energetics in Hirundines and other birds. *Comparative Biochemistry & Physiology*, **63**A, 581–5.

Hails, C. J. & Bryant, D. M. (1979). Reproductive energetics of a free-living bird. *Journal of Animal Ecology*, **48**, 471–82.

Heusner, A. A. (1982). Energy metabolism and body size. 1. Is the 0.75 mass exponent of Kleiber's equation a statistical artefact? *Respiratory Physiology*, **48**, 1–12.

Jungas, R. L. (1968). Fatty acid synthesis in adipose tissue incubated in tritiated water. *Biochemistry (American Chemists' Society)*, **Pa6**, 3708–17.

King, J. R. & Farner, D. S. (1961). Energy metabolism, thermoregulation, and body temperature. In A. J. Marshall, *Biology and comparative physiology of birds*, ed. pp. 215–88. Academic Press: New York.

King, W. W. & Hadley, N. F. (1979). Water flux and metabolic rates of free-roaming scorpions using the doubly-labelled water technique. *Physiological Zoology*, **52**, 176–89.

Klein, P. D., James, W. P. T., Wong, W. W., Irving, C. S., Murgatroyd, P. R., Cabrera, M., Dallosso, H. M., Klein, E. R. & Nichols, B. L. (1984). Calorimetric validation of the doubly-labelled water method for determination of energy expenditure in man. *Human Nutrition: Clinical Nutrition*, **38**C, 95–106.

Lifson, N., Gordon, G. B., Visscher, M. B. & Nier, A. O. (1949). The fate of utilised molecular oxygen of respiratory carbon dioxide studied with aid of heavy oxygen. *Journal of Biological Chemistry*, **180**, 803–11.

Lifson, N. & McClintock, R. (1966). Theory and use of turnover rates of body water for measuring energy and material balance. *Journal Theoretical Biology*, **12**, 46–74.

Luz, B., Kolodny, Y. & Horowitz, M. (1984). Fractionation of oxygen isotopes between mammalian bone phosphate and environmental drinking water. *Geochimica Cosmochimica Acta*, **48**, 1689–93.

Maren, T. H. (1967). Carbon anhydrase: chemistry, physiology, and inhibition. *Physiological Reviews*, **47**, 597–768.

Masman, D. & Klaassen, M. (1987). Energy expenditure for free-flight in trained and free-living Kestrels *Falco tinnunculus*, *Auk* **104**, 603–16.

Masman, D., Daan, S. & Beldhuis, J. A. (1987). Ecological energetics of the Kestrel: Daily energy expenditure throughout the year based on time-energy budget, food intake, and doubly-labelled water methods. *Ardea* **76**, 64–81.

McClintock, R. & Lifson, N. (1958*a*). CO_2 output of mice measured by $D_2^{18}O$ under conditions of isotope re-entry into the body. *American Journal of Physiology*, **195**, 721–5.

McClintock, R. & Lifson, N. (1958*b*). Determination of the total carbon dioxide outputs of rats by the $D_2^{18}O$ method. *American Journal of Physiology*, **192**, 76–8.

Mugaas, J. N. & King, J. R. (1981). Annual variation of daily energy expenditure by the Black-billed Magpie: A study of thermal and behavioural energetics. *Studies in Avian Biology*, No. 5, pp. 1–78, Cooper Ornithological Society.

Nagy, K. A. (1980). CO_2 production in animals: analysis of potential errors in the doubly-labelled water method. *American Journal of Physiology*, **238**, R466–73.

Nagy, K. A. (1983). The doubly-labelled water method: a guide to its use. *UCLA Publ.*, No. **12–1417**, 45 pp., Los Angeles.

Nagy, K. A. & Costa, D. P. (1980). Water flux in animals: analysis of potential errors in the tritiated water method. *American Journal of Physiology*, **238**, R454–65.

Pianka, E. R. & Parker, W. S. (1975). Age specific reproductive tactics. *American Naturalist*, **109**, 453–64.

Prentice, A. M., Coward, W. A. & Whitehead, R. G. (1984). Measurement of energy expenditure by the doubly-labelled water method. *Annual Report of the Nestlé Foundation*, ed. B. Schurch, pp. 61–9.

Roberts, S. B., Coward, W. A., Schlingenseipen, K. H., Norhia, V. & Lucas, A. (1986). Comparison of the doubly-labelled water ($^2HH^{18}O$) method with indirect calorimetry and a nutrient balance study for simultaneous determination of energy expenditure, water intake, and metabolisable energy intake in preterm infants. *American Journal of Clinical Nutrition*, **44**, 315–22.

Sackett, W. M. (1978). Carbon and hydrogen isotope effects during thermocatylic production of hydrocarbons in laboratory simulation experiments. *Geochimica Cosmochimica Acta*, **42**, 571–80.

Schoeller, D. A., Van Santen, E., Peterson, D. W., Dietz, W., Jaspan, J. & Klein, P. D. (1980). Total body water measurements in humans with

^{18}O and ^{2}H labelled water. *American Journal of Physiology*, **238**, R545–65.

Schoeller, D. A. (1983). Energy expenditure from doubly-labelled water: some fundamental considerations in humans. *American Journal of Clinical Nutrition*, **38**, 999–1005.

Schoeller, D. A. & Webb, P. (1984). Five day comparison of the doubly-labelled water method with respiratory gas exchange. *American Journal of Clinical Nutrition*, **40**, 153–8.

Schoeller, D. A., Leitch, C. A. & Brown, C. (1986*a*). Doubly-labelled water methods: *in vivo* oxygen and hydrogen isotope fractionation. *American Journal of Physiology*, **251**, R1137–43.

Schoeller, D. A. Ravussin, E., Schutz, Y., Acheson, K. J., Baertschi, P. & Jequier, E. (1986*b*). Energy expenditure by doubly-labelled water: Validation in humans and proposed calculation. *American Journal of Physiology*, **250**, R823–30.

Schoeller, D. A., Minagawa, M., Slater, R. & Kaplan, I. R. (1986*c*). Stable isotopes of carbon, nitrogen, & hydrogen in the contemporary North American food web. *Ecology of Food and Nutrition*, **18**, 159–70.

Speakman, J. R. (1986). Calculation of CO_2 production in doubly-labelled water studies. *Journal of Theoretical Biology*, **126**, 101–4.

Speakman, J. R. & Racey, P. A. (1987). The equilibrium concentration of oxygen-18 in body water: Implications for the accuracy of the doubly-labelled water technique and a potential new method of measuring RQ in free-living animals. *Journal of Theoretical Biology*, **127**, 79–95.

Tatner, P. & Bryant, D. M. (1986). Flight cost in a small passerine measured using doubly-labelled water: implications for energetics studies. *Auk*, **103**, 169–80.

Tatner, P. (1988). A model of the natural abundance of oxygen-18 and deuterium in the body water of animals. *Journal of Theoretical Biology*, **133**, 267–80.

Taylor, C. B. & Hulston, J. R. (1972). Measurement of oxygen-18 ratios in environmental waters using the Epstein-Mayeda technique. Part 2: Mathematical aspects of the mass spectrometry measurements, the equilibration process, correction terms, and computer calculation Delta-SMOW values. *Institute for Nuclear Sciences Publications*, **557**, pp. 29, Low Hutt, New Zealand.

Taylor, C. B. (1973). Measurement of oxygen-18 ratios in environmental waters using the Epstein-Mayeda technique. Part 1: Theory and experimental details of the equilibrium technique. *Institute for Nuclear Sciences Publications*, **556**, pp. 24, Low Hutt, New Zealand.

Terwilliger, D. T. (1977). An improved mass spectrometric method for determining the hydrogen abundance ratio. *International Journal Mass Spectrometry and Ion Physics*, **25**, 393–9.

Travis, J. (1983). A method for the statistical analysis of time energy budgets. *Ecology*, **63**, 19–25.

Walsberg, G. E. (1983). Avian ecological energetics. In: *Avian Biology*, **7**, ed. Farner, D. S., King, J. R. & Parkes, K. C., pp. 161–220. New York: Academic Press.

Weathers, W. W. & Nagy, K. A. (1980). Simultaneous doubly-labelled water ($^3HH^{18}O$) and time budget estimates of daily energy expenditure in *Phainopepla nitens*. *Auk*, **97**, 861–7.

Westerterp, K. R. & Bryant, D. M. (1984). Energetics of free-existence in swallows and martins (Hirundinidae) during breeding: a comparative study using doubly labeled water. *Oecologia*, **62**, 376–81.

Williams, J. B. (1985). Validation of the doubly-labelled water technique for measuring energy metabolism in starlings and sparrows. *Comparative Biochemistry & Physiology*, **80**A, 349–53.

Wong, W. W. & Klein, P. D. (1986). A review of techniques for the preparation of biological samples for mass spectrometric measurements of hydrogen-2/hydrogen-1 and oxygen-18/oxygen-16 isotope ratios. *Mass Spectrometry Review*, **5**, 313–42.

Wong, W. W., Lee, L. S. & Klein, P. D. (1987). Oxygen isotope ratio measurements on carbon dioxide generated by reaction of microliter quantities of biological fluids with guanidine hydrochloride. *Analytical Chemistry*, **59**, 690–3.

Wood, R. A., Nagy, K. A., MacDonald, N. S., Wakakuwa, S. T., Beckman, R. J. & Kaaz, H. (1975). Determination of oxygen-18 in water contained in biological samples by charged particle activation. *Analytical Chemistry*, **47**, 646–50.

E. GNAIGER, J. M. SHICK and J. WIDDOWS

Metabolic microcalorimetry and respirometry of aquatic animals

Introduction

Cellular energy transformations are accompanied by heat changes indicative of the physiological and metabolic activities of living organisms. The thermodynamic approach by metabolic (direct) calorimetry reveals the integrated sum of all enthalpy changes occurring within the experimental chamber, that is the living open system and its environment. The term microcalorimetry is loosely defined, indicating that heat flux is measured in the range of 1 to 1000 μW, equivalent to 0.0022 to 2.2 nmol O_2 s^{-1} or 8 to 8000 μmol O_2 h^{-1} for aerobic metabolism. Due to recent technological advancements, metabolic microcalorimetry has become a sensitive, non-invasive method for the study of respiration and complex metabolic processes in animals (Gnaiger, 1983a; Pamatmat, 1983; Shick, Gnaiger, Widdows, Bayne & de Zwaan, 1986; Widdows, 1987).

In physiological energetics, three main aspects of microcalorimetric investigations are distinguished: (i) Studies of environmentally induced transitions between different metabolic states, especially aerobic–hypoxic–anoxic transitions and aerobic recovery. (ii) Energetic studies of animal behaviour, associated with locomotion, gas exchange, and biological rhythms. (iii) Energy balance studies, combining metabolic calorimetry with simultaneous or parallel measurements of respiration and biochemical changes. The classic method of indirect calorimetry depends on the calculation of heat changes from measured oxygen consumption and theoretical oxycaloric equivalents. This indirect method is extended by including anaerobic metabolite changes in the calculation of theoretical heat changes. Comparison of direct and indirect calorimetry constitutes the thermodynamic energy balance method, the crucial test for a complete biochemical description of net processes under various metabolic states. Thermochemical interpretation of biochemical and calorimetric data is not only required in energy balance studies, it is important for understanding the functional significance of heat flux in physiological energetics (Gnaiger, 1983a). Specifically, calorespirometry – the simultaneous measurement of

Table 1. *The oxycaloric equivalent,* $\Delta_k H_{O_2}$, *as a function of the proportion of catabolic substrates, given as mass fractions; K – carbohydrate, L – lipid, P – protein; calculated from RQ [CO_2/O_2] and NQ [N/O_2]. The deviation of the oxycaloric equivalent [%] is relative to a generalized oxycaloric equivalent of* -450 *kJ. mol*$^{-1}$. *The last line shows an example of carbohydrate to lipid conversion, associated with a high respiratory quotient and oxycaloric equivalent. The last column lists the oxyenthalpic or combustion equivalent of oxygen,* $\Delta_c H_{O_2}$, *combining respiration and excretion, R + U, in terms of the bomb calorimetric enthalpy of combustion of catabolized substrates (after Gnaiger, 1983b).*

Substrate fractions			RQ	NQ	$\Delta_k H_{O_2}$ kJ/mol	deviation %	$\Delta H_{C O_2}$ kJ/mol
K	L	P					
-1.00	–	–	1.00	0.00	-478	$+6.1$	-473
–	-1.00	–	0.72	0.00	-445	-1.1	-441
–	–	-1.00	0.97	0.27	-451	$+0.2$	-528
–	–	-1.00^a	0.84	0.27	-443	-1.6	-528
-0.22	-0.51	-0.27	0.80	0.05	-450	–	-461
-0.11	-0.39	-0.50	9.83	0.10	-450	–	-475
-0.06	-0.25	-0.69	0.87	0.15	-450	–	-491
-0.02	-0.14	-0.84	0.91	0.20	-450	–	-506
-0.81	$+0.18$	-0.09	1.20	0.05	-497	$+10.4$	-507

^{a} For urea as excretory product; all other values are for ammonia.

heat flux and oxygen flux in an open flow or perfusion system – enables the partitioning of total heat flux into aerobic and anaerobic components.

Aerobic energy balance
Oxycaloric equivalent and calorimetric-respirometric ratio
The heat dissipated in respiration or aerobic catabolism (k), $_k\dot{Q}$ [$\mu W = \mu J\cdot s^{-1}$], is calculated from the oxygen flux, $\dot{N}_{O_2}$ [nmol $O_2\cdot s^{-1}$],

$$_k\dot{Q} = \dot{N}_{O_2} \times \Delta_k H_{O_2} \tag{1}$$

The conversion factor, $\Delta_k H_{O_2}$ [kJ·mol^{-1} O_2 = $\mu W/(nmol\ O_2\cdot s^{-1})$], is the theoretical *oxycaloric equivalent*, the catabolic enthalpy change per mol O_2 (Table 1). This thermochemically derived conversion factor can be tested directly by calorespirometry, the simultaneous measurement of total heat flux, $_t\dot{Q}$ [μW], and oxygen flux, $\dot{N}_{O_2}$, which yields the *calorimetric–respirometric ratio* or CR ratio, [kJ·mol^{-1} O_2],

$$CR\ ratio = \Delta_t Q_{O_2} = {}_t\dot{Q}/\dot{N}_{O_2} \tag{2}$$

In metabolic calorimetry the animal is usually allowed to move freely within the calorimeter chamber. Any free energy temporarily conserved in locomotory work is ultimately dissipated within the chamber if the experimental design does not provide a mechanical energy transducer. Theoretically, the experimental CR ratio, $\Delta_t Q_{O_2}$, and the oxycaloric equivalent, $\Delta_x H_{O_2}$, should match. As a prerequisite for this aerobic energy balance, catabolism must be fully aerobic and dissipative (see below), the theoretical oxycaloric equivalent must be correct, and the two experimental terms in equation (2) must be accurate, not merely reproducible or precise.

Calculation of oxycaloric equivalents for aquatic animals involves assumptions on the relative proportions of catabolic substrates, or measurement of the respiratory quotient, RQ [mol $CO_2 \cdot mol^{-1}$ O_2], and nitrogen quotient, NQ [mol N excreted mol^{-1} O_2],

$$\Delta_k H_{O2} = -360 - 118\,RQ + 85\,NQ \tag{3}$$

(Taken from equation 8b in Gnaiger, 1983b; here a larger sample of amino acid compositions was used; see Gnaiger & Bitterlich, 1984). If not only ammonia (equation 3) but also urea is excreted, the expression $+85\,NQ$ is replaced by $+(85 - 28\,x_{urea})\,NQ$. The nitrogen quotient accounts for the total nitrogen excreted in ammonia and urea, and X_{urea} is the fraction of nitrogen excreted as urea, $(x_{ammonia} + x_{urea} = 1)$.

A general estimate of the oxycaloric equivalent in water in the biological temperature range is $-450\,kJ \cdot mol^{-1}$ O_2, e.g. for $NQ = 0.05$ and $RQ = 0.80$ (corresponding to a proportional respiratory substrate loss of -22% carbohydrate, -51% lipid, and -27% protein; equation 2 in Gnaiger, 1983b; Table 1).

Carbon dioxide and ammonia in water

Side reactions of dissociation and buffering occur in the aqueous environment, and the corresponding heat effects must be accounted for. The gases are in the dissolved state, 81% of carbonic acid is dissociated as bicarbonate at pH 7 (pK$' = 6.37$, apparent dissociation constant, in pure water at 25 °C), and ammonia is in the form of ammonium ion. The proton quotient or production of protons per mol O_2 consumed, H^+/O_2, is approximately (Gnaiger, 1983b) in a closed system.

$$H^+/O_2 = RQ\,\frac{1}{1 + \exp(pK' - pH)} - NQ \tag{4}$$

Again, $-NQ$ in equation (4) is replaced by $-(x_{ammonia}\,NQ)$ if urea is excreted simultaneously. At steady state pH under constant physiological conditions all protons generated (equation 4) must be excreted. Natural environmental buffers have a low enthalpy of neutralization, usually not more than -8 kJ·mol^{-1}H$^+$. This is incorporated in the oxycaloric equivalent (equation 3 and Table 1). For carbohydrate and protein (ammoniotelic) at pH 7, H$^+$/O$_2$ is 0.81 and 0.52, respectively. Aerobic buffer effects are small, -6.5 and -4.2 kJ·mol^{-1}O$_2$. If O$_2$ and CO$_2$ are exchanged between a gaseous and aqueous phase, then the oxycaloric equivalent for glycogen is -469, 2% less than -478 kJ·mol^{-1}O$_2$ (Table 1). Irrespective of the transport mechanism of ammonia in the form of NH$_3$(aq) or NH$_4^+$(aq) across respiratory surfaces, ammonium ion is the end-product in water under normal conditions. Endproducts but not transient transport forms have to be considered in thermochemical energy budget calculations. This was possibly misunderstood by Brafield (1985); his oxycaloric value is 5% too low for ammoniotelic protein catabolism in aquatic animals.

Metabolic and bomb calorimetry: direct and indirect

Another conceptual problem arises in the context of aerobic energy balance studies. The oxycaloric equivalent, $\Delta_k H_{O_2}$, is calculated for cellular and aqueous conditions to compare oxygen consumption and actually measured metabolic heat flux (equations 2 and 3). In ecophysiological energetics, however, indirect calorimetry is used to calculate the scope for growth,

$$P = C + F + R + U \qquad (5)$$

where the energy change in biomass, P, the energy input, C, and energy loss in faeces, F, are obtained by bomb calorimetry. Analogous to respirometers in indirect metabolic calorimetry, automatic CHN analyzers can be used in 'indirect bomb calorimetry' to obtain the combustion enthalpy of ash-free dry organic matter, $\Delta_c H$ [kJ/g $_{af}W$], from measured mass fractions of organic carbon, ω_C [g C/g $_{af}W$], and organic nitrogen, ω_N [g N/g $_{af}W$] (Gnaiger & Bitterlich, 1984).

$$\Delta_c h = 11.2 - 66.27\, \omega_C - 4.44\, \omega_N \qquad (6)$$

With careful homogenization, ashing and avoidance of moisture contamination of the dry material, reproducibility is better than 3%, and inaccuracies due to variable residual water contents are <2% (± 0.5 kJ·g^{-}1; Gnaiger & Bitterlich, 1984).

To account for the total catabolic loss in respiration and excretion, $R+U$, in terms of the combustion enthalpy, the *respiratory combustion equivalent*, $\Delta_c H_{O_2}$ has to be used (equation 7.1; Table 1). Alternatively, excretory loss, U, is calculated separately as a product of ammonia excretion, $\dot{N}_{NH_3}$ [nmol·s^{-1}], and the combustion equivalent of ammonia, $\Delta_c H_{NH_3}$ [kJ·mol^{-1} NH$_3$] (equation 7.2),

$$\text{R} + U = \dot{N}_{O_2} \times \Delta_c H_{O_2} \tag{7.1}$$

$$\text{R} + U = \dot{N}_{O_2} \times \Delta_k H_{O_2} + \dot{N}_{NH_3} \times \Delta_c H_{NH_3} \tag{7.2}$$

Based on the equality of the two expressions of equation (7), the combustion equivalents of oxygen consumption and ammonia are related as

$$\Delta_c H_{O_2} = \Delta_k H_{O_2} + NQ \times \Delta_c H_{NH_3} \tag{8}$$

Using the examples given in Table 1 and solving for the combustion equivalent for ammonia (equation 8), we obtain approximately $\Delta_c H_{NH_3} = -280\text{kJ mol}^{-1}\text{NH}_3$. Necessarily, this 'apparent' equivalent is different from the enthalpy of combustion of aqueous or gaseous ammonia (-380 and -345 kJmol^{-1}, respectively). The correction factors for the transition from the cellular state of R to the standard state of combustion are lumped in $\Delta_c H_{NH_3}$, to ensure consistency of thermodynamic state in the energy balance equation. For applications of these concepts see Hawkins *et al.* (1985) and Zamer & Shick (1987).

Calorespirometry: simultaneous heat and oxygen measurements

Errors in calculating oxycaloric equivalents have to be put into perspective relative to experimental accuracy. In 6 simultaneous calorespirometric experiments with the aquatic oligochaete *Lumbriculus variegatus*, the CR ratio was -451 ± 38 kJ·mol^{-1} O$_2$ (S.D.) (Gnaiger & Staudigl, 1987). If errors are $\pm 5\%$ for each measurement, then the error of the CR ratio is $\pm \sqrt{5^2 + 5^2} = \pm 7\%$ or ± 32 kJ·mol^{-1} O$_2$. This embraces the range of oxycaloric equivalents for different catabolic substrates (Table 1), emphasizing the need of high accuracy in calorimetric and respirometric energy balance studies.

In nonsimultaneous measurements of heat and oxygen flux, different states of activity of the animals in the calorimeter and respirometer may explain the large range of CR ratios, -420 ± 60 kJ mol^{-1} (S.D.), calculated from 11 mean values for different species. However, a significantly lower variability, -460 ± 13 kJ mol^{-1}, was calculated from simultaneous calorespirometric studies, again using 11 mean values reported for various aquatic euryoxic and stenoxic animals (Gnaiger & Staudigl, 1987). This difference underscores the need for simultaneous calorespirometry, thus eliminating any variability due to different metabolic states of animals in nonsimultaneous measurements. The important conclusion drawn from these results is that several anoxia-tolerant (oligochaetes, bivalves) and intolerant species (Crustacea, salmonid fish larvae) alike are fully aerobic under normoxic, stress-free conditions. The measured CR ratios and expected aerobic oxycaloric equivalents (Table 1) agree. Aerobic energy balance studies, therefore, provide a methodological baseline for calorespirometry that can then be applied to studies of metabolically complex situations involving anoxia or net biosynthesis.

The most versatile applications of calorespirometry are offered by microcalorimeters with perfusion (open-flow) chambers, connected to an open-flow respirometer (Fig. 1).

Perfusion microcalorimeters

Instrument design. Principles of design and the range of micro-calorimeters used in 'studies of biochemical processes and of cellular

Fig. 1. An open-flow calorespirometer: Combination of the LKB-2107 flow sorption microcalorimeter (left) and the Cyclobios Twin-Flow respirometer (right). Left inset; operation principle of the heat flow calorimeter: A, $0.5\ cm^3$ pyrex animal chamber; D, heat detector; E_d detector heat exchanger; E_e; external heat exchanger; E_i, internal heat exchanger; H, heat sink; R, reference heat detector; S, signal amplifier; T_d, detector thermopile; T_r, reference thermopile. Right inset; Twin-Flow principle; the inflow and outflow of the perfusion medium through gold or stainless steel capillaries connecting to the calorimeter is indicated by arrows: DS, drive shaft for simultaneous switching of the 4-way valves; EC, electrode cable of the POS; EH, stainless steel POS sleeve; MF, magnetic stirrer; POS, polarographic oxygen sensor; PP, peristaltic pump; RM, rotating magnet; SC, right stirring chamber in calibration position; V, 4-way valve; WR, water reservoir (after Gnaiger, 1983c; Foto: P. Flöry, Cyclobios).

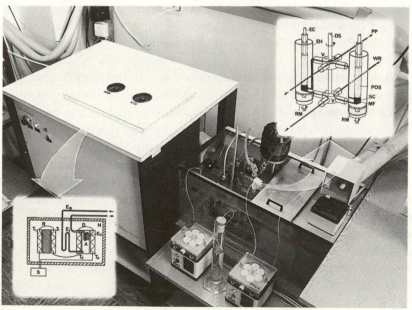

systems such as micro organisms and tissue cells, etc.', has been reviewed recently by Wadsö (1987). Here we are exclusively concerned with studies of 'etc.', living animals. Adiabatic calorimeters ideally exchange no heat with the surroundings, and the observed temperature increase is directly proportional to the heat flux within the reaction chamber. An adiabatic calorimeter with continuous aeration for studies of fish up to 100 g is described by Smith, Rumsey & Scott (1978). The chamber must be closed to avoid heat exchange and disturbances in sensitive adiabatic microcalorimeters, which excludes an open-flow mode. By simultaneously measuring the decline in P_{O_2} with a polarographic oxygen sensor, an initial phase of progressive hypoxia can be distinguished from anoxia when depletion of dissolved oxygen is completed (Hammen, 1983).

This strictly unidirectional experimental regime is avoided by continuous perfusion of the animal chamber. High-precision thermistors are used to measure the temperature difference between water flowing through a thermally insulated animal chamber and an identical twin chamber serving as a reference (Lock & Ford, 1983). An entirely different principle is employed in heat conduction calorimeters where the temperatures of outflow and inflow water are essentially identical. This is achieved either by efficient heat exchangers adjacent to the animal chamber (Fig. 1) or by a counter current principle (Fig. 2). In contrast to adiabatic instruments, a heat conduction calorimeter maintains nearly constant temperatures within the reaction chamber, which is surrounded by semiconductor Peltier elements. These thermopiles provide good thermal contact between the reaction chamber and a constant-temperature heat sink, for rapid conduction of heat at minute temperature differences, ΔT, between chamber and heat sink. In heat conduction microcalorimeters, ΔT is in the order of 10^{-6} °C μW^{-1}, that is <0.0001 °C in a typical physiological experiment. The recorded signal, however, is not temperature but the thermopile voltage which, in turn, is proportional to ΔT (Calvet & Prat, 1963).

The thermocouples of the thermopiles surrounding one chamber are connected in series, but in opposition to the thermopiles of an identical twin or reference chamber. The twin principle (Fig. 1) serves to cancel out any uniform temperature disturbances of the heat sink, provided that the sensitivities and response times of the twin halves are identical. If possible, it is advantageous to monitor both, the differential twin-signal and the single mode signal of the reference chamber; then abnormal baseline disturbances can be detected and corrected by calibrated mathematical models (Kaufmann & Gnaiger, 1981). In order to achieve a stable experimental baseline it is advisable to site the calorimeter in a dry, temperature controlled room

120   *E. Gnaiger, J. M. Shick and J. Widdows*

Fig. 2. Perfusion vessels of the LKB-Thermometrics microcalorimeter,
(*a*) Module with 3.5 cm^3 chambers; twin system (after Görman
Nordmark *et al.*, 1984): a, stainless steel capillaries connecting to the
respirometer; b, outer stainless steel tube; c, heat exchanger brass bolts;
d, e, animal chamber, the magnification shows the 'nested chamber' and
flow directions; f, steel cylinder, immersed in the thermostated water
bath; g, heat sinks; h, reference chamber; 1, inflow; 2, outflow as
counter current heat exchanger; 3, nested animal chamber; 4, 100 μm
mesh bottom. (*b*) Module with a single 25 cm^3 perfusion chamber (after
Widdows, 1987): 1, temperature controlled water jacket; 2, inflow from
water reservoir; 3, heat sink; 4, thermostat; 5, counter current heat
exchanger; 6, outflow 150 μm filter; 7, thermopile; 8, outflow to
polarographic oxygen sensor and peristaltic pump; 9, 25 cm^3 stainless
steel chamber; 10, experimental animal.

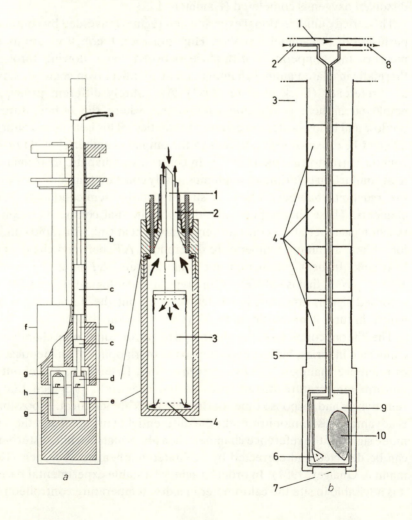

(within ±1 °C), to avoid direct sunlight and to provide efficient air circulation with additional fans.

The choice of a particular calorimeter primarily depends on the size of the experimental animals and of the budget. The lower limit of animal size is set by the sensitivity and long-term stability of the instrument, availability of large numbers (e.g. of meiofauna), and tolerable crowding effects. With a stability in the order of ±0.5 μW, 0.1 mg $_d W$ would represent the minimum dry biomass, but under anoxia the minimum is rather 1 mg due to 70–95% reduced anoxic heat flux (Gnaiger, 1983d). Long-term (>24 h) stabilities better than 2 μW, required at the lower size limit, can only be achieved in heat conduction microcalorimeters with small reaction chambers in a twin arrangement (Figs. 1 and 2*a*) and at low perfusion rates (Suurkuusk & Wadsö, 1982). The upper limit of animal size depends on the geometry and dimensions of the calorimeter chamber, and on the feasability of high perfusion rates sufficient for maintaining aerobic conditions. With increasing size of the chamber (Fig. 2*b*) the twin configuration becomes less efficient due to temperature inhomogeneities in space, but with larger biomass, less rigorous standards are set with respect to sensitivity and stability, hence single mode long-term stabilities of 3 μW are sufficient (Gnaiger, 1983d; Widdows, 1987).

Static and dynamic calibration. Electrical power dissipated across a resistor is instantaneously and completely transformed into an equivalent rate of heat dissipation, and can be accurately regulated, <±0.1%. Therefore, electrical calibration is most convenient, although for some applications cross-calibration is required by chemical methods using the high enthalpies of some buffer reactions (Chen & Wadsö, 1982). Commercial microcalorimeters are equipped with built-in precision resistance heaters (e.g. 50Ω). It is important to place the calibration resistor into a representative position in thermal contact or better within the animal chamber, especially at high perfusion rates and for dynamic calibration. Electric wires leading from an external power source into the chamber may cause thermal leaks. The most elegant solution is achieved by an internal power source sealed into a calibration capsule containing an electronic circuit for switching on and off a known power at fixed time intervals, e.g. two hours (Widdows, 1987). The internal calibration capsule is powered by three 1.5 V silver oxide microbatteries (RM47) which produce a background heat flux of ca. −6 μW due to the timer circuit and the enthalpy changes associated with the chemical reaction in the battery. Comparison of the built-in external calibration in the ThermoMetric microcalorimeter and the internal calibration capsule agree within <1%.

At steady state, a constant rate of heat dissipation is balanced by an equal rate of heat flow. Then the steady deflection of the voltage across the thermopiles is directly proportional to the heat flux, resulting in a strictly linear calibration. The inverse of the sensitivity is the static calibration constant, typically between 3.3 $\mu W/\mu V$ (Suurkuusk & Wadsö, 1982) and 25 $\mu W/\mu V$ (Pamatmat, 1983). When the calibration current is switched on or off, the output signal follows exponentially from an initial to a final steady-state, due to the inertia of the system. If the average signal during time intervals Δt [s], is $\dot{Q}$, following a change of the apparent signal by ΔQ, then the total heat flux is calculated as

$$_t\dot{Q} = \dot{Q} + \tau(\Delta Q / \Delta t) \tag{9}$$

τ [s] is the exponential time constant, the ratio of heat capacity to thermal conductivity of the calorimeter chamber and thermopiles (Calvet & Prat, 1963). With increasing size of biomass and calorimeter chamber, the heat capacity increases and heat conduction decreases, the latter due to longer thermal diffusion paths and a decreased surface to volume ratio. Therefore, the time constant of heat conduction calorimeters increases with increasing size of the animal chamber, from ca. 120–150 s for the 0.5 and 3.5 cm^3 chamber filled with water (Fig. 1 and 2), to 600 s for the 150 cm^3 thin-walled glass chamber of the Pamatmat calorimeter (Fig. 6). The time constant obtained from dynamic calibrations should be tested for accurate reconstruction of entire transition periods. With increasing chamber size and decreasing time intervals, e.g. <600 s, second-order exponential time constants become increasingly significant for accurate time corrections (Randzio & Suurkuusk, 1980; Suurkuusk & Wadsö, 1982). In studies with active animals, however, these second-order effects are frequently obscured by variable heat exchange due to changing locomotion and ventilation.

An entirely different dynamic aspect predominates the initial period of an experiment, after placing the animal in the calorimeter chamber. Thermal disturbances of the heat sink cannot be avoided during insertion of the chamber, but can be minimized by thermal pre-equilibration and always very slowly lowering the chamber into the calorimeter. A standardized procedure of filling the chamber must be imitated without actually adding an animal, to determine the stabilization time in blank runs, e.g. 1 h for the 3.5 cm^3 perfusion chamber (Fig. 2a), but 4 to 5 h in a calorimeter with a 150 cm^3 chamber (Pamatmat, 1983).

Perfusion rate. High perfusion or flow rates exert disturbing effects on the baseline and distort the sensitivity or calibration constant. Both effects are minimized by a counter-current principle achieved by two concentric stainless steel tubes (Fig. 2). Since a larger volume of water is contained in the outer than the inner capillary, the retention and equilibration time is longer if the inflow is directed through the outer capillary. Reversed flow can be used initially to eliminate air bubbles through the outer capillary (Görman Nordmark *et al.*, 1984). With reversed flow one obtains the option of a 'nested chamber' design with improved water exchange characteristics (Fig. 2*a*). This is particularly important for homogeneous perfusion of sedimenting particles, such as batches of *Artemia* embryos (Hand & Gnaiger, 1988).

The change between inflow and outflow concentration of oxygen, ΔC_{O_2} [mmol $O_2 \cdot dm^{-3}$ = nmol·mm^{-3}], is a function of biomass, $_dW$ [g dry weight], weight specific oxygen flux, $\dot{N}_{O_2}$ (nmol $O_2 \cdot g^{-1} \cdot s^{-1}$], and perfusion rate, $\dot{V}_W$ [mm$^3 \cdot s^{-1}$],

$$\Delta C_{O2} = {_dW} \times \dot{n}_{O_2} / \dot{V}_W \tag{10}$$

The perfusion rate must be adjusted to the metabolic requirements, to maintain outflow oxygen sufficiently high relative to the inflow oxygen concentration, $C{in}_{O2}$,

$$\dot{V}_W = ({_dW} \times \dot{n}_{O_2}) / (C{in}_{O_2} \times R_{O_2}) \tag{11}$$

R_{O_2} is the oxygen reduction ratio, that is the signal change relative to the calibration value (Gnaiger, 1983*c*),

$$R_{O_2} = \Delta C_{O_2} / C{in}_{O_2} = \text{signal change / inflow signal} \tag{12}$$

When aerobic heat flux is converted to oxygen flux (Table 1), the calorimetric signal can serve, in conjunction with equation (11), as a control for appropriate perfusion rates. At air saturation concentrations between 0.41 and 0.21 nmol $O_2 \cdot mm^{-3}$ (4 and 25 °C fresh water and sea water, respectively), every μW heat flux must be multiplied by 0.04–0.07 mm$^3 \cdot s^{-1}$ or 0.13–0.25 *ml.h*$^{-1}$ to obtain the perfusion rate required for maintaining R_{O_2} at 0.15 (outflow concentration at 85% of inflow).

For example, calculate the perfusion rate required to set R_{O_2} at 0.15 in an experiment with 0.014 g oligochaetes at 20 °C (Gnaiger & Staudigl, 1987): At a total heat flux of 76 μW, $\dot{n}_{O_2}$ is 12 nmol $O_2 \cdot g^{-1} \cdot s^{-1}$, $C{in}_{O_2}$ is 0.280 nmol O_2 mm^{-3} in air-saturated fresh water near sea level (101 kPa), then the flow rate must be 4.0 mm$^3 \cdot s^{-1}$ or 14 ml·h^{-1}. At 10 °C the perfusion rate can be reduced to 1.5 mm^3 s^{-1} or 5.3 ml h^{-1}, assuming $Q_{10} = 2.2$. With a 1 g marine mussel at a respiratory activity of 3 nmol s^{-1} at 15 °C (Fig. 5), a perfusion rate of 40 mm$^3 \cdot s^{-1}$ or 145 ml·h^{-1} yields a R_{O_2} of 0.30.

Simultaneous respirometry: the Twin-Flow respirometer

The above calculations apply equally for designing open-flow respirometric experiments, where polarographic oxygen sensors (POS) are

employed for continuous monitoring of inflow and outflow oxygen concentrations. As the oxygen reduction ratio, R_{O_2}, is a relative value (equation 12), it can be calculated from the experimental signal recorded
arbitrarily as mm or percentage chart recorder deflection, voltage or current
of the POS. Oxygen flux, $\dot{N}_{O_2}$ [nmol·s^{-1}], at steady state is (equations 11
and 12),

$$\dot{N}_{O_2} = R_{O_2} \times Cin_{O_2} \times \dot{V}_W = \Delta C_{O_2} \times \dot{V}_W \tag{13}$$

The Twin-Flow respirometer and its applications in calorespirometry
(Fig. 1) were described in detail (Gnaiger, 1983c; Cyclobios, 1985). Cin_{O_2} is
kept constant by continuous equilibration of water with air (or a gas
mixture). The water reservoir and the entire respirometer are simply
immersed into a water bath regulated at constant temperature ($\pm$ 0.05 °C)
accurately at or slightly above experimental temperature of the calorimeter.
Otherwise supersaturation of heated water results in gas bubble formation
in the perfusion system. Moreover, the water reservoir should be located at
the highest level of the perfusion system to avoid negative pressure and
hence bubble formation.

Accurate calibration of the oxygen sensors is essential in a perfusion
respirometer. If the variability of oxygen measurements in the inflow and
outflow amounts to only 0.5%, then the error of oxygen consumption is
4.4% at R_{O_2} of 0.15. Frequent calibrations of the POS in a fixed position
improve its accuracy to better than 0.5%, which is possible with the
Twin-Flow principle (Gnaiger, 1983c). Calibration with inflow water at
Cin_{O_2} is made when the POS is intersected between the water reservoir and
animal chamber. By a system of two 4-way valves the sensor is switched into
measuring position at the outflow. The difference in the signal is directly
proportional to R_{O_2} and $\dot{N}_{O_2}$ (equation 13). During calibration the record of
oxygen flux is interrupted, but this loss of signal is avoided by using two POS
switched at intervals into alternate positions (Fig. 1, inset). As a prerequisite for simultaneous calorimetry, the direction of flow remains
unchanged despite the switching between measuring and calibration positions of the POS.

A peristaltic pump is connected to the outflow for regulating perfusion at
a constant rate ($\pm$0.5%). Frequent measurements of the perfusion rate,
$\dot{V}_W$, are required to correct for gradual changes due to ageing of the
tubings. The measurements are easily made by using a calibrated 1 or 2 ml
pipette and a stop watch. Inaccuracies affect the respirometric results and
the CR ratio, as the calorimetric signal is independent of the accurate value
of $\dot{V}_W$. The response time of the respirometer is an inverse function of the
perfusion rate and may be very different from that of the calorimeter.

Therefore, the accuracy of instantaneous CR ratios depends on the quality of dynamic corrections of both systems (Gnaiger, 1983c). In cases of changing ventilatory and locomotory patterns, the apparent fluctuations of oxygen consumption may not correlate with changes in metabolic flux but merely reflect external oxygen exchange.

Artefacts of oxygen consumption and CR ratios can also arise from oxygen diffusion or microbial and chemical oxygen consumption in the flow lines. All flow lines and connections from the water reservoir to the outflow POS must be essentially leak tight and diffusion free. Teflon or silicone tubings are not feasible. Gold capillaries with 1 mm inner diameter are flexible and inhibit bacterial growth at the surface. Where glass capillaries are too fragile, stainless-steel capillaries can be applied. Non-biological blank oxygen consumption commonly occurs when using stainless steel. The problem of corrosion can be minimized by coating the inner surfaces with an organosilane concentrate (Prosil–28). Blank oxygen uptake is determined after removing the animals from the system and replacing the chamber without cleaning precautions. Microbial blank oxygen uptake reaches steady state after several hours of contact with the animal (Dalla Via, 1983). The blank flux varies with P_{O_2} (Gnaiger, 1983c) and must be subtracted from the measured flux. Heat dissipation in the empty animal chamber is automatically indicated in the experimental baseline of the calorimeter hence no further corrections are required.

Closed-system calorimetry and respirometry: *parallel* heat and oxygen measurements

The continuous depletion of the small amount of oxygen dissolved in water limits the usefulness of closed 'batch-type' calorimeters in the study of aerobic aquatic animals. In 150 cm^3 air-saturated water a 40 mm long specimen of *Mytilus edulis* is already experiencing hypoxia by the time thermal equilibrium of the calorimeter is re-established. Prolonged aerobic conditions in water can be maintained by including a gas phase in the closed chamber and stirring the water for equilibration (Görman Nordmark *et al.*, 1984). The ventilatory activity of some animals such as *Mytilus edulis* provides sufficient stirring, >80% air saturation in 40 cm^3 of seawater placed in the 150 cm^3 calorimeter chamber (Fig. 6).

Use of a two-phase system has permitted measurement of heat flux by isolated ovaries at oxygen levels that approximate those prevailing *in vivo* (Bookbinder & Shick, 1986). Parallel measurements of oxygen flux in gently agitated Gilson respirometer flasks indicate that the majority of the energy metabolism in these massive, unperfused organs is anaerobic. The unstirred boundary layer adjacent to the ovaries in the static calorimeter chamber may be larger than *in vivo*, and open-flow calorimetry seems better suited for such studies. A large anaerobic component was postulated from parallel microcalorimetric and respirometric measurements of the isolated snail heart (Herold, 1977). However, the average CR ratio was *lower* than the oxycaloric equivalent (less heat instead of additional anaerobic heat was observed), but calculation of the linear slope between calorimetric and respirometric rates at various P_{O_2} yields a regression-CR ratio of -459 kJ·mol^{-1} (r = 0.98). This agreement of the regression-CR ratio with the oxycaloric equivalent, in combination with an intercept different from zero, usually indicates a problem with the baseline position.

The 20 to 50-fold higher concentration of oxygen in air compared to water, makes closed-chamber calorimeters more suitable for studying air-breathing animals, e.g. insects (Coenen-Stass, Schaarschmidt & Lamp-recht, 1980; Dunkel, Wensman & Lovrien, 1979; Peaking, 1973). Prolonged measurement of heat flux by, e.g. the terrestrial slug *Arion hortensis*, does not subject the animal to hypoxia in a closed 150 cm^3 chamber (Fig. 3). Such long-term measurements are likely to detect all phases of activity and associated metabolism (Pamatmat, 1983), and are far less tedious than the manual manipulations involved in traditional volumetric measurements of oxygen. In the present example, oxygen consumption by *A. hortensis* was measured every 0.25 h for six hours in a Gilson constant pressure

Fig. 3. Continuous recording of heat flux in air by a terrestrial slug, *Arion hortensis* at 15 °C. Minimum and maximum hourly values of heat flux from this recording and minimum and maximum hourly oxygen flux for the same specimen measured on the day following the calorimetric experiment yield estimates of the CR ratio.

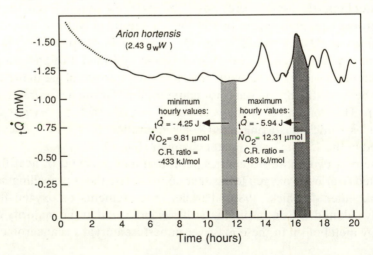

respirometer. Comparison of the minimum and maximum hourly values of $\dot{N}_{O_2}$ measured periodically, with those for $_t\dot{Q}$ measured continuously reveal CR ratios similar to the theoretical oxycaloric equivalent (Fig. 3). Direct calorimetry thus seems particularly well suited to monitor daily patterns of activity and energy flux in such animals.

Intertidal animals experience intermittently terrestrial conditions and then rely on aerobic and anaerobic energy metabolism to varying extents. Although total heat flux can be measured under simulated intertidal conditions, the method does not distinguish between aerobic and anaerobic sources of heat, and parallel or sequential indirect measurements are required to do so. Neither form of indirect calorimetry (respirometry, or biochemical measurement of anaerobic end-products) is sufficient to quantify total metabolism. Studies on *Mytilus edulis*, which generally closes its shell valves during exposure to air, indicate that 0 to 45% of the heat dissipated during exposure derives from aerobic respiration (Shick *et al.*, 1986), whereas *Cardium edule*, which gapes and breathes air when exposed, remains fully aerobic (Widdows & Shick, 1985), having a mean CR ratio of -433 kJ·mol^{-1} O_2. Likewise, intertidal specimens of the sea anemone *Anthopleura elegantissima* have a mean CR ratio of -473 kJ·mol^{-1} O_2 during 12 hours of aerial exposure (Shick & Dykens, 1984). Subtidally acclimatized specimens of the sea anemone *Actinia equina* become active and have high heat flux during acute intertidal exposure, although their oxygen flux declines over the same period; this indicates an increasing

Fig. 4. Heat flux of *Mytilus edulis* (0.4 g $_dW$; 15 °C) during aerial exposure, measured in the 28 cm^3 closed ampoule of the LKB-ThermoMetrics microcalorimeter. On reimmersion the mussel released a gas bubble that was measured volumetrically. Consumption of all the O_2 in this bubble and the heat flux integrated above the minimum rate yield the net CR ratio, indicating a fully aerobic peak due to air gaping (after Shick *et al.*, 1986).

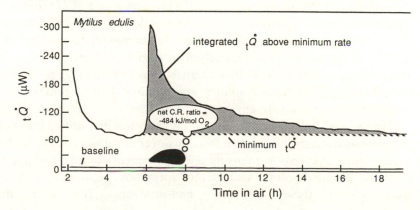

reliance on anaerobic metabolism as their hydrostatically supported body collapses, which increases diffusion distances and reduces gas exchange surface area (Shick, 1981).

In intertidal bivalves, air-gaping in particular can cause an individual's heat flux to increase 5 to 8-fold, as metabolism switches from being predominantly anaerobic to largely aerobic during periods of high rate (Fig. 4). Parallel oxygen and heat measurements confirm the largely aerobic nature of such peaks; the net CR ratio for maximum flux in air-exposed *M. edulis* is -467 kJ·mol^{-1} O$_2$ (Shick *et al.*, 1986). Reassuringly, we also observed agreement in the rates of aerial heat flux of *M. edulis* measured in the ThermoMetric calorimeter (-0.305 ± 0.065 mW g^{-1}; N=12) and the Pamatmat calorimeter (-0.288 ± 0.047 mW·g^{-1}; N=9) (mean $\pm$ S.E.; Shick *et al.*, 1986).

Microcalorimetry, and fundamental aspects of aerobic and anoxic metabolism

Quantification of the extent of 'metabolic shutdown' following aerobic–anoxic transitions require highly sensitive methods. Using the modular 4-channel ThermoMetric microcalorimeter, we studied the anoxic rate of *Mytilus edulis* in the size range of 0.016 to 1.0 g dry weight (Fig. 5), with 4 or 3 specimens (18 or 25 mm long) in 5 cm^3 ampoules, and 3 or 1 specimens (35 or 45 mm) in 28 cm^3 ampoules. Confirming Pamatmat's preliminary result, the anoxic mass exponent is not different from 1.0. This linear function of anoxic heat flux and body mass is fundamentally different from the exponential or power function of aerobic metabolism (Fig. 5; see also Widdows, 1987). According to a dimensional analysis of the metabolic power function by Heusner (1985), the $\dot{N}_{O_2}/mass^{2/3}$ ratio is the 'mass-independent metabolism'. Consequently, the deviation of the aerobic mass exponent of 0.73 (Fig. 5) from the value of $2/3 = 0.67$ would indicate deviation from biological similitude with increasing size of *M. edulis*. This conclusion cannot be drawn from the direct calorimetric measurements, as the constant $_t\dot{Q}/mass$ ratio indicates strict linear mass-independence under anoxia. Clearly, the functional analysis of aerobic and anoxic biochemical mechanisms can make a positive contribution towards a general physiological theory, but further studies are required before contrasting the Aerobic Surface Law with an 'Anoxic Mass Law'.

Furthermore, the meaning of metabolic rate must be clarified for relating distinct mechanisms such as heat flux, turnover of carbon chains, oxidative metabolism, glycolysis to pyruvate and subsequent operation of pyruvate oxidases, succinate-propionate-acetate formation, and accumulation or excretion of these anaerobic end-products. The term direct

Fig. 5. Anoxic heat flux in N_2 gas and oxygen flux in aerobic seawater as a function of body size of *Mytilus edulis* (March, 12 °C), plotted on equivalent scales (-450 μW nmol^{-1} O_2/s^{-1}]. The exponential mass exponents, b, are significantly different (N = 18). Anoxic heat flux is a linear function of dry body mass.

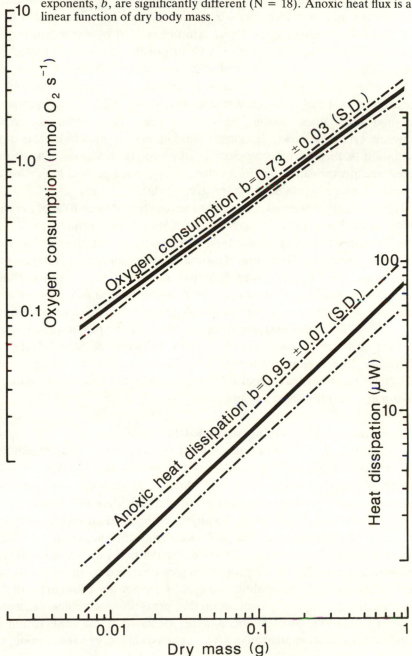

calorimetry has led to the contention, that this method provides the general approach to measure metabolic rate 'directly', without further explanation of the functional significance of heat flux. It must be pointed out that the calorimetric and the biochemical measures of metabolic flux, that is heat flux and ATP turnover respectively, are not related by a constant coefficient: the heat dissipated per unit ATP turnover is ca. -80 kJ$\cdot$mol^{-1} for fully coupled aerobic metabolism, but less than -40 kJ$\cdot$mol^{-1} for the propionate-acetate pathway (Gnaiger, 1983a). The direct calorimetric measure of heat flux is identical to catabolic power (Gibbs energy change per unit time) under aerobic conditions, but a large difference exists between catabolic power and heat flux in anoxic metabolism due to significant isothermal entropy changes of glycolytic reactions. Therefore, metabolic flux measured by direct calorimetry is not a general indication of the rate of entropy production (Gnaiger, 1983a).

The frequently observed reduction of anoxic heat flux to 10% of aerobic values would be equivalent to a reduction of anoxic ATP turnover to 20% of aerobic values calculated on the basis of propionate-acetate excretion and oxygen consumption. Biochemical estimates of metabolic flux under anoxia are frequently lower (de Zwaan & Wijsman, 1976), indicating that either unknown metabolic pathways contribute to the total anoxic flux, or that oxidative processes based on residual oxygen under 'anaerobic' conditions (0.5 to 2% air saturation) are detected by the non-specific calorimetric method, but not by biochemical analyses (Gnaiger & Staudigl, 1987). Therefore, careful energy balance studies, based on enthalpy changes of all metabolic reactions, are required for extending the concept of indirect calorimetry into the anoxic domain.

Anoxic and anaerobic energy balance

Shick, de Zwaan and de Bont (1983) measured, simultaneously, the anoxic heat flux and the accumulation of metabolites in *Mytilus edulis*, testing Gnaiger's (1980) claim that direct and indirect (biochemical) measurements of anoxic energy flux in euryoxic invertebrates do not agree. Thermochemical interpretation (Gnaiger, 1983a) of their results revealed the expected exothermic 'anoxic gap', as a significant fraction of the anoxic heat flux could not be explained by known anoxic pathways. This result is in line with a gap in the anoxic proton balance observed by intracellular pH measurements and metabolite changes (Graham & Ellington, 1985). Additional support for the contention that previous biochemical estimates of anoxic catabolic flux are incomplete, stems from the observation of anoxic ammonia accumulation in *Mytilus edulis*, contrary to assumptions on ammonia fixation to account for the accumulated alanin (Bitterlich, Gnaiger & Widdows, unpublished).

Departures from the anoxic metabolic arrest of invertebrates include the cockle *Cardium edule* (40% anaerobic/aerobic $_t\dot{Q}$; Pamatmat, 1980), which is a rather aerobically-poised bivalve that does not tolerate anoxia well. The clam, *Mulinia lateralis* maintains anoxic heat flux from 5–30% (Gnaiger & Shick, unpubl.) to 97% of aerobic values (Shumway, Scott & Shick, 1983). The large 'anaerobic'/aerobic ratio (40%) in the oligochaete, *Lumbriculus variegatus* (Gnaiger, 1980), was obtained under perfusion conditions without simultaneous proof of complete oxygen removal, and the strictly 'anoxic'/aerobic ratio is lower (16%; Gnaiger & Staudigl, 1987). The problem of oxygen leakage and diffusion is obviously enhanced in perfusion systems, and cannot be neglected when using closed chambers sealed with inappropriate materials (e.g. silicone stoppers). Even an ideal oxygen conformer maintains at 1–2% air saturation an aerobic rate of 1–2% of the normoxic value. This condition is frequently referred to as 'anaerobic'. At a measured anaerobic/aerobic ratio of 20%, therefore, 5–10% of the 'anaer-obic' rate would be supported by oxygen consumption, which is not only important for energy flux considerations (Gnaiger & Staudigl, 1987), but critical for anaerobic redox balance calculations and estimations of total 'anaerobic' ATP turnover.

Aerobic recovery from hypoxia and efficiency of biosynthesis

During aerobic recovery from anoxia the conservation of heat in glyconeogenic processes results in CR ratios lower than the oxycaloric equivalent for dissipative metabolism, indicating the internal thermodyna-mic efficiency of coupled metabolism. Therefore, the 'overshoot' of heat flux during aerobic recovery (Fig. 6) is less than the corresponding oxygen flux (Famme & Knudsen, 1983; Shick *et al.*, 1986). The relatively large chamber of the Pamatmat calorimeter has proved useful in studying the hypoxic–aerobic transition when aerially-exposed 'intertidal' mussels are reimmersed. The experimental mussel is suspended in air above the water by a nylon monofilament cemented to its shell (Fig. 6). The glass chamber is then sealed inside a larger epoxy-coated tin canister, to which a second monofilament is attached. The entire vessel is finally placed between the thermopiles of one half of the double-twin calorimeter. The mussel is reimmersed at simulated 'high tide' by rotating the chamber 90° using the second monofilament (Fig.6). This minimum contact with the vessel during the momentary opening of the calorimeter presents a minimum of thermal disturbance. Re-equilibration time as determined in blank experiments is consistently 0.75 to 0.83 hours, and the reproducible blank re-equilibration curve allows calculation of aquatic heat flux for all but the initial 0.25 h of recovery.

A quantitative understanding of the efficiency of anabolism and of maintenance is a prerequisite for modern analyses of energy allocation and growth efficiency (Hawkins, Bayne & Day, 1986). Anabolism during recovery from anoxia and hypoxia, as studied by direct calorimetry and biochemistry, can serve as a model for thermodynamic efficiencies in anabolic ATP utilization and growth. Today, students of ecological and physiological energetics are increasingly prepared to ponder over the complexities involved in this thermodynamic approach that is fundamental in physiological energetics, comparable to the aspects of efficiency of catabolic ATP production and metabolic regulation (Gnaiger, 1987). An experimental and theoretical analysis, comparing the biochemical and thermodynamic efficiency of aerobic recovery metabolism after accumulation of succinate and lactate, is available (E. Gnaiger, M. Shick & J. Widdows, in preparation).

Fig. 6. Heat flux of *Mytilus edulis* (0.32 g $_{d}W$; 15 °C) during aerial exposure and aquatic recovery, measured in the 150 cm^3 closed chamber (insets) of the Pamatmat calorimeter. The sharp deflection at the start of recovery and exponential re-equilibration as measured without an animal results from opening the calorimeter to rotate the sample chamber and immerse the mussel. Parallel measurement of oxygen flux yields low CR ratios corrected for re-oxygenation of body fluids and mantle cavity water, indicating the high thermodynamic efficiency of recovery metabolism (after Shick *et al.*, 1986).

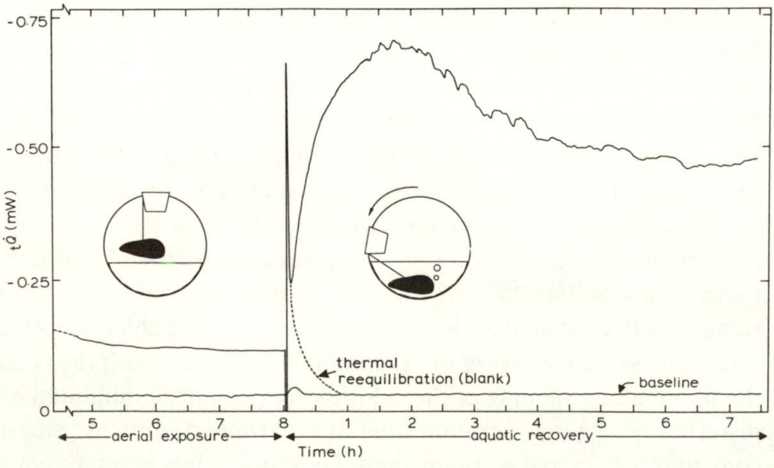

Acknowledgments
Supported by FWF Austria, project no. JO187B (EG) and NATO
research grant 27181. We dedicate this work to Kjell Johansen, Viking &
Physiologist – 'beija flor is very much alive'.

References
Bookbinder, L. H. & Shick, J. M. (1986). Anaerobic and aerobic energy
metabolism in ovaries of the sea urchin *Strongylocentrotus droebachien-
sis*. *Marine Biology*, **93**, 103–10.
Brafield, A. E. (1985). Laboratory studies of energy budgets. In *Fish
energetics: new perspectives*, ed. P. Tytler & P. Calow, pp. 257–81.
London & Sydney: Croom Helm.
Calvet, E. & Prat, H. (1963). *Recent progress in microcalorimetry*. Oxford:
Pergamon Press.
Chen, A. & Wadsö, I. (1982). A test and calibration process for microcalorimeters used as thermal power meters. *Journal Biochemical Biophysical Methods*, **6**, 297–306.
Coenen-Staß, D., Schaarschmidt, B. & Lamprecht, I. (1980). Tempera-
ture distribution and calorimetric determination of heat production in
the nest of the wood ant, *Formica polyctena* (Hymenoptera, Formici-
dae). *Ecology*, **61**, 238–44.
Cyclobios (1985). Twin-Flow respirometry. *Cyclobios Newsletter*, **1**, 1–4.
Dalla Via, G. J. (1983). Bacterial growth and antibiotics in animal
respirometry. In *Polarographic Oxygen Sensors. Aquatic and Physiological Applications*, ed. E. Gnaiger & H. Forstner, pp. 202–18. Berlin,
Heidelberg, New York: Springer.
De Zwaan, A. & Wijsman, T. C. M. (1976). Anaerobic metabolism in
bivalvia (Mollusca). Characteristics of anaerobic metabolism. *Comparative Biochemistry Physiology*, **54B**, 313–24.
Dunkel, F., Wensman, C. & Lovrien, R. (1979). Direct calorific heat
equivalent of oxygen respiration in the egg of the flour beetle *Tribolium
confusum* (Coleoptera: Tenebrionidae). *Comparative Biochemistry
Physiology*, **62A**, 1021–9.
Famme, P. & Knudsen, J. (1983). Transitory activation of metabolism,
carbon dioxide production and release of dissolved organic carbon by
the mussel *Mytilus edulis* L., following periods of self-induced anaerobiosis. *Marine Biology Letters*, **4**, 183–92.
Gnaiger, E. (1980). Energetics of invertebrate anoxibiosis: direct calorimetry in aquatic oligochaetes. *FEBS Letters*, **112**, 239–42.
Gnaiger, E. (1983*a*). Heat dissipation and energetic efficiency in animal
anoxibiosis: economy contra power. *Journal Experimental Zoology*,
228, 471–90.
Gnaiger, E. (1983*b*). Calculation of energetic and biochemical equivalents
of respiratory oxygen consumption. in *Polarographic Oxygen Sensors.
Aquatic and Physiological Applications*, ed. E. Gnaiger & H. Forstner,
pp. 337–45. Berlin, Heidelberg, New York: Springer.
Gnaiger, E. (1983*c*). The twin-flow microrespirometer and simultaneous

calorimetry. In *Polarographic Oxygen Sensors. Aquatic and Physiological Applications*, ed. E. Gnaiger & H. Forstner, pp. 134–66. Berlin, Heidelberg, New York: Springer.

Gnaiger, E. (1983*d*). Microcalorimetric monitoring of biological activities: ecological and toxicological studies. *Science Tools*, **30**, 21–6.

Gnaiger, E. (1987). Optimum efficiencies of energy transformation in anoxic metabolism. The strategies of power and economy. In *Evolutionary physiological ecology*, ed. P. Calow, pp. 7–36, London: Cambridge Univ. Press.

Gnaiger, E. & Bitterlich, G. (1984). Proximate biochemical composition and caloric content calculated from elemental CHN analysis: a stoichiometric concept. *Oecologia*, **62**, 289–98.

Gnaiger, E. & Staudigl, I. (1987). Aerobic metabolism and physiological reactions of aquatic oligochaetes to environmental anoxia. Heat dissipation, oxygen consumption, feeding and defaecation. *Physiological Zoology*, **60**, 659–677.

Görman Nordmark, M., Laynez, J., Schön, A., Suurkuusk, J. & Wadsö, I. (1984). Design and testing of a new microcalorimetric vessel for use with living cellular systems and in titration experiments. *Journal of Biochemical Biophysical Methods*, **10**, 187–202.

Graham, R. A. & Ellington, W. R. (1985). Phosphorus nuclear magnetic resonance studies of energy metabolism in molluscan tissues: intracellular pH change and the qualitative nature of anaerobic end products. *Physiological Zoology*, **58**, 478–90.

Hammen, C. S. (1983). Direct calorimetry of animals entering the anoxic state. *Journal of Experimental Zoology*, **228**, 397–403.

Hand, S. C. & Gnaiger, E. (1988). Anaerobic dormancy quantified in *Artemia* embryos: a calorimetric test of the control mechanism. *Science*, **239**, 1425–7.

Hawkins, A. J. S., Bayne, B. L. & Day, A. J. (1986). Protein turnover, physiological energetics and heterozygosity in the blue mussel, *Mytilus edulis*: the basis of variable age-specific growth. *Proceedings Royal Society London B*, **229**, 161–76.

Hawkins, A. J. S., Salkeld, P. N., Bayne, B. L. Gnaiger, E. & Lowe, D. M. (1985). Feeding and resource allocation in the mussel *Mytilus edulis*: evidence for time-averaged optimization. *Marine Ecology Progress Series*, **20**, 273–87.

Herold, J. P. (1977). Advantage of microcalorimetric investigations in cardiac energetic physiology: determination of oxidative efficiency in the isolated snail heart. *Comparative Biochemistry Physiology*, **58**A, 251–54.

Heusner, A. A. (1985). Body size and energy metabolism. *Annual Reviews of Nutrition*, **5**, 267–93.

Kaufmann, R. & Gnaiger, E. (1981). Optimization of calorimetric systems: Continuous control of baseline stability by monitoring thermostat temperatures. *Thermochimica Acta*, **49**, 63–74.

Lock, M. A. & Ford, T. W. (1983). Inexpensive flow microcalorimeter for measuring heat production of attached and sedimentary aquatic microorganisms. *Applied Environmental Microbiology*, **46**, 463–67.

Pamatmat, M. M. (1979). Anaerobic heat production in bivalves (*Polymesoda caroliniana* and *Modiolus demissus*) in relation to temperature, body size, and duration of anoxia. *Marine Biology*, **53**, 223–9.

Pamatmat, M. M. (1980). Facultative anaerobiosis of benthos. In *Marine Benthic Dynamics*, Belle W. Baruch Symposium Marine Science 11, ed. K. R. Tenore & B. C. Coull, pp. 69–90. Columbia: Univ. South Carolina Press.

Pamatmat, M. M. (1983). Simultaneous direct and indirect calorimetry. In *Polarographic Oxygen Sensors. Aquatic and Physiological Applications*, ed. E. Gnaiger & H. Forstner, pp. 167–75. Berlin, Heidelberg, New York: Springer.

Peaking, G. J. (1973). The measurement of the costs of maintenance in terrestrial poikilotherms: A comparison between respirometry and calorimetry. *Experientia*, **29**, 801–2.

Randzio, S. L. & Suurkuusk, J. (1980). Interpretation of calorimetric thermograms and their dynamic corrections. In *Biological Microcalorimetry*, ed. A. E. Beezer, pp. 311–41. London: Academic Press.

Shick, J. M. (1981). Heat production and oxygen uptake in intertidal sea anemones from different shore heights during exposure to air. *Marine Biology Letters*, **2**, 225–36.

Shick, J. M., De Zwaan, A. & De Bont, A. M. T. (1983). Anoxic metabolic rate in the mussel *Mytilus edulis* L. estimated by simultaneous direct calorimetry and biochemical analysis. *Physiological Zoology*, **56**, 56–63.

Shick, J. M., Gnaiger E., Widdows, J., Bayne, B. L. & De Zwaan, A. (1986). Metabolism and activity in the mussel *Mytilus edulis* L. during recovery from intertidal exposure. *Physiological Zoology*, **59**, 627–42.

Shumway, S. E., Scott, T. M. & Shick, J. M. (1983). The effects of anoxia and hydrogen sulphide on survival, activity and metabolic rate in the coot clam, *Mulinia lateralis* (Say). *Journal of Experimental Marine Biology Ecology*, **71**, 135–46.

Smith, R. R., Rumsey, G. L. & Scott, M. L. (1978). Net energy maintenance requirements of salmonids as measured by direct calorimetry: Effect of body size and environmental temperature. *Journal of Nutrition*, **108**, 1017–24.

Suurkuusk, J. & Wadsö, I. (1982). A multiple channel modular microcalorimeter. *Chimica Scripta*, **20**, 155–63.

Wadsö, I. (1987). Calorimetric techniques. In *Thermal and energetic studies of cellular biological systems*, ed. A. M. James, pp. 34–67. Bristol: John Wright.

Widdows, J. (1987). Application of calorimetric methods in ecological studies. In *Thermal and energetic studies of cellular biological systems*, ed. A. M. James, pp. 182–215. Bristol: John Wright.

Widdows, J. & Shick, J. M. (1985). Physiological responses of *Mytilus edulis* and *Cardium edule* to aerial exposure. *Marine Biology*, **85**, 217–32.

Zamer, W. E. & Shick, J. M. (1987). Physiological energetics of the intertidal sea anemone *Anthopleura elegantissima*. II. Energy balance. *Marine Biology*, **93**, 481–91.

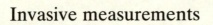

Invasive measurements

P. J. BUTLER AND A. J. WOAKES

Telemetry

Introduction

Telemetry, measurement at a distance, is not a single, easily definable technique, but encompasses a wide range of experimental methods. In general, telemetry equipment has to be specifically designed for a particular experiment. The number of data channels, type of transducers, size of animal, required lifetime and range are all likely to vary markedly between different experimental situations. This review can therefore only give basic information about the technique, and has been split into two sections, the first reviewing recent studies *using* telemetry, in the hope that here can be found references to experimental protocols that are similar to the reader's requirements. The second section will introduce basic techniques used in telemetry, and includes detailed descriptions of how to build a simple transmitter for monitoring electrocardiograms (ECG) and body temperature.

Biotelemetry can be a difficult technique for those with little or no experience of electronic design or construction, for, unlike the majority of subjects covered in this book, most of the equipment cannot be purchased. Furthermore, the devices are expected to operate in hostile environments, so a great deal of perseverance is also required if each failure mode is to be identified and eliminated. However, the use of biotelemetry equipment allows measurements to be made that would not otherwise be possible, particularly in those fields where:

(i)   disturbance of the experimental animal may well invalidate the acquired data;
(ii) information has to cross some form of barrier;
(iii) the environment is hazardous or difficult for the experimenter to operate in;
(iv) physiological data are required from naturally behaving animals.

In these cases, the application of biotelemetry will amply repay the effort expended in acquiring expertise in the technique.

The transmitters may be small, with a volume of less than 1 ml, or large, implanted, ingested or mounted externally, carry their own batteries or depend on radio power beamed into them; operate using ultrasonics as the transmission medium at a few tens of kHz, or infra-red radiation at 10^{14} Hz, modulate this carrier by variation in frequency, amplitude or some parameter of a stream of pulses, and even reverse the principle, or trigger internally placed stimulators by an externally mounted command transmitter.

Whatever method of transmission is used, there must be a signal to transmit and the most easily obtained signals related to physiological variables are those that either originate in the animal itself, e.g. electrocardiogram (ECG), electromyogram (EMG), and merely require amplification before transmission, or those for which the sensor does not have to be in direct contact with the tissues of the animal and requires very little power to energise it, e.g. temperature. Not surprisingly, therefore, these are the most regularly telemetered signals. More useful physiological variables such as respiratory air flow, blood flow, blood pressure and partial pressure of gases in respiratory air and blood have been determined much less frequently by telemetric techniques.

A large number of variables may be transmitted simultaneously. Such multichannel systems are exceptionally complex and most make use of digital techniques. The basic principles of these systems have been adequately reviewed by Kimmich (1975) and those wishing to progress to these devices are advised to consult this reference. More modest systems can be constructed by those with only a limited knowledge of electronics, and the transmitter described later in this paper should be capable of a good performance if correctly built.

Often the animal carrying the telemetric device is in a confined (even if largely unrestrictive) space and its behaviour can be regularly monitored and related to the physiological data. It is becoming more common, however, to release animals with telemetric equipment into their natural environment, in which case it is desirable to have some means of monitoring their behaviour. Thus, using transmitters to track animals may become an essential part of physiological studies, and the recent use of satellites such as the NIMBUS-6 and ARGOS systems (Jennings & Gandy, 1980 – dolphins; Priede, 1984 – basking sharks), makes this an even more powerful tool. Systems are being developed that enable the position of relatively small animals such as golden eagles, *Aquila chrysaetos*, and Andean condors, *Vultur gryphus*, to be determined (Strikwerda *et al.*, 1986; French, 1986). It is also intended to incorporate the transmission of physiological signals to the satellite (Howey *et al.*, 1984).

Recent telemetric studies of animal physiology

Telemetry has been used to study a wide range of physiological variables in a number of different animals ranging from crabs (Walcott, 1980; Graham, 1981) to polar bears (Øritsland, Stallman & Jonkel, 1977) and much important data, related to respiratory and cardiovascular physiology, have been obtained by monitoring body temperature alone from animals such as tuna, lamnid sharks, alligators, tits, emperor penguins, muskrats, ground squirrels and northern elephant seals under natural or semi-natural conditions (Carey *et al.*, 1971; Smith, 1975; Boyd & Sladen, 1971; MacArthur, 1979; Wang, 1978; McGinnis & Southworth, 1971; Reinertsen & Haftorn, 1986). However, the present review will only attempt to cover the various applications of telemetry during one particular activity state in animals, namely, exercise, partly because this is an area in which the authors themselves are involved but also because this is an aspect of animal physiology where telemetry clearly comes into its own. Ideally, the animal needs to be free and not tethered to recording equipment. Most of the studies on exercise involving telemetric techniques have been confined to fish, birds and mammals, so it is these three groups of vertebrates that will be considered.

Fish

Activity. An important aspect of studying exercise in any animal is to be aware of its normal level of activity. An understanding of the migratory behaviour of commercially important marine fish is of particular importance if an optimum balance between exploitation and conservation of stocks is to be achieved, and acoustic telemetry has proved most useful in this respect (Harden Jones & Arnold, 1982). Monitoring displacement alone, however, does not give an indication of active expenditure of energy in marine fish. In fact the simultaneous recordings of tail beat frequency and depth have indicated that the eel actively swims downwards during its vertical migration and does not, as had been suggested, glide downwards as a result of being negatively buoyant (Westerberg, 1984). Most fish, however, may not be very active for most of their life (Holliday, Tytler & Young, 1972/73).

Indirect indicators of metabolic rate. Once the animals' normal levels of activity are known, these can then be simulated in water channels, wind tunnels or on treadmills, and a number of physiological variables can be recorded, using either conventional hard-wired or telemetric techniques. The relationship between heart rate and aerobic metabolism has been studied in a number of species in such conditions in an attempt to use heart rate as an indicator of metabolism in wild free-living animals. The relationship between heart rate and oxygen uptake is given by the Fick equation:

$$\dot{V}_{O_2} = HR \times SV \,(Ca_{O_2} - C\bar{v}_{O_2})$$

where $\dot{V}_{O_2}$ = rate of oxygen uptake
  HR = heart rate
  SV = cardiac stroke volume
  Ca_{O_2} = oxygen content of arterial blood
  $C\bar{v}_{O_2}$ = oxygen content of mixed venous blood

It is clear that variations in SV and $(Ca_{O_2} - C\bar{v}_{O_2})$ will modify the relationship between $\dot{V}_{O_2}$ and HR, and that in poikilotherms the 'calibration' temperature must be the same as the field temperature. Priede & Tytler (1977) found that there was not a good linear relationship between these two variables in a number of fish species. Priede & Young (1977), however, were able to use the data to determine maximum oxygen consumption for a given heart rate telemetered from wild brown trout in a

Fig. 1. Cumulative frequency distribution curves of time at different heart rates for two brown trout (*Salmo trutta* L.) in a Scottish loch. Heart rate (beats min^{-1}) was determined from the ECG transmitted by an ultrasonic transmitter. Heart rate is also expressed as a percentage of the assumed active (maximum) heart rate (90) and indicates the maximum possible oxygen consumption (Max $\dot{V}O_2$) which is shown on the third ordinate scale. (Redrawn from Priede & Young, 1977.)

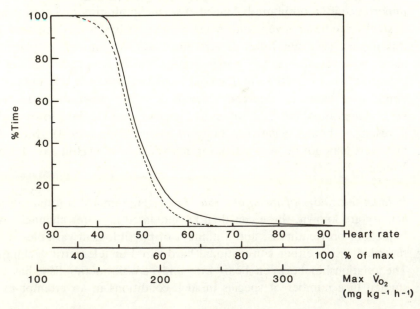

lock and to estimate maximum mean oxygen uptake over the recording period (Fig. 1).

Clear linear relationships have been obtained from rainbow trout, swimming in a water channel, between telemetered EMG signals obtained from the 'mosaic' (white) epaxial muscle mass and oxygen uptake (Weatherley *et al.*, 1982), and between average EMG activity from the *levator arcus palantini* muscle and oxygen uptake (Rogers & Weatherley, 1983). It is apparent from these studies, however, that much care is required when trying to use an indirect indicator of oxygen uptake, and that it is not always justified to extrapolate directly from the situation in the laboratory to the field. Fairly extensive laboratory studies have been performed on pike, *Esox lucius*, and the evidence suggests that heart rate measured by telemetry would provide a good indication of metabolic rate in the field (Armstrong, 1986).

Birds

Flight. Unlike the situation with fish there is little or no commercial pressure for improving our understanding of migration, and thus the physiology of flight, in birds. It is merely a desire to find out the mechanisms involved in these activities that has driven scientists to develop a number of ingenious techniques for use in such studies. Migration routes and the distances covered have largely been determined by ringing studies and radar but, as already indicated, the use of telemetry, particularly via

Fig. 2. Traces of electromyograms from the pectorialis muscles of a pigeon flying at 10 m s^{-1} in a wind tunnel (*a*) and during a free spontaneous flight (*b*). During the latter the EMG was transmitted by an FM transmitter attached to the back of the bird. The time marker is in seconds. (From Butler *et al.*, 1977).

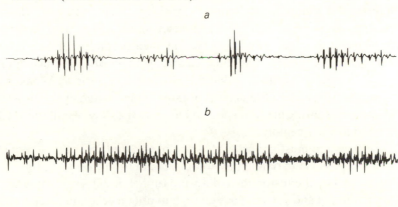

satellite links, is becoming more feasible (Strikwerda *et al.*, 1986; French, 1986).

Most data on the physiological adjustments to flight have been obtained from birds flying in wind tunnels, but in one such study (Butler, West & Jones, 1977) it was noted that the flight pattern was somewhat different from that in freely flying birds from which EMG of the flight muscles was telemetered (Fig. 2). Telemetry is an obvious technique to use with flying birds, but unfortunately it can have its limitations. Hart & Roy (1966) published the first comprehensive set of respiratory data obtained by radiotelemetry from flying birds (pigeons). To prevent the loss of the bird and transmitter and, presumably, to keep the transmitter within range of the receiver, the birds were restrained by a nylon line tied to a harness so that flights were, on average, of only 9 s duration. Thus, the birds had barely taken off and were nowhere near a steady state, so that all the measured values were considerably different from what they would have been after several minutes of flight (see Butler *et al.*, 1977). Similar criticisms can be levelled at the subsequent papers by these authors (Berger, Roy & Hart, 1970*a*; Berger, Hart & Roy, 1970*b*). A long range (80 km) transmitter was attached externally to herring gulls, *Larus argentatus*, and used to monitor heart rate during flights of up to 20 km in distance (Kanwisher *et al.*, 1978). This is clearly a great advance over the earlier systems as far as the duration of the flights is concerned, but external mounting and the possible lack of information on the bird's behaviour during such long flights are still defects of the technique.

It has recently been demonstrated, however, that implantable, short-range transmitters can be used with freely flying birds engaged in long flights (Butler & Woakes, 1980). Barnacle geese, *Branta leucopsis*, were raised from eggs, hatched in a laboratory incubator, imprinted (Lorenz, 1970) on a human and trained to fly behind a pick-up truck containing their foster parent. A two-channel, FM radiotransmitter (Fig. 3) was completely implanted into the trained geese and measurements of heart rate and respiratory frequency were obtained during flights of an average duration of 14.4 min and at a mean air velocity of 18.7 m s^{-1} (Butler & Woakes, 1980). A perfectly clean ECG signal was obtained (Fig. 4) and by taking a cine film of the geese during flight, respiratory activity (i.e. lung ventilation) could be related to wing beating.

Perhaps the most important feature of the study was the demonstration that it is possible to obtain physiological data from free flying birds that are unstressed by the restrictions of a wind tunnel or externally mounted leads or equipment and yet are close enough to obtain accurate measurements of their air speed and behaviour.

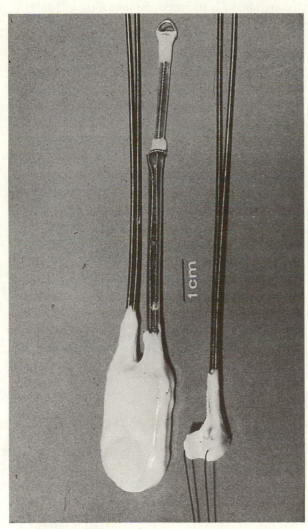

Fig. 3. Encapsulated, 2-channel FM radiotransmitter used for transmitting respiratory frequency and heart rate from freely flying geese. The bipolar ECG electrode was placed underneath the sternum, the transmitter was placed in the abdominal cavity and the thermistor was placed in the lumen of the trachea. The silicone 'wings' were secured around the outside of the trachea.

Swimming and diving. Although flight is the predominant form of exercise in birds, it is by no means the only form, and another aspect of the behaviour of air-breathing vertebrates that has fascinated physiologists for many decades is the ability of the aquatic species, particularly the homeotherms, to remain submerged under water for extended periods while feeding. The current view as to how this is achieved is largely the result of data obtained via telemetry from freely diving birds. Millard, Johansen & Milsom (1973) used radiotransmitters to record heart rate and blood flow in a few major arteries of adélie and gentoo penguins, *Pygoscelis adeliae* and *P. papua*, respectively. Radiotelemetry has since proved a useful technique for monitoring heart rate from freely diving birds and it is now clear that, as indicated by Millard *et al.* (1973), the cardiac response during most natural dives is completely different from that obtained during involuntary submersion and may be more like that seen during exercise in air.

Using totally implanted pulse interval modulate (PIM) transmitters

Fig. 4. Traces from a ♀ barnacle goose showing ECG, heart rate and respiratory frequency obtained from an implanted 2-channel radiotransmitter (see Fig.3) before, during and after a flight of 11 min 52 s duration. The bird was imprinted and trained to fly behind a moving truck. (*a*) take off; (*b*) steady flapping flight at 22 m s^{-1}, 5 min after take-off; (*c*) 3 min after landing. The vertical dashed lines in (*a*) indicate when (i) the truck starts to move; (ii) the bird begins to run and flap its wings; (iii) the bird is airborne. (From Butler & Woakes, 1980).

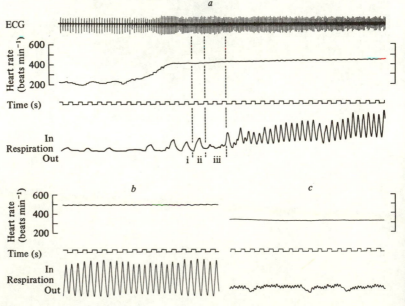

(Fig. 9) Butler & Woakes (1979) demonstrated that, contrary to the 'classical' hypothesis, there is no maintained bradycardia during voluntary dives in pochards and tufted ducks, *Aythya ferina* and *A. fuligula* respectively (Fig. 5). Monitoring heart rate by telemetry has subsequently demonstrated that there is no diving bradycardia during voluntary submersion in cormorants *Phalacrocorax auritus* (Kanwisher, Gabrielsen & Kanwisher, 1981), Humbold penguins, *Spheniscus humboldti* (Butler & Woakes, 1984) and redhead ducks, *A. americana* (Furilla & Jones, 1986). Thus, the absence of this bradycardia during voluntary submersion raised the question as to whether all of the other adjustments characteristic of the 'classical' response (widespread peripheral vasoconstriction, increased anaerobiosis, reduced aerobic metabolism) are present during natural dives.

Further studies on tufted ducks and Humboldt penguins have shown that there is no reduction in oxygen uptake during normal voluntary dives in these animals. In fact, in the ducks it is 3.5 times the resting value and not significantly different from that during maximum sustained swimming speed at the surface, whereas in the penguins it is 25% above (but not significantly different from) the resting value (Woakes & Butler, 1983; Butler &

Fig. 5. Mean changes in heart rate ($\pm$ SE) in tufted ducks, *Aythya fuligula* during voluntary feeding dives to a depth of 1.9 m. R = resting heart rate. Heart rate was determined from the ECG transmitted from an implanted PIM radiotransmitter (see Fig. 9). Times of submersion and surfacing are indicated by the vertical lines. (From Butler, 1987).

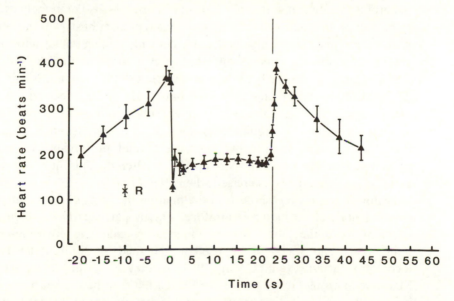

Woakes, 1984). It appears from these values and from knowledge of usable oxygen stores and normal dive durations that these (and presumably other birds that dive) normally remain submerged for periods that are well within their aerobic limits. They seem to metabolise complete aerobically when submerged and to replace the oxygen stores during the period at the surface between dives. There is no evidence to suggest that under normal circumstances they resort to anaerobiosis and accumulation of lactic acid (Butler & Stephenson, 1987). It has been proposed, therefore, that the cardiovascular and metabolic responses to voluntary diving in these animals are normally similar to those seen during exercise in air (see Butler, 1987).

Continued studies on tufted ducks with implanted radiotransmitters have demonstrated that under certain circumstances (e.g. if travelling long horizontal distances or if disoriented and unexpectedly trapped under ice), there is a progressive or more rapidly developing bradycardia (Stephenson, Butler & Woakes, 1986). This suggests that these animals are able to switch, under such circumstances, to the 'classical' oxygen conserving response. Clearly these data are only indicative of what may be happening; no definite conclusions can be drawn. There is no doubt, however, that information obtained via telemetry from freely diving birds has completely changed our understanding of the physiological adjustments to diving in these animals.

There is a close linear relationship between heart rate and oxygen uptake in tufted ducks during steady-state swimming at different velocities (Woakes & Butler, 1983) indicating that heart rate may be a useful measure of aerobic metabolism under natural conditions. It is also extremely useful to be able to measure heart rate, via telemetry, from birds with no other form of instrumentation attached to obtain truly resting values and to be able to determine the effect of different types of instrumentation.

It is clear from the study by Woakes & Butler (1986) that although the direct recording techniques they used have little or no effect on oxygen uptake (provided the animals are given sufficient time to settle after attaching the recording equipment), they do cause an elevation of heart rate. Thus the relationship between heart rate (and, presumably, the other measured variables) and oxygen uptake is altered. This indicates that extreme caution must be exercised when attempting to use heart rate as an indicator of aerobic metabolism in the field on the basis of one particular experimental set up in the laboratory. This is particularly so when the relationship in the laboratory is obtained by changing environmental temperature and yet is to be used in the field at different levels of locomotory activity (e.g. Owen, 1969; Wooley & Owen, 1977; Ferns, Macalpine-Leny & Goss-Custard, 1979). It is felt that, before heart rate can be usefully used as an indicator of aerobic metabolism in the field, more exhaustive studies on the relationship between these two variables in a number of different situations must be performed. Calibration in the field against the doubly-labelled water method (see Tatner & Bryant – this volume) should also be attempted.

Mammals

Running. Mammals are generally larger, and some at least can be trained to carry external packages more readily, than birds. It has therefore been possible to implant pressure and flow transducers into a number of larger mammals and then to telemeter the data via a transmitter either implanted in, or carried in a backpack by the exercising animal. Thus, it has been possible to obtain valuable information on the cardiovascular adjustments to exercise in freely running animals. Unlike the situation in humans (Wade *et al.*, 1956), there is no reduction in mesenteric and renal blood flows during severe, prolonged exercise in Alaskan sled dogs or in mongrels trained to run beside a van containing the receiving equipment (Van Citters & Franklin, 1969; Vatner *et al.*, 1971). However, if the increase in heart rate (and therefore in cardiac output) during severe exercise was drastically reduced, by producing a complete AV block in the dogs, there would then be substantial increases in resistance in the iliac and mesenteric vascular beds. Similar studies with similar results have also been performed on baboons, *Papio anubis* (Vatner, 1978).

Perhaps the most impressive athlete of all mammalian species is the horse in which maximum oxygen uptake ($\dot{V}_{O_2max}$) may be 40 times the resting value (Snow & Harris, 1985). Systemic blood pressure, foot fall and a number of respiratory variables, including oxygen uptake, have been telemetered from exercising horses (Hörnicke, Engelhardt & Ehrlein, 1977; Hörnicke, Meixner & Pollmann, 1983).

An interesting feature of running, in horses, is the tight 1:1 relationship between step frequency and respiratory frequency during the canter and gallop (Fig. 6). This forces the horse to ventilate its lungs at frequencies up to 140 breaths min^{-1}. It has been suggested that this may facilitate ventilation. It is equally possible, however, that the frequency of limb movements may impose a restriction on respiratory frequency and be a contributory cause of the hypoxaemia that occurs in horses during exercise (Bayly *et al.*, 1983; Woakes, Butler & Snow, 1987). The use of telemetric techniques has much more to offer in the attempts to understand the factors that limit performance, particularly under racing conditions, in these supreme athletes.

Diving. The monitoring of heart rate alone via telemetry in unrestrained and freely diving mammals (including humans – Butler & Woakes, 1987) has provided much important information and, as is the case with birds, has altered our views of the physiological adjustments that occur during normal diving (Harrison, Ridgway & Joyce, 1972; Jones *et al.*, 1973; Fedak, Pullen & Kanwisher, 1987). The first of these studies indicated that in grey seals, *Halichoerus grypus*, in a holding tank, heart rate during trained dives (i.e. when the seals dived on command, signalled by an underwater light) is higher than during involuntary dives. In the second study there were variable cardiac responses during spontaneous feeding dives of harbour seals, *Phoca vitulina*, in a tank, with one seal exhibiting no bradycardia at all during 20% of such dives. In the third study, heart rate was measured (via acoustic transmitters) from grey seals at different swimming speeds in a swimming flume and in harbour seals while free at sea. While ventilating their lungs heart rate was 110–120 beats min^{-1} and when the seals were swimming (or resting) the under-water heart rate was approximately 35 beats min^{-1}. Neither of these values varied in any systematic way with oxygen uptake (or swimming speed; Fig. 7), although mean heart rate (together with oxygen uptake) did increase with swimming speed as the seals spent a greater proportion of their time at the surface, at higher velocities. Interestingly, much lower heart rates (<10 beats min^{-1}) were

Fig. 6. The relationship between respiratory frequency and step frequency in five horses during the walk, trot and gallop. Mean and range of respiratory frequency in the horses while standing is also given. All data were transmitted telemetrically. The dashed line is the line of equality between respiratory and step frequencies. (Redrawn from Hörnicke *et al.*, 1983.)

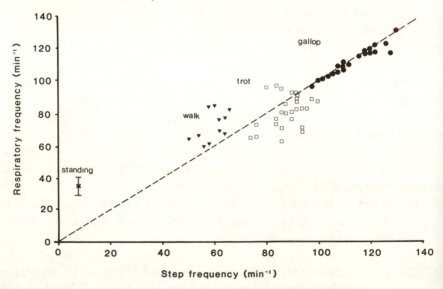

obtained when the seals were prevented from surfacing by closing off access
to the respirometer in the flume. Similar patterns of heart beat were seen in
the free range harbour seals. Unusually low heart rates were observed only
on two occasions, when one seal engaged in aggressive behaviour with
another. There were anticipatory increases in heart rate upon surfacing. It is
interesting to note that Hill *et al.* (1983) recorded a gradual reduction in
heart rate during longer (up to 52 min) dives in Weddell seals, *Leptony-
chotes weddelli*, and that this is similar to the situation in tufted ducks
swimming long horizontal distances underwater for food (e.g. under ice).

It has been suggested (probably with some justification) that the events in
the rest of the cardiovascular system and the type of metabolism seals have,

Fig. 7. The relationship between heart rate and oxygen
consumption in a grey seal in a swimming flume. ($\triangle$), heart rates during
dives; ($\triangledown$), heart rates while at the surface, breathing; ($\blacktriangledown$), heart rates
over complete dive-breathing cycles. Heart rate was obtained from an
acoustic transmitter attached to the seal. (Redrawn from Fedak, 1986.)

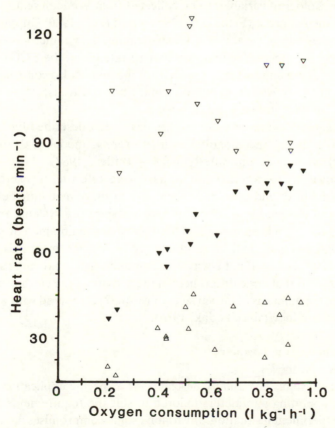

can be deduced from the mean heart rate (Fedak *et al.*, 1987). However, heart rate will only ever be an *indicator* of other events. Clearly these animals may engage in high levels of activity when under water and 75–85% of their time is spent below water when engaged in their normal diving behaviour. To a far greater extent than birds, these marine mammals lead a sub-aquatic existence, with brief visits to the surface to replenish the oxygen stores (and to remove CO_2). It might be more appropriate to consider the increase in heart rate accompanying these visits to the surface, as a ventilation *tachycardia* rather than the lower heart rate, when submerged, being seen as a bradycardia (cf. Belkin, 1964 for turtles). The normal method of exercising in seals is while under water, and the important question is whether there are differences in the cardiovascular response to swimming near the surface when there is easy access to air (or during dives of short duration) compared with that which occurs during deep (and, therefore, longer) dives.

Recent use of solid state electronics (Hill, 1986) has enabled a number of different physiological variables to be collected from Weddell seals, diving naturally in the Antarctic (Falke *et al.*, 1985; Qvist *et al.*, 1986; Guppy *et al.*, 1986). Attached to each seal was a peristaltic pump and a microprocessor. The latter not only measured and stored heart rate (from the ECG), body temperature, depth and swimming velocity of the seal, it also controlled the pump that not only withdrew blood samples but could also inject radio labelled metabolites in freely diving seals.

Data obtained in some of these studies have been described by Guppy *et al.* (1986) as being 'incompatible with an exercise model of diving' (as proposed from studies on tufted ducks – Butler, 1982). That rather depends, however, on whether the authors were referring to exercise in fully terrestrial birds and mammals when, as a result of matching between the delivery of oxygen by the circulatory *and* respiratory systems and the demands of the exercising muscles, blood gases are maintained at steady levels, or whether they had in mind exercise in marine mammals when ventilation is invariably (if not always) intermittent. There are large swings in blood gases, so that large differences in the activity of the cardiovascular system when the animal is at the surface (or resting?) compared with when it is submerged are inevitable (Fedak, 1986).

Techniques in telemetry
Basic principles
Biotelemetry systems are usually designed for a specific application, with the various operating parameters tailored to suit the requirements of the particular experimental situation. Invariably, some compromises have to be

made, and only the individual experimenter can decide on the order of the priorities involved. However, the main limitation in biological research is usually based on the maximum allowable size of the transmitter. This should be as small as possible so as to produce the least interference with the behavioural or physiological response being investigated. This in turn restricts the available power by limiting the number or size of the batteries, and one then has to balance the requirements of transmission range and operational lifetime, both of which make demands on the fixed amount of energy stored in the batteries. To some extent various techniques exist that can partially overcome these restrictions. The system described in detail later in this section uses a pulse mode of transmission, with the information being encoded as a variation in the time between each pulse. The radio frequency transmitter is then switched on only for the limited duration of each pulse and is switched off between pulses. This saves a great deal of power and can easily extend the lifetime of a battery pack by a substantial amount. The actual improvement depends on both the on-off ratio and the increase in power required to compensate for the lower efficiency of this pulse mode of transmission.

Extending the operational life of the powerpack is only justified if one is sure that the mechanical integrity of the transmitter will also survive for this period. Transducers and their leads are invariably the weakest link in a biotelemetry system, and the illustrated transmitter reduces this potential problem by restricting the parameter to that requiring no complex transducer. Even so, electrode leads are often the source of most failures. The transmitter described later has been designed as an integral unit, with the electronic package mounted rigidly alongside the replaceable battery. A separate antennae is not used, the RF power being radiated from the tuning coil that surrounds the electronic package, thus giving the largest possible radiating area. The actual radio frequencies used by these transmitters are determined to a major extent by legal restrictions so care should be taken to ensure that the law is not inadvertently broken by operating outside the frequency bands that are allocated to biomedical telemetry devices. Operators outside of the United Kingdom are advised to contact their own regulating agency for information on these bands. Details are given with the transmitter of the component changes required to modify the operating frequency, which will otherwise meet the UK regulations. Note that certain countries, including the UK also require each operator to be licensed.

Components
Yet more restrictions are imposed on the construction of bio-telemetry transmitters by the limited range of suitable electronic com-

Table 1. *Resistors*

Type no.	Construction	Temp. Coef. ppm/deg.C.	Resistance Range ohms	Dimensions L × w × t or L × dia. (mm)	Manufacturer	U.K. Supplier
SFR16T	Metal film	100	10R – 3M	3.7 1.7	Mullard	STC
RCO1	Surface mount 1206 size	200	1R – 10M	3.2 1.6 0.6	Mullard + others	STC + others
CR21	Surface mount 0805 size	300	10R – 10M	2.0 1.25 0.45	Kyocera	Unitel
RSX-O	Carbon film axial leads	400	1R – 4M7	3.2 1.0	Roederstein	Steatite
RSX–00	Carbon film axial leads	400	1R – 1M	2.2 0.7	Roederstein	Steatite
KLM-2	Metal film axial leads	100	100R – 33K	2.2 0.8	Siegert	Lemo
RKL-2	Carbon film axial leads	700	10R – 22M	2.2 0.8	Siegert	Lemo

ponents that are available, and one often has to design a system using this as a major consideration. The suitability of a component for use in biotelemetry will depend on the parameters discussed in the previous section, but in general, passive devices must be as small as possible while active components must also be able to operate on the minimum of supply currents and voltages. Fortunately, these qualities are increasingly required by consumer electronics manufacturers and not only is the range of suitable devices increasing but their price is also falling rapidly. The various tables give the important parameters of a range of components while their suppliers are listed at the end of the chapter. All of the items needed to construct the ECG transmitter can be found in these tables but they are not intended as an exhaustive survey of the field, and further manufacturers and distributors can also be found in the trade press.

These form the bulk of components used in circuit design, and so any reduction in their size can significantly reduce the volume of the final product. The SFR16T series is the standard resistor used in full size construction and can also be used where transmitter volume is not critical.

Ceramic capacitors are generally small components, and standard packages can be used. The grade of ceramic used in their construction not only determines their capacity:size ratio, but also their temperature coefficients and stability. Where large values of capacitance are required, electrolytic capacitors will have to be used. Those based on solid tantalum have exceptional capacitance for their size, but even so, they tend to be too large for telemetry use at values above 47 μF. Variable capacitors may be used to

Table 2. *Capacitors, fixed and variable*

Type no.	Construction	Capacitance Range F.	Dimensions L × w × t or L × dia. (mm.)	Manufacturer	U.K. Supplier
Fixed					
EDPT	Disc ceramic radial leads	470p–1n	1.7 1.7 1.0	Stettner	Steatite
SMD 0805	Surface mount ceramic	.47p–22n	2.0 1.25 0.5 min 1.3 max	Mullard + others	STC + others
SMD 1206	Surface mount ceramic	.47p—100n	3.2 1.6 0.5 min 1.6 max	Mullard + others	STC + others
TAJ	Surface mount tantalum	0.1μ–100μ	3.2 1.6 1.6 7.3 4.3 2.9	STC, Kemet	STC
Minitan MTZ	Cylindrical axial lead	0.33μ–47μ	3.2 1.8 6.4 3.8	Corning	Corning
Tantalex 189D	Cylindrical axial leads	0.47μ–15μ	2.3 1.3 6.4 2.0	Sprague	Sprague
Variable					
9401/2	Open	0.2/0.6p–5.0/20p	1.0 3.5	Johansen	Tekelec
3S-Triko	Open	1.0/2.5—1.0/22p	1.25 3.2	Stettner	Steatite
3018	Sealed	1.4/4.0—0.8/44p	4.6 4.1 3.1	Stettner	Steatite
TZB04	Sealed	1.4/3.09—8.5/40p	4.5 4.0 3.0	Murata	ACM

tune the radio frequency sections of transmitters. Most consist of ceramic discs, one of which can rotate relative to the other and, by changing the area of overlap of metalised portions of the discs, so adjust the capacitance.

A wide range of transistors and diodes are available in the SOT-23 package. Practically all CMOS (complementary metal oxide semiconductor) digital integrated circuits can be found in SO (surface outline) packs, and are suitable for low-power applications where more than 3 V is available. The range of analogue integrated circuits suitable for telemetry applications is more limited. Considerable savings in space can be made by using dual or quad devices mounted in a single package.

Mercury and silver oxide cells are widely used as power sources, with mercury having a better power:volume ratio, and silver oxide a higher voltage, with both giving a stable operating voltage throughout the discharge period. A range of cells based on lithium are now available, providing voltages from 1.2 to 3.6 V. In larger sizes, these have considerably greater capacities than other forms of battery. A limited range of zinc-air cells are also now being produced. These have approximately twice the capacity of equivalent mercury cells, but do require access to air, and

Table 3. *Analogue integrated circuits: low power, miniature packages*

Type no.	Function	Minimum voltage V	Operating current μA	Package	Manufacturer	U.K. Supplier
TLC251	Op. amplifier programmable	1	10–1000	SO8	Texas	STC, Surtech
TLC25L2	Dual op. amp.	1	20	SO8	Texas	do.
TLC25L4	Quad op. amp.	1	40	SO14	Texas	do.
LP339	Quad comparator	2	60	SO14	National Semiconductor	Hi-tek
7660DY	Voltage doubler	1.5	170	SO8	Siliconix	Siliconix

The following devices are not, at the time of writing, available in small package sizes.

LM4250	Op. amp, also dual & quad	2	7.5–100	—	National Semiconductor	Hi-tek
ICL8021	Op. amp. also dual & triple	2	10–300	—	Intersil	Intersil
ICL7611	Op. amp, also dual and quad	1	10–1000	—	do.	do.

Package sizes: L × t × W (W + leads) mm.

SOT23 (transistor)	2.9	0.9	1.3	(2.5)
SO8	5.1	1.7	4.0	(6.2)
SO14	10.2	1.7	4.0	(6.2)

Table 4. *Batteries*

Type no.	Chemistry	Voltage V	Capacity mA.hr.	Dimensions dia. × h mm	Weight g	Manufacturer	U.K. Supplier
LD319	Silver oxide	1.5	15	5.7 2.7	0.3	Ucar	Short
SR48	Silver oxide	1.5	75	7.9 5.4	1.1	—	All Batteries
R13H	Mercury	1.4	85	7.9 5.4	1.0	Duracell	do.
MP675H	Mercury	1.4	210	11.6 5.4	2.6	do.	do.
DA675H	Zinc–air	1.4	400	11.6 5.4	1.9	do.	do.
LC01	Lithium CuO	1.5	500	14.1 12.0	4.5	Saft	do.
BEL	Lithium thionyl chloride	3.6	370	22.5 7.0	5.0	Tadiran	Bauch
LS3	do.	3.6	850	14.5 25.0	8.5	Saft	All Batteries
AL125	do.	3.6	1400	32.0 10.0	19.0	Altus	do.

are not therefore suitable for use in implants. The table illustrates only a small selection of the wide range of batteries that are available.

A long life biopotential transmitter

This transmitter (Figs. 8, 9), originally developed by Fryer (1970) and modified by the authors, uses a pulse interval modulation (PIM) technique and transmits both ECG and temperature data in the VHF

Fig. 8. Circuit diagram of the pulse interval modulated transmitter. The component values are:

Resistors (ohms)

R1	270 K	R6	Note 1	R12	39 K
R2	100 K	R8	100 K	R13	15 K
R3	10 K	R9	3.9 M	R14	68 K
R4	330 K	R10	28 K	R15	100 (Note 2)
R5	1 M	R11	27 K		

Capacitors (farads)

C1	2.2 μ	C5	1 n	C9	10 p
C2	1 n	C6	1 n	C10	22 p
C3	2.2 μ	C7	1 n	C11	Note 3
C4	1 n	C8	15 p		

Transistors

T1, 2, 3, 5 BCW32R   T4, 6 BCW29R   T7 BFS17R
All transistors in SOT23 packages.

Notes
(1) Resistor R6 should be chosen to give the desired pulse frequency. A value of 20 K will give a frequency of approximately 400 Hz. This resistor can be replaced by a thermistor (47 K) if temperature information is required. For accurate frequency adjustment, a suitably selected, high value resistor may be taken from the junction of R6:T3 to the positive supply (see R7, Fig. 10).
(2) Resistor R15 determines the RF output power of the transmitter and therefore its power consumption. With the value given, the range will be of the order of 20 m, with a current drain of 40 μA.
(3) Capacitor C11 is used to adjust the RF frequency of the transmitter and can either be a fixed or variable component. It should not exceed the value of C10.

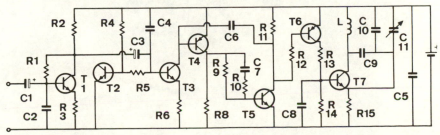

frequency band. It is capable of giving an excellent performance with both a low-power consumption and a range adequate for most biological experiments. A decoder is an essential partner to this transmitter as the pulse frequency is far too high to be counted manually (see section: Decoders and receivers). As the ECG can contain frequency components up to 100Hz, a pulse frequency (effectively the sampling rate) of 400 Hz has been chosen for this example. EMG signals will require appropriate changes to this parameter. The mean sampling rate varies with temperature, at approximately 40 Hz/°C. The pulse duration is approximately 40 μs, this being the shortest that the authors' receiver (Sony CRF5090) can follow without distortion. This gives a duty cycle of approximately 2%, and while over 1 mA is taken by the radio frequency section during the pulse, the average current is less than 40 μA for a range of over 20 m. Powered by a 675 mercury cell, this transmitter will operate for approximately six months. Built using standard components, the transmitter should weigh 5.0 g including the batteries, but excluding encapsulation. By using the smallest possible components and an LD 319 cell, the transmitter has been built down to a weight of 0.72 g. Details of the operation of this circuit can be found in Fryer (1970) and Woakes (1980).

Fig. 9. The major components of the PIM transmitter. The antenna is cast into the epoxy resin frame, which will also hold the electronic package and the R13 mercury cell shown positioned by the frame.

Transmitter construction

If the very smallest dimensions are required for the final transmitter, a number of special tools are necessary. A temperature controlled soldering iron with the smallest possible tip is absolutely essential, for any attempt to use an unsuitable iron will result in bridged contacts and ruined components. The recommended values are a tip temperature of 300 °C and a 1 mm maximum tip diameter. Most of the construction will have to be done under a binocular microscope, but this, along with fine forceps, springbow scissors (for cutting fine copper wire) and scalpels should be to hand in most physiology laboratories. Assorted files, a small hacksaw, hand drill and a set of needlefiles complete the list.

The normal 'cordwood' construction method uses components in matched packages, closely packed together and mounted between printed circuit end boards. However, this can be modified by using one board, for mechanical rigidity only, and bringing all the component leads to the opposite surface, where they may be interconnected with fine wire soldered into place between the pins and encapsulated in epoxy resin to form a solid component block (Fig. 9). Figure 10 illustrates the high packing density that can be achieved by this method, and also acts as a guide for the actual layout of the ECG transmitter although the plan may require modification according to the component size. The actual dimensions depend on the battery being used to power the transmitter. The cell determines the width (remembering to allow for the added bulk of the coil) and height of the package, the length alone being adjustable to allow some flexibility in the construction phase.

Take each component, as it is required, from the test jig. This will ensure that the correct component is always used, and prevent the loss of any of these small devices. Preform the component leads so that they all exit from one end of the package or, in the case of the SOT23 transistors, bend their leads down and around the underside of the package, and prepare to mount the components with epoxy, upright onto the base board. This may be thin plastic or glass, or stripboard filed or machined down to 0.3 mm thickness, but retaining an area of copper strip for applying power and test signals during construction. Firstly, epoxy the battery contact, a strip of thin (0.1 mm) stainless steel 2 mm wide and 10 mm long, onto the board and insulate it from the components, again with a thin coat of epoxy. Use standard 2-component epoxy adhesive, and heat to approximately 80 °C to shorten the curing time. If you are not following the guides, then take a careful note of where every component is placed so that each lead may be later identified. Furthermore, try to arrange the components so that their leads do not touch except where required, and that the interconnecting

Fig. 10. The arrangement of the components on the baseboard. The subminiature components are held in place by epoxy resin onto the baseboard along with the negative battery contact strip, to which R15 is directly soldered. The lower portion of the figure illustrates the manner in which the components are interconnected. The arrowed connections are made to the transistor bases, otherwise hidden below their emitter contacts along the side of the component package. Using RSX-0 resistors, SOT23 transistors and MTZ tantalum capacitors, the base should measure 7.5 × 6.5 mm, suitable for use with the R13 cell. Larger components would require a 11 × 9 mm baseboard and be suitable for use with the 675 series cell.

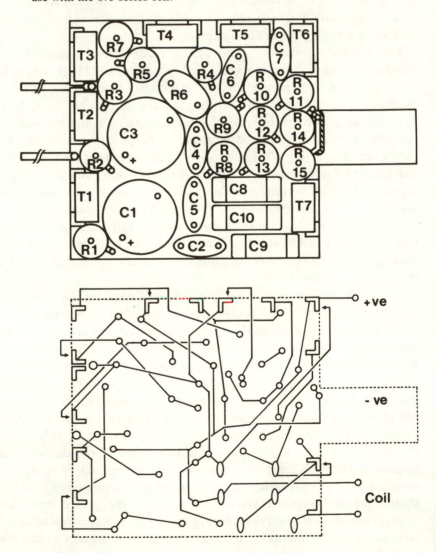

wires will be as short as possible and do not cross unnecessarily. If fitted to the circuit, the tuning capacitor will require treatment to allow the rotating sections to operate normally when embedded in epoxy encapsulation. If machining facilities are available, a plastic case can be made up to fit around the capacitor, leaving only the tuning exposed. Fully enclosed variable capacitors are available, but with some size penalty. Otherwise, pack the rotating parts with silicon grease and then coat thinly with epoxy. Allow this to set and test the capacitor for free operation before mounting with the other components. Where accurate tuning is not required, this variable capacitor can either be replaced with a suitable fixed component, chosen to give approximately the correct frequency or left out completely (Fig. 10). Final tuning can then take place by adjusting the configuration of the coil.

The components can now be wired together. Use the finest possible copper wire, of 40–100 μm diameter (Goodfellow Metals). It is best to use uninsulated wire, as stripping off the insulation can be difficult with this fine gauge. Work from the signal input end of the transmitter, wiring one node at a time, and referring to the component plan at each step to ensure the correct leads are being used. Wrap one turn of wire around each component lead and when the node is complete, use fine resin cored solder, or a dab of solder cream, to solder the wire to the leads. With experience, it is quicker to coat each component lead with solder cream, take the wire to the lead and solder into place without the necessity of winding the wire around the lead. Take lengths of wire to the stripboard to allow connection of the power lines, input signals and output test points. Test for correct operation as each stage is finished, and, if necessary, apply a light coat of epoxy to secure and insulate the wiring on this tested section. The output coil is best attached directly to the capacitors that bridge it, using their leads for mechanical strength. A dummy test coil can be temporarily soldered in place, and the complete transmitter tested again. The actual output coil can either be wound around the electronic block or embedded in an outer frame. This latter arrangement allows some degree of uniformity to be achieved, of particular importance if a number of identical transmitters are being built at the same time. Make up a plastic former of the same width and length as the completed electronic package. Apply a thin coat of silicon grease as a release agent, and lightly epoxy two turns of insulated wire (0.25 mm, single core) around the former, spacing the turns a few millimetres apart.

When the epoxy has set, slide the now rigid coil off the former, place it over the electronic block and solder it in position. Check the operating frequency with a cell pushed up close to the outside of the coil (as the presence of this mass of metal will affect the measurement) and with the variable capacitor set to the centre of its range. A grid dip meter is a particularly useful, and relatively inexpensive, instrument for measuring this resonant frequency. If the frequency is too low, build another coil with fewer turns, if too high, with less turns. Using quick-set epoxy, this trial and error technique need take little time and is only necessary for the first transmitter, all succeeding coils being built with the same number of turns as the first correct one.

File excess epoxy off the coil and make up a female mould to the required outer dimensions of the frame. This can be fairly simple, and use a teflon or silicon rubber cylinder, to form a concave recess in the frame for the cell. Coat the mould with silicon grease and epoxy the coil into the mould, packing all spaces with epoxy except the central cavity. Remove all traces of silicon grease from the coil before trying to add further epoxy. When set, remove the frame, file off excess epoxy, drill holes for the two input pins and for access to the variable capacitor, and file recesses for the battery contacts (Fig. 9). Make up two input pins from stainless steel wire and epoxy them into place, then lightly epoxy the electronics block into position, add the remaining battery contact and wire up the block to these connections. Test, then top up the frame with electronic encapsulating epoxy, forming a solid, mechanically strong, transmitter. Glass, in the form of appropriately sized microscope cover slips, can be epoxied onto the major faces of the transmitter. This will help in preventing water vapour penetrating the package and will considerably extend the operation life of the device. The cell is soldered to both contacts, thus holding it rigidly in place, and the ECG lead soldered to the input pins. When soldering to stainless steel, orthophosphoric acid may be used as an etching flux, but it must be thoroughly cleaned from the soldered joint to prevent subsequent corrosion. This flux is also suitable for soldering the leads to the cells, as it allows the joint to be made very quickly. Do not apply the iron to the cell for more than one second. Overheating the cell may, at least, cause a reduction in capacity, and possibly lead to an explosion.

Input leads

The design and correct positioning of biopotential electrodes will vary greatly between animals and so cannot be covered in detail by this section, but it is hoped that each experimentor will have some experience of hard wired recording of these signals. This should give a good idea of the problems likely to be encountered with his own experimental animals. The following points, some of them of a more general nature should be noted:

(i)   The biopotential transmitter requires for ECG work, inputs of between 1 and 10 mV, from fairly low impedance electrodes (less than 100 K).

(ii) The animal is probably free to exercise vigorously. The electrodes will therefore require careful positioning if movement artefact and interference is to be avoided.

(iii) Similarly the animal may be able to pull at or rub off externally mounted equipment. It should either be firmly attached, or internally implanted.

(iv) It is possible that the presence of external electrodes, leads and transmitter cannot be tolerated by the animal, or may interfere with its behaviour. Full implantation will then be necessary.

(v) The design of the electrodes and leads should be suitable for the duration of the experiment. Long term investigations will require materials resistant to both body fluids and to the mechanical stress imposed by movement of the subject.

Stainless steel (preferably EN58J or AISA 316) has proved a perfectly good material for implanted ECG electrodes, and has been used by the authors for extended periods with no sign of degradation. Leads should preferably be made of the same material and teflon coated multistranded wire can be used where little movement is expected. Where the leads may be continually twisted, bent or stretched then the only material that the authors have found that will survive for more than a few weeks is a wire formed from 3 or 4 stainless steel wires wound into a close coil, about 1 mm in diameter (APC Medical). Two of these leads have lasted over five years in one experimental animal and the wire is also widely used in the medical field. It may be insulated by sleeving with medical grade silicon rubber tubing using a two part heat curing SR compound as both a lubricant during this process and as a sealant. By injecting the compound into the sleeved wire and curing the rubber, a solid, void free lead is formed that is resistant to both body reactions and to mechanical stress (Fig. 3). Alternatively, the silicon rubber can be expanded by immersion in toluene, and the stainless steel lead(s) inserted. The tubing will contract down tightly onto the lead as the solvent evaporates. The leads can then be soldered to the input pins, the soldered joint insulated with epoxy and allowed to set. The silicon rubber sleeving can then be pulled back down over the joint to form a tight, waterproof seal, ready for final encapsulation.

Mounting the transmitter

Finally, the transmitter must be mounted in or on the animal. For implantation, one must accept that there is no certain way of protecting the transmitter from body fluids except by the use of hermetic techniques. If a more limited life of about one to two years is acceptable, then dip the transmitter in a just molten mix of five parts paraffin wax to one part beeswax. This will leave a thin, slightly flexible layer that is impermeable to body fluids. It is not, however, a good adhesive and failure occurs at leads and other breaks in the coating. A top coat of silicon rubber provides a more robust surface, easily handled and cleaned and into which suture threads and loops can be embedded to secure the transmitter during the implantation procedure. For animal applications, medical grade silicon rubber is not essential. Dip coating of the transmitters can either be with a flowing compound (734 RTV, Dow Corning), or a non-slump rubber thinned down with toluene. Note that these devices must never be heat sterilized, a mild gas or chemical method must be used.

Mounting on the exterior of the animal can either be by adhesive or by harness. If a suitable mounting cannot be devised then implantation, even with its associated increase in unreliability, is the only answer.

Decoders and receivers

Telemetry transmitters will require receivers that not only cover the correct radio frequency bands, but are also able to demodulate the form of modulation being used. The ECG transmitter uses a mode of modulation that is not catered for on most domestic quality VHF receivers. Where regulations allow, the transmitter frequency may be shifted above 108 MHz into the air communications band. As amplitude modulation is used in this band, any receiver that has this extra facility will be able to correctly demodulate the pulsed signal from this transmitter. If this frequency shift is not possible, then a communications receiver should be obtained as these have facilities to handle most forms of modulation. Alternatively, if the chosen receiver has an air band, then it may be possible to arrange for the AM detector used on this band to be operational on the FM band as well. Such a modification should be fairly straightforward but will depend on the actual receiver in question. It is possible that, with this pulse form of modulation, various receiver facilities may not be operational, especially automatic frequency control and possibly the signal strength meter.

The ECG transmitter will require a decoder in order to re-create the input signal from the pulse-interval modulated output from the receiver. Full details of a suitable circuit are given in Woakes (1980), or are available from the authors.

Advanced systems

The ECG transmitter used to illustrate the construction methods developed by the authors can be further modified, either by measuring more complex and difficult parameters, by pushing one dimension of biotelemetry (size, lifetime, range, etc.) to its limit or by increasing the number of channels telemetered by the system. The ECG transmitter may be matched to any form of transducer and pre-amplifier, as long as its output is in the form of a suitable current drawn from the base of transistor is Q3. This parameter controls the modulation of the pulse interval through the time taken to discharge the timing capacitor, C4. The transducer will, of course, have to meet the usual requirements of most biotelemetry equipment: small size, low power consumption, low voltage requirements etc., that will drastically reduce the chances of being able to utilize normal laboratory devices. Furthermore, the simple circuitry of this transmitter is not stable enough, either with time or variations in temperature, for monitoring variables accurately over extended periods.

The transmitter may be modified in range and lifetime. These factors will, in turn, change the power consumption and influence the desired size of battery. Although these are all interrelated variables, the pulse systems do allow a certain amount of interdependence through adjustment of the pulse parameters and details are given in the relevant sections.

Most multichannel systems use a time multiplex form of encoding the data, with each channel represented in a particular time slot within a transmitter cycle, or frame. A large proportion of the electronics can therefore be common to all channels in both the transmitter and decoder with a consequent saving in size, power consumption and cost. Balanced against this saving, extra components must be used to sample each channel, to control the sequence of events and to provide a means of synchronization between encoder and decoder. However, most of these functions are available in CMOS digital integrated circuits, and at about four channels and above, this method becomes much more economical than multiple subcarrier systems. This time multiplex method can also be used in conjunction with a pulsed RF output, leading to further economies in battery power and size. The most advanced systems convert the output from each channel into a digital number and transmit a stream of data bits. This is perhaps the ultimate in sophistication as the data is relatively immune to degradation once it has been encoded. It does, however, require considerable complexity at the transmitter, but this is mainly digital circuitry and fast A/D converters with serial output and built in channel multiplexers are now available in one silicon chip (Texas TLC540 series).

Further advances will undoubtedly make use of integrated circuits built specifically for biotelemetry, and first steps have already been made in this direction. Silicon chips containing a number of transistors, resistors and capacitors can be connected together by a top metallization coat. By varying the design of this last stage, a vast number of different circuits can be obtained. Unfortunately, this form of custom-made integrated circuit is extremely expensive to build in small numbers and it is unlikely that this sort of device can be used in physiological research, unless a standard design can be agreed upon. This is a great pity, for the ECG transmitter, apart from one or two components in the RF section, could be contained within a few cubic millimetres, and this would open up wide avenues of fruitful research. However, this equipment is being increasingly applied in the medical field, for patient monitoring and sport physiology, and, with long production ruins, the price should drop to a reasonable level. Biotelemetry may then become common place, and even used on a disposable basis, thrown out when the batteries expire.

Conclusion

There is no doubt that telemetric techniques have already given us a glimpse of the physiology of untethered and unrestricted animals, sometimes in semi-natural conditions. Solid state electronics and satellites are already enabling scientists to obtain information simultaneously on the behaviour and physiology of animals in their natural environment. We are, therefore, entering an exciting era. At last we are able to free ourselves and our animals from the constraints of the laboratory. Given the right conditions the next ten years will see telemetry, in the widest sense, play a dominant role in helping us understand how animals, including humans, cope with their environment.

References

Armstrong, J. D. (1986). Heart rate as an indicator of activity, metabolic rate, food intake and digestion in pike, *Esox lucius. Journal of Fish Biology*, **29** (Supplement A), 207–21.

Bayly, W. M., Grant, B. D., Breeze, R. G. & Kramer, J. W. (1983). The effects of maximal exercise on acid-base balance and arterial blood gas tension in thoroughbred horses. In *Equine Exercise Physiology*, ed. D. H. Snow, S. G. B. Persson & R. J. Rose, pp. 400–7. Cambridge: Burlington Press.

Belkin, D. A. (1964). Variations in heart rate during voluntary diving in the turtle *Pseudemys concinna, Copeia*, 321–30.

Berger, M., Roy, O. Z. & Hart, J. S. (1970*a*). The co-ordination between respiration and wing beats in birds. *Zeitschrift für vergleichende Physiologie*, **66**, 190–200.

Berger, M., Hart, J. S. & Roy, O. Z. (1970*b*). Respiration, oxygen consumption and heart rate in some birds during rest and flight. *Zeitschrift für vergleichende Physiologie*, **66**, 201–14.

Boyd, J. C. & Sladen, W. J. L. (1971). Telemetry studies of the internal body temperatures of adélie and emperor penguins at Cape Crozier, Ross Island, Antarctica. *The Auk*, **88**, 366–80.

Butler, P. J. (1982). Respiration and cardiovascular control during diving in birds and mammals. *Journal of Experimental Biology*, **100**, 195–221.

Butler, P. J. (1987). The exercise response and the 'classical' diving response during natural submersion in birds and mammals. *Canadian Journal of Zoology*, **66**, 29–39.

Butler, P. J. & Stephenson, R. (1987). Physiology of breath-hold diving: a bird's eye view. *Science Progress, Oxford*, **71**, 439–58.

Butler, P. J., West, N. H. & Jones, D. R. (1977). Respiratory and cardiovascular responses of the pigeon to sustained, level flight in a wind-tunnel. *Journal of Experimental Biology*, **71**, 7–26.

Butler, P. J. & Woakes, A. J. (1979). Changes in heart rate and respiratory frequency during natural behaviour of ducks, with particular reference to diving. *Journal of Experimental Biology*, **79**, 283–300.

Butler, P. J. & Woakes, A. J. (1980). Heart rate, respiratory frequency and wing beat frequency of free flying barnacle geese *Branta leucopsis*. *Journal of Experimental Biology*, **85**, 213–26.

Butler, P. J. & Woakes, A. J. (1984). Heart rate and aerobic metabolism in Humboldt penguins, *Spheniscus humboldti*, during voluntary dives. *Journal of Experimental Biology*, **108**, 419–28.

Butler, P. J. & Woakes, A. J. (1987). Heart rate in humans during underwater swimming with and without breath-hold. *Respiration Physiology*, **64**, 387–99.

Carey, F. G., Teal, J. M., Kanwisher, J. W., Lawson, K. D. & Beckett, J. S. (1971). Warm-bodied fish. *American Zoologist*, **11**, 135–45.

Falke, K. J., Hill, R. D., Qvist, J., Schneider, R. C., Guppy, M., Liggins, G. C., Hochachka, P. W., Elliott, R. E. & Zapol, W. M. (1985). Seal lungs collapse during free diving: evidence from arterial nitrogen tensions. *Science*, **229**, 556–7.

Fedak, M. A. (1986). Diving and exercise in seals: a benthic perspective. In *Diving in Animals and Man*, ed. A. O. Brubakk, J. W. Kanwisher & G. Sundnes, pp. 11–32. Trondheim: Tapir Publishers.

Fedak, M. A., Pullen, M. R. & Kanwisher, J. (1987). Circulatory responses of seals to periodic breathing: heart rate and breathing during exercise and diving in the laboratory and open sea. *Canadian Journal of Zoology*, **66**, 53–60.

Ferns, P. N., MacAlpine-Leny, I. H. & Goss-Custard, J. D. (1979). Telemetry of heart rate as a possible method of estimating energy expenditure in the redshank *Tringa totanus* (L.). In *A Handbook on Biotelemetry and Radio Tracking*, ed. C. J. Amlaner & D. W. MacDonald, pp. 595–601. Oxford: Pergamon Press.

French, J. (1986). Tracking animals by satellite. *Electronics & Power*, May 1986, 373–76.

Fryer, T. B. (1970). *Implantable biotelemetry systems*. NASA SP-5094. Washington: NASA Technology Utilization Publications.

Furilla, R. A. & Jones, D. R. (1986). The contribution of nasal receptors to the cardiac response to diving in restrained and unrestrained redhead ducks (*Aythya american*). *Journal of Experimental Biology*, **121**, 227–38.

Graham, J. M. (1981). Development of ultrasonic and electromagnetic tags for heart rate monitoring in crabs. *Marine Physics Group Report 17, Scottish Marine Biological Association*, 1–17.

Guppy, M., Hill, R. D., Schneider, R. C., Qvist, J., Liggins, G. C., Zapol, W. M. & Hochachka, P. W. (1986). Microcomputer-assisted metabolic studies of voluntary diving of Weddell seals. *American Journal of Physiology*, **250**, R175–87.

Harden Jones, F. R. & Arnold, G. P. (1982). Acoustic telemetry and the marine fisheries. *Symposia of the Zoological Society, London*, **49**, 75–93.

Harrison, R. J., Ridgway, S. H. & Joyce, P. L. (1972). Telemetry of heart rate in diving seals, *Nature*, **327**, 280.

Hart, J. S. & Roy, O. Z. (1966). Respiratory and cardiac responses to flight in pigeons. *Physiological Zoology*, **38**, 291–306.

Hill, R. D. (1986). Microcomputer monitor and blood sampler for free-diving Weddell seals. *Journal of Applied Physiology*, **61**, 1570–6.

Hill, R. D., Schneider, R. C., Liggins, G. C., Hochachka, P. W., Schuette, A. H. & Zapol, W. M. (1983). Microprocessor controlled recording of bradycardia during free diving of the antarctic Weddell seal. *Federal Proceedings*, **42**, 470.

Holliday, F. G. T., Tytler, P. & Young, A. H. (1972/73). Activity levels of trout (*Salmo trutto*) in Airthrey Loch, Stirling, and Loch Leven, Kinross. *Proceedings of R.S.E.*, **74**, 315–31.

Hörnicke, H., Engelhardt, W. V. & Ehrlein, H. J. (1977). Effect of exercise on systemic blood pressure and heart rate in horses. *Pflugers Archives*, **372**, 95–9.

Hörnicke, H., Meixner, R. & Pollmann, U. (1983). Respiration in exercising horses. In *Equine Exercise Physiology*, ed. D. H. Snow, S. G. B. Persson & R. J. Rose, pp. 7–16. Cambridge: Burlington Press.

Howey, P. W., Witlock, D. R., Fuller, M. R., Seegar, W. S. & Ward, F. P. (1984). A computerized biotelemetry receiving and datalogging system. In *Biotelemetry VIII*, ed. H. P. Kimmich & H. J. Klewe, pp. 442–6.

Jennings, J. G. & Gandy, W. F. (1980). Tracking pelagic dolphins by satellite. In *A Handbook on Biotelemetry and Radio Tracking*, ed. C. J. Amlaner & D. W. MacDonald, pp. 753–9. Oxford: Pergamon Press.

Jones, D. R., Fisher, H. D., McTaggart, S. & West, N. H. (1973). Heart rate during breath-holding and diving in the unrestrained harbor seal (*Phoca vitulina richardi*). *Canadian Journal of Zoology*, **51**, 671–80.

Kanwisher, J. W., Gabrielsen, G. & Kanwisher, N. (1981). Free and forced diving in birds. *Science*, **211**, 717–19.

Kanwisher, J. W., Williams, T. C., Teal, J. M. & Lawson, K. O. (1978). Radiotelemetry of heart rates from free-ranging gulls. *The Auk*, **95**, 288–93.

Kimmich, H. P. (1975). Multichannel biotelemetry. *Biotelemetry*, **2**, 207–55.

Lord, R. D., Bellrose, F. C. & Cochran, W. W. (1962). Radiotelemetry of the respiration of a flying duck. *Science*, **137**, 39–40.

Lorenz, K. (1970). *Studies in Animal and Human Behaviour*, Vol. 1. Translated by R. Martin. London: Methuen.

MacArthur, R. A. (1979). Seasonal patterns of body temperature and activity in free-ranging muskrats (*Ondatra zibethicus*). *Canadian Journal of Zoology*, **57**, 25–33.

McGinnis, S. M. & Southworth, T. P. (1971). Thermoregulation in the Northern elephant seal, *Mirounga angusstirostris*. *Comparative Biochemistry and Physiology*, **40**A, 893–8.

Millard, R. W., Johansen, K. & Milsom, W. K. (1973). Radiotelemetry of cardiovascular responses to exercise and diving in penguins. *Comparative Biochemistry and Physiology*, **46**A, 227–40.

Øritsland, N. A., Stallman, R. K. & Jonkel, C. J. (1977). Polar bears: heart activity during rest and exercise. *Comparative Biochemistry and Physiology*, **57**, 139–41.

Owen, R. B. (1969). Heart rate, a measure of metabolism in blue-winged teal. *Comparative Biochemistry and Physiology*, **31**, 431–6.

Priede, I. G. (1984). A basking shark (*Cetorhinus maximus*) tracked by satellite together with simultaneous remote sensing. *Fisheries Research*, **2**, 201–16.

Priede, I. G. & Tytler, P. (1977). Heart rate as a measure of metabolic rate in teleost fishes; *Salmo gairdneri*, *Salmo trutta* and *Gadus morhua*. *Journal of Fish Biology*, **10**, 231–42.

Priede, I. G. & Young, A. H. (1977). The ultrasonic telemetry of cardiac rhythms of wild brown trout (*Salmo trutta L.*) as an indicator of bio-energetics and behaviour. *Journal of Fish Biology*, **10**, 299–318.

Qvist, J., Hill, R. D., Schneider, R. C., Falke, K. J., Liggins, G. C., Guppy, M., Elliot, R. L., Hochachka, P. W. & Zapol, W. M. (1986). Hemoglobin concentrations and blood gas tensions of free-diving Weddell seals. *Journal of Applied Physiology*, **61**, 1560–9.

Reinertsen, R. E. & Hafthorn, S. (1986). Different metabolic strategies of northern birds for nocturnal survival. *Journal of Comparative Physiology* B, **156**, 655–63.

Rogers, S. C. & Weatherley, A. H. (1983). The use of opercular muscle electromyograms as an indicator of the metabolic costs of fish activity in rainbow trout, *Salmo gairdneri* Richardson, as determined by radiotelemetry. *Journal of Fish Biology*, **23**, 535–47.

Smith, E. N. (1975). Thermoregulation of the American alligator, *Alligator mississippiensis*. *Physiological Zoology*, **48**, 177–94.

Snow, D. H. & Harris, R. C. (1985). Thoroughbreds and greyhounds: biochemical adaptations in creatures of nature and of man. In *Circulation, Respiration, and Metabolism*, ed. R. Gilles, pp. 227–39. Berlin: Springer-Verlag.

Stephenson, R., Butler, P. J. & Woakes, A. J. (1986). Diving behaviour and heart rate in tufted ducks (*Aythya fuligula*). *Journal of Experimental Biology*, **126**, 341–59.

Strikwerda, T. E., Black, H. D., Levanon, N. & Howey, P. W. (1986). The bird-borne transmitter. *Johns Hopkins APL Technical Digest*, **6**, 60–7.

Tatner, P. & Bryant, D. M. (1988). The doubly-labelled water technique for measuring energy expenditure. (This volume).

Van Citters, R. L. & Franklin, D. L. (1969). Cardiovascular performance of Alaska sled dogs during exercise. *Circulation Research*, **24**, 33–42.

Vatner, S. F. (1978). Effects of exercise and excitement on mesenteric and renal dynamics in conscious, unrestrained baboons. *American Journal of Physiology*, **234**, H210–4.

Vatner, S. F., Higgins, C. B., White, S., Patrick, T. & Franklin, D. (1971). The peripheral vascular response to severe exercise in untethered dogs before and after complete heart block. *The Journal of Clinical Investigation*, **50**, 1950–60.

Wade, O. L., Combes, B., Childs, A. W., Wheeler, H. O., Cournand, A. & Bradley, S. E. (1956). The effect of exercise on the splanchnic blood flow and splanchnic blood volume in normal man. *Clinical Science*, **15**, 457–63.

Walcott, T. G. (1980). Heart rate telemetry using micropower integrated circuits. In: *A Handbook on Biotelemetry and Radio Tracking*, ed. C. J. Amlaner & D. W. MacDonald W., pp. 279–86. Oxford: Pergamon.

Wang, L. C. (1978). Energetic and field aspects of mammalian torpor: the Richardson's ground squirrel. In *Strategies in Cold: Natural Torpidity and Thermogenesis*, ed. L. C. H. Wang & J. W. Hudson W., pp. 109–45. New York: Academic Press.

Weatherley, A. H., Rogers, S. C., Pincock, D. G. & Patch, J. R. (1982). Oxygen consumption of active rainbow trout, *Salmo gairdneri* Richardson, derived from electromyograms obtained by radiotelemetry. *Journal of Fish Biology*, **20**, 479–89.

Westerberg, H. (1984). Diving behaviour of migrating eels studies by ultrasonic telemetry. In *Biotelemetry VIII*, ed. H. P. Kimmich & H. J. Klewe, pp. 367–70. Nijmegen: Braunschweig.

Woakes, A. J. (1980). Biotelemetry and its application to the study of avian physiology. Ph.D. Thesis. University of Birmingham, Birmingham, U.K. British Library no. D 35979/81.

Woakes, A. J. & Butler, P. J. (1983). Swimming and diving in tufted ducks, *Aythya fuligula*, with particular reference to heart rate and gas exchange. *Journal of Experimental Biology*, **107**, 311–29.

Woakes, A. J. & Butler, P. J. (1986). Respiratory, circulatory and metabolic adjustments during swimming in the tufted duck, *Aythya fuligula*. *Journal of Experimental Biology*, **120**, 215–31.

Woakes, A. J., Butler, P. J. & Snow, D. H. (1987). The measurement of respiratory airflow in exercising horses. In *2nd International Conference on Equine Exercise Physiology*, ed. J. R. Gillespie, N. E. Robinson pp. 194–205 Davis, California: ICEEP Publications.

Wooley, J. B. & Owen, R. B. (1977). Metabolic rates and heart rate-metabolism relationships in the black duck (*Anas rubripes*). *Comparative Biochemistry and Physiology*, **57**A, 363–7.

N. H. WEST

Methods for measuring blood flow and distribution in intermittently-ventilating and diving vertebrates

Introduction

The measurement of cardiovascular function in those airbreathing vertebrates which dive, or exhibit intermittent ventilation, has been a source of interest to comparative physiologists for a quarter of a century (Johansen, 1963). The ability to measure blood flow is crucial to our understanding of these animals, for not only is cardiac output influenced by the alternation of ventilatory and non-ventilatory episodes, but also the distribution of blood flow may be radically affected by the ongoing ventilatory activity. In those vertebrates (mammals and birds) with a completely divided ventricle this may be reflected in a redistribution of cardiac output to systemic tissues and organs. In amphibians and reptiles however, in which intracardiac shunting is made possible by no, or incomplete, division of the ventricle, the potential exists for differential perfusion of the pulmonary and systemic circulations. Intermittent ventilation and diving are behavioral as well as physiological phenomena, so that their biological significance, as well as that of the associated changes in blood flow may be best assessed in behaving, unrestrained animals. This leads inexorably to experimental situations that contain conflicting elements: on the one hand animals must be free to exhibit normal or near normal behaviour, while on the other hand the techniques we have at our disposal for the measurement of blood flow are by their very nature invasive. Any such instrumentation will therefore tend to mitigate against 'normal' physiological function, and perhaps alter, or even abolish, the phenomena of interest (Burggren, 1987).

The aim of this chapter is to discuss the limitations and advantages of some of the large number of techniques available for the measurement of blood flow (Noordergraaf, 1978) currently in use by comparative physiologists studying diving or intermittently ventilating species.

Flow measurement in vessels

Electromagnetic flow transducers

The principle of operation of the electromagnetic flow transducer is a straightforward application of Faraday's law of magnetic inductance (1832). This states that an electromotive force is generated and a current will flow if a conductor, in this case blood, moves through a magnetic field. The field and the direction of motion of the conductor have to be perpendicular in order to maximize the induced current. Under these conditions, the voltage developed between the detecting electrodes is proportional to magnetic field strength (dependant on the coil current), the current path length in the moving conductor (the diameter of the vessel), the mean instantaneous velocity of blood flow, and therefore volume flow (velocity * cross sectional area of vessel). For a comprehensive review of the theory of blood flow measurement by magnetic inductance, see Wyatt (1984). For a discussion of the historical development of DC and AC electromagnetic flow meters, see Woodcock (1975).

In order to detect flow in an unopened artery, the artery is slotted into a channel in the head of a flow probe that contains an electromagnet (Fig. 1). Energization of the coil produces a magnetic field across the direction of

Fig. 1. Size comparison between an electromagnetic flow probe (Zepeda) and an ultrasonic crystal used in the construction of a pulsed Doppler probe. The diameter of both the crystal, and the probe lumen, is 1 mm. (From Burggren 1987.)

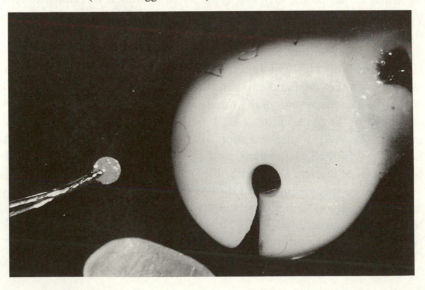

blood flow. Two detecting electrodes, which closely contact the vessel wall in a plane at right angles to the field, sense the flow-dependant voltage, which is delivered by connector wires to the appropriate circuits for amplification, demodulation and ultimately, display of the signal.

Unfortunately, the application of electromagnetic flow probes to the measurement of arterial blood flow in ectothermic vertebrates is difficult, particularly if results from chronic preparations are desired. As pointed out by Cameron (1986), chronic implantation of flow probes in ectotherms rarely results in tissue encapsulation, which serves to stabilize the probe and to ensure electrical continuity with the blood vessel. Consequently, their use has been limited in ectothermic vertebrates, and typically probes function for a matter of days, rather than weeks, after implantation (West & Burggren, 1984; Butler, Milsom & Woakes, 1984). A major limitation in the absence of encapsulation, and therefore stabilization of the flow head, is that the bulk of the probe lies in a plane at 90° to the long axis of the vessel to which it is applied, and the probe leads are stiff. Therefore, the position of the probe tends to be sensitive to torsional forces, so that the task of keeping the probe close to the optimal 90° angle relative to blood flow is difficult, particularly if the vessel is small.

An additional limitation of the usefulness of electromagnetic probes in ectotherms is the irreducible size of the electromagnet (Spencer & Barefoot, 1968) and therefore flow probe, relative to the size of the experimental animals and the perivascular space available. In experiments in which pulmonary and cutaneous arterial blood flow was measured in *Bufo marinus*, only in the largest animals were we able to place probes on these arteries ipsilaterally (West & Burggren 1984). A problem in these experiments was the length of extrinsic pulmonary artery available for probe placement compared to probe width, about 5 mm. Similar difficulties have been encountered by those investigators attempting to measure cardiac output in the ventral aorta of fish (Cameron, 1986).

It is often desirable to measure blood flow by two probes on adjacent vessels (West & Burggren, 1984). In this situation, noise may be generated by the interaction between the AC magnetic fields generated by neighbouring probes. Very often, however a combination of excitation frequencies may be chosen that minimises such interaction. A related problem may be encountered in experiments in which flow is recorded while afferent nerves are electrically stimulated (Van Vliet & West, 1986). For example, we recently investigated the effects of stimulating the recurrent laryngeal nerves in conscious toads, while recording pulmocutaneous or aortic pressures and flows (Van Vliet & West, 1986). In order to minimise interference with the flow signals during stimulation a stimulus isolation unit was used, only the contralateral nerve was stimulated, and we adjusted the relationship between the axes of the stimulating electrode and flow probe for minimum interference. In spite of these precautions, only a minority of trials were interference-free.

It is necessary to determine zero flow in order to provide a zero calibration point for the electromagnetic flow meter. Unfortunately, although many models provide 'magnetic zero', i.e. a zero level corresponding to zero excitation current, in many circumstances this will be found to be different from 'occlusion zero', i.e. zero determined by occluding the vessel downstream from the probe with magnet excitation on. This is due to the presence of stray capacitive impedance and/or the presence of a 'transformer voltage' that can alter the level of the zero flow signal (Woodcock, 1975). It is usually possible to adjust magnetic zero to occlusion zero, for example during probe implantation, but capacitive impedance may vary over time. Ideally, a miniature hydraulic cuff occluder should be fitted downstream of the probe on implantation, to enable a periodic determination of occlusion zero. Unfortunately, this is rarely feasible in small non-mammalian preparations. Workers using diving and ectothermic vertebrates in particular have often resorted to considering end diastolic flow levels to represent zero, particularly under conditions of bradycardia associated with submersion, Ach infusion or depressor nerve stimulation (Kirkebo, 1968; Shelton & Burggren, 1976; Van Vliet & West, 1986). This is not an entirely satisfactory arrangement. Most volume flow occurs near zero during the long diastole characteristic of ectotherms, so that unfortunately, even a small zero error has the potential of producing large errors in the estimation of volume flow.

It is dangerous to become fixated on recording pulsatile flows, to the exclusion of mean flow recordings. If mean flows are wanted, this ultimately leads to the daunting and time-consuming task of calculating mean flow by means of area integration (West & Burggren, 1984). Both the instantaneous velocity and mean flow signals contain information that it is difficult, or impossible, to derive from each other. Ideally, both should be recorded. A convenient way of doing this, if funds permit, is to record instantaneous velocities on magnetic tape, and to replay through low-pass filters to obtain mean flow.

Volume flow ($\dot{Q}$) calibration of electromagnetic flow probes (or any other type of flow probe) is best carried out *in vivo*. During human surgery, for example, it is often possible to doubly cannulate an artery upstream from the flow probe, forming a bypass loop of cannula. Blood is then withdrawn into a syringe from the upstream vessel *via* a T-fitting in the loop of cannula, the upstream segment of cannula clamped, and this blood ejected at known rates downstream past the flow probe. Unfortunately, the length of available artery precludes such an approach in most comparative work, so that in many cases not only is calibration not carried out *in vivo*, but it may not even be possible to calibrate *in situ*. *In vitro* calibration introduces several unknown variables, including a different electrode contact if the probe is removed and then replaced on the vessel, local changes in wall resistivity and increases in thickness (up to 50%) in excised vessels, alterations in the resistivity of the perivascular fluid, and thermal effects (Brunsting and ten Hoor, 1968). It is highly desirable therefore, but not always possible, to leave the artery to be calibrated *in situ*. Calibration flow may be provided by an infusion pump, gravity feed or an air displacement system and vessel distension controlled with a pressure reservoir or variable downstream resistance, in order to maintain normal transmural pressures and electrode-vessel wall contact (Meisner, Messmer and Hagl, 1968). The calibration medium should be heparinized blood of normal haematocrit, preferably drawn from the animal before sacrifice. A normal haematocrit is necessary to avoid resistance errors because the perivascular fluid, the vessel wall and the intravascular fluid provide, in practice, alternate paths in parallel for induced current flow (Gessner, 1961). The resistivity of blood increases almost exponentially with increasing haemoglobin concentration. Therefore the flow-dependant voltage at the sensing electrodes will be larger for a given volume flow if isotonic saline rather than blood of normal haematocrit is used as the calibration medium (Dedichen & Schenk, 1968). The critical factor appears to be the resistivity of the blood compared to that of the shunt pathways, vessel wall and perivascular fluid, because cannulating flow probes are much less affected by haemoglobin concentration (Brunsting and ten Hoor, 1968).

Permanent magnet flood flow transducers

In spite of the early rejection of blood flow detection using DC magnetic fields (Kolin, 1936), based on the instability and polarization of the electrodes, both cuff and cannulating permanent magnet flow transducers are currently commercially available. Being passive devices, their output can be recorded with a DC amplifier and is essentially an analogue of instantaneous flow rate integrated over the cross section of the vessel. The detecting electrodes are composed of sintered silver-silver chloride, the presence of silver chloride promoting a low electrode noise, low impedance interface with a small half cell potential and hence offset potential. DC offset potential is claimed to be typically 200 microvolts, and electrode noise to be in the order of 1.5 microvolts peak to peak. Unfortunately, the offset potential drift rate is some 200 microvolts/hour and it is not possible to provide a magnetic zero. However, the transducers possess the advantage of being relatively easy to interface with a simple telemetry transmitter, as no transducer excitation voltage is required. Therefore, they may have a potential role to play in the investigation of relative flows in unrestrained small vertebrates.

Pulsed Doppler flow transducers

The availability of miniature pulsed Doppler flow probes, utilising ultrasound (Hartley & Cole, 1974) has presented new opportunities to comparative physiologists, including the ability, for the first time, to chronically measure flow in vessels with diameters of <1 mm, of measuring dimensional changes in small arteries (Van Vliet & West, 1987) and of continuously monitoring regional blood flow distribution in small, conscious vertebrates (Haywood *et al.*, 1981). Miniature Doppler probes have many advantages for comparative work: their simplicity (and relative cheapness), their small size (Fig. 1), facilitating flow measurement in small vessels such as the arteries of chick embryos (Clark & Hu, 1982), the use of fine unshielded signal leads, which are easily implantable subcutaneously, and the presence of a reliable internal zero flow reference. Furthermore, the physical characteristics of the cuff-type miniature Doppler probe approach those of the 'ideal' probe, which is softer than the vessel wall to allow changes in diameter, but also geometrically stable. They possess one major disadvantage for the worker interested in volume flow ($\dot{Q}$): strictly speaking they are velocimetric devices, that is to say their output is a function of a sampling of peak velocity of blood flow rather than a measurement of flow velocity integrated over the cross-sectional area of the blood vessel (see Peronneau *et al.*, 1972; Nakamura *et al.*, 1986).

In the miniature pulsed Doppler flow probe a piezoelectric crystal is opposed to the wall of the vessel, typically by being molded into a Silastic cuff. The crystal face is oriented at some 45° to the axis of blood flow and is excited to emit short bursts of ultrasound energy (typically 8 cycles, 20 MHz) that are directed into the moving stream of blood, impinge on the blood cells, and are reflected back to the crystal with an altered frequency (KHz) due to their velocity (Hartley *et al.*, 1984). The high frequency of the transmitted signal maximizes the acoustic power of the reflected waves, as the acoustic power of reflections from blood is proportional to f^4, while absorption is proportional to f^2. Echoes of the ultrasound pulses are delayed in their return to the crystal by a time proportional to the distance from the site from which they are reflected (Fig. 2).

$$Df = 2fo.V./c.\cos \text{Theta}$$

where
Df = Doppler shift
fo = original frequency
V = fluid velocity (meters.s^{-1})
c = velocity of sound in blood ($\sim$1.58 mm.μs^{-1})
Theta = angle between the flow and acoustic axes

By using the same crystal as a receiver, and by adjusting a receiving 'gate' with a 0.2 μs width to transduce Doppler shifted returns over a time delay range of 1–15 μs it is, in effect, possible to sample the velocity of blood flow at a distance of $\sim$1–10 mm into the vessel. The sample volume is about 0.5 mm^3, its size and shape being a function of the band widths of the transducer and receiving amplifier, the transmitter pulse and sampling pulse widths, and the pattern of the acoustic beam (Baker & Yates, 1973). In

Fig. 2. Instantaneous arterial blood flows recorded by the pulsed Doppler technique in a conscious toad, *Bufo marinus*. Doppler flow probes were implanted two weeks before the recording was made. Event marker indicates the introduction of 10% CO_2 into the experimental chamber. (West, unpublished.)

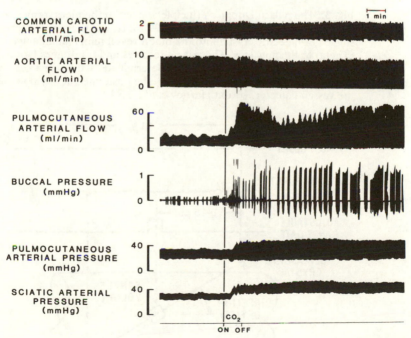

practice the gate is set to sample velocities around maximum instantaneous velocity by maximizing signal voltage with the aid of an audio monitor.

The small size of the probes, the orientation of the cuff along the long axis of the vessel, their tolerance to tissue fluids, and their fine leads make them practical for chronic implantation in small animals. It is also possible to customize probes. At the simplest level, preformed silastic cuffs may be trimmed to length with a scalpel to match the available vessel length. If crystals and the attached leads are purchased they may be fitted to cuffs constructed from silastic or polyethylene tubing. The tubing wall is notched at 45° to the long axis, and a wedge of tubing removed sufficient to accommodate a crystal. The crystal is attached with cyanoacrylate, and the joint strengthened with silicone sealant. Finally, a longitudinal cut is made in the tubing diametrically opposite the crystal, and the cuff cut to length.

It is possible to simultaneously pulse several crystals, thereby avoiding electrical interference between closely adjacent probes (Haywood *et al.*, 1981). Such interference can occur between independently pulsed flow meters even if these are situated some 5 meters apart (Burggren, pers. com.) Acoustic interference between probes appears to rarely be a problem if probes are fitted with acoustic baffles of polyurethane foam. For example,

Fig. 3. The pulsed Doppler principle applied to the measurement of blood velocity. Bursts of ultrasound (20 MHz) are transmitted from a single peizoelectric crystal into the blood vessel lumen at an angle (Theta) to the flow axis. The crystal receives and transduces Doppler shifted echoes after a variable delay determined by the range gate, which in effect, enables the sample volume to be positioned at any point across the vessel diameter. (From Hartley *et al.*, 1984.)

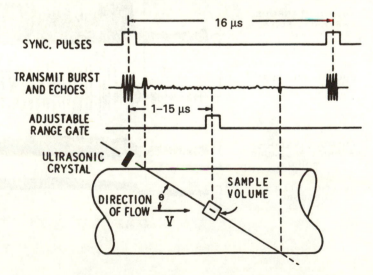

it has proved possible to implant miniature Doppler probes on the common carotid, aorta, and pulmocutaneous aorta in *Bufo marinus* in order to investigate the factors influencing the distribution of cardiac output (Fig. 3). Flows were recorded some two weeks after implantation, the limiting factor being the patency of the accompanying pressure cannulas.

It is important when implanting Doppler probes to ensure that good acoustical coupling occurs between the probe and the wall of the vessel. In particular, air bubbles under the crystal will attenuate the acoustic returns. A useful test is to swish the probe in a beaker of saline before implantation. The addition of a small amount of detergent reduces surface tension and bubble tenacity. This also provides an opportunity to weed out any non-functional probes, while a muffling of the audio signal indicates the persistence of air bubbles. The inside diameter of the cuff should closely approximate the outside diameter of the artery to which it is applied. It is usually convenient to pass a pair of small curved forceps under the cleared segment, and to allow them to spring open sufficiently to straighten the artery while the flexible cuff is applied, and secured with a half-knot. At this point peak flow may be tuned with the audio output, and the flow signal inspected. Persistent air bubbles may be dealt with by the introduction of a blunt-ended 32 g needle between the cuff and vessel wall, and this space forcibly irrigated with saline. A securing knot may then be tied. Flow should be monitored, because it is easy to partially constrict the flexible cuff in an excess of enthusiasm. Provided air bubbles are cleared on implantation, acoustic coupling rarely appears to be a problem in chronic preparations.

It may not be necessary to provide an occlusion zero either on implantation, or by means of an implanted occluder, as one commercially available model of miniature pulsed Doppler meter provide internal calibrations in terms of KHz Doppler shift, including 0 KHz.

Calibration of miniature pulsed Doppler probes may be performed *in situ* or *in vitro* using gravimetrically or volumetrically measured volume flows. In practice, it is convenient to calibrate all traces in terms of Doppler shift (KHz) at the time of recording, using the flowmeter's internal calibration, and to then measure 4 or 5 levels of volume flow in terms of KHz shifts upon final calibration. In order to perform volume flow calibrations as accurately as possible it is important to take the following into consideration:

(i) Under laminar flow conditions, in which the velocity profile is parabolic, peak velocity is twice the mean velocity across the lumen, and there is usually an acceptably reproducible empirical relationship between Doppler shift (KHz) and volume flow. It has been shown, for example, that miniature Doppler probe output provides an acceptable estimate of volume flow in rat arteries (Haywood *et al.*, 1981). Scans of the velocity profile in

Fig. 4(*a*). Flow velocity profiles at peak systole (filled circles) and end diastole (triangles), expressed in terms of frequency shift, and recorded across the diameter of the extrinsic pulmonary artery of a conscious Green turtle (*Chelonia mydas*). High heart rates occurred during ventilatory periods, and low heart rates during voluntary submergence. The profiles were constructed over several beats. The transmitted frequency was 20 MHz. (West, unpublished.)

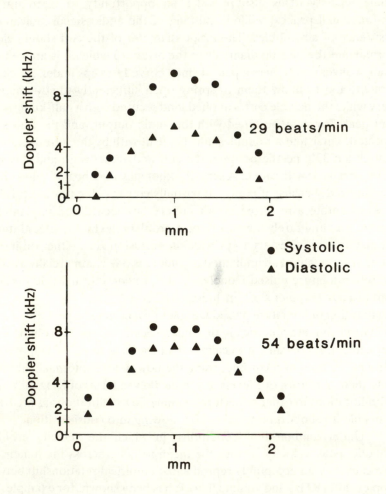

a   *Chelonia mydas*

(*b*) The relationship between Doppler shift (KHz) and the angle between the flow axis and the ultrasound beam, assuming a transmission frequency of 5 MHz. (From Klepper, 1983.)

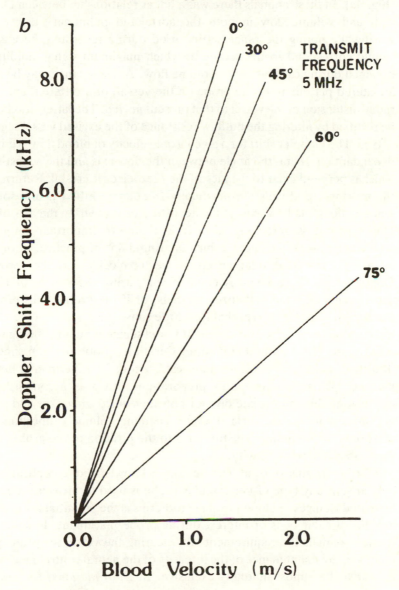

the extrinsic pulmonary artery of the turtle *Chelonia mydas* indicate that this is parabolic, or near parabolic, over a wide range of heart rates (Fig. 4*a*). In these animals there was a linear relationship between Doppler shift and volume flow in both the aorta and pulmonary artery. It is worthwhile noting the range setting used during recording, because this should correspond to the setting at which maximum signal amplitude is obtained during calibration for volume flow. A discrepancy may be due to an altered placement of the probe on the vessel, or a different amount of radial distension of the vessel to that present *in vivo*. The latter situation can be rectified by altering the outflow resistance of the excised vessel segment.

(ii)   The Doppler shift for a given axial velocity of blood flow is critically dependant on Theta, the angle between the flow axis and the acoustic axis, which is perpendicular to the face of the piezoelectric crystal. Returns from the moving blood cells are maximally frequency shifted if ultrasound is emitted along the flow axis. If the acoustic axis is at 90° to the axis of flow, Doppler shift is zero (Fig. 4*b*). It is therefore important to prevent significant changes in Theta, which is around 45° in practice, in order to maintain a constant relationship between velocity and the associated Doppler shift. Such changes could occur during the course of chronic experiments on unanaesthetized animals, or if the probe is disturbed on calibration. This factor is probably of little consequence as a source of error if the probe is implanted on a straight vessel segment but may be significant if the vessel has a natural curvature. Certainly, probe position should be disturbed as little as possible upon calibration. In experiments on *Bufo marinus* (West and Smits, in preparation) this was accomplished by cannulating the appropriate divisions of the *truncus arteriosus* and providing an outflow *via* the atria, thereby enabling volume calibration to be performed with minimum disturbance to the position of the probes within the thoracoabdominal cavity.

A little recognised, or at least seldom acknowledged, complication that may arise if any type of perivascular probe is used to measure reflexively mediated changes in blood flow in ectotherms is the potential for denervating the vascular beds of interest during probe placement. In some ectothermic animals the requirements for clearing the vessel for probe placement and the maintenance of the integrity of the neural control mechanisms appear to be almost mutually exclusive. In *Bufo marinus*, for example, positioning a perivascular probe on the pulmocutaneous artery (Smits, West & Burggren, 1986) requires great care in order to avoid damaging the recurrent laryngeal nerve with which it is closely associated. The recurrent laryngeal nerve contains afferents from baroreceptors within the pulmocutaneous artery, which are important for cardiovascular regulation in the anuran (Van Vliet & West, 1987). Similarly, the pulmonary vagus lies close to the extrinsic pulmonary artery in anurans and chelonians (West and Burggren, 1984). In the case of miniature Doppler probes it is possible to incorporate the vagus into the probe lumen (West *et al.*, unpublished). An additional complication arises from the fact that the extrinsic pulmonary artery undergoes vagally controlled changes in calibre (Milsom *et al.*, 1977), necessitating a careful choice of probe diameter.

Measurement of blood flow distribution
Radioactive microspheres

The injection of solid foreign particles into the central arterial circulation in order to measure the distribution of cardiac output, organ blood flow and cardiovascular shunts has been a widely used technique since the work of Pohlman (1909). A significant advance in technology was the ability to label the indicator particles with gamma-emitting nuclides such as strontium and caesium, enabling their quantification in tissue to be performed with relative ease and accuracy (Heymann *et al.*, 1977). Currently the most commonly used indicator particles are plastic microspheres with diameters of 15–35 μm, and a specific gravity of 1.3.

To use microspheres successfully as indicators of organ blood flow it is important that they are well mixed and evenly distributed within the central circulation. To this end the spheres should be visualized before injection to check for clumping. Various workers have used ultrasonification, vortex mixing, treatment with Tween 80 detergent, suspension in 10% Dextran and vibration during injection as precautions to prevent aggregation and ensure an even distribution of suspended spheres (Millard, Baig & Vatner, 1977; Jones *et al.*, 1979). In order to ensure an even distribution within the systemic blood leaving the heart, spheres should ideally be introduced into the left atrium or pulmonary vein, or if this is not practical, the left ventricle (Jones *et al.*, 1979). Many workers have used the equality of flow to bilaterally paired organs, e.g. kidneys, or muscles, as an index of even distribution in the systemic cardiac outflow (Jones *et al.*, 1979, Laughlin *et al.*, 1982).

The diameter of microspheres should be chosen carefully with an eye to what is to be measured. If the spheres are large enough to be trapped both by tissue capillaries and a–v anastomoses then their distribution represents total organ blood flow. The spheres must be trapped on the first pass through the organ, and must remain trapped until counted. Although 15 μ spheres are satisfactory in mammals, it is unwise to assume that this is the case for ectothermic vertebrates. For example, although 15 μ spheres are satisfactory in Varanid lizards, there was 7–15% recirculation of these spheres in the turtle *Chrysemys*, that necessitated a switch to 25μ spheres (Heisler & Glass, 1985). In principle, smaller spheres which are trapped in capillaries, but not in a–v anastomoses, could be used to measure both nutritive and shunt flow if both tissue and venous blood samples are taken.

Injection of the volume containing microspheres should have no effect on the cardiovascular variables, particularly if serial injections with different labels are to be performed (Jones *et al.*, 1979). In practice, few authors have reported on the effects of microsphere injection on such variables as blood pressure, although these could influence the distribution of the measured flow, particularly in circulations with intracardiac shunts.

Quantification of cardiac output and organ blood flow is commonly performed using the 'reference sample' or 'artificial organ' technique. Blood is withdrawn from an artery at a constant known rate before, during and after microsphere injection. Then:

(i) Cardiac Output (ml. min^{-1}) =

$$\frac{\text{'reference sample' withdrawal rate (ml. min}^{-1})}{\text{fraction of injected spheres in 'reference sample'}}$$

(ii) Organ flow (ml. min^{-1}) = Cardiac Output (ml. min^{-1}) × fraction of injected spheres in organ

The microsphere method is attractive for its ability to provide a measurement of blood flow distribution in a relatively undisturbed conscious animal in two or more different physiological states, for example during apnea and air breathing. However, it must be recognized that the precision of the method in such repeated determinations is mainly determined by the number of spheres trapped in the reference blood sample or organ. Buckberg *et al.* (1971)· investigated the non-random variability of the distribution of microspheres in dogs and sheep by injecting the spheres into the left atrium or ventricle, and simultaneously collecting reference samples of two different volumes/unit time (V1 and V2). The ratio of spheres/ volume in each sample was then measured to determine if the difference in concentration between samples was greater than that expected by chance. The main factor determining the percentage difference between micro- sphere concentrations was the total number of spheres in the two samples. Error decreased as the total number of spheres increased. Variability was least for any total sphere number if the two samples were equal in volume (f = 0.5, Fig. 5). Disparities in sample volume produced the greatest effect when total sphere number was low. The authors concluded that a total number of spheres close to 400 in either the reference sample or organ of interest would provide sufficient precision for practical purposes, while 384 spheres/sample provided 10% precision at the 95% confidence level. More recently it has been determined experimentally in the rat that the reference sample must contain at least 200 spheres for an accurate determination of cardiac output. Cardiac output was over-estimated with less than 200 spheres/sample (Ishise *et al.*, 1980). Large numbers of injected spheres and large reference samples are therefore optimal provided that these are consistent with a lack of haemodynamic disturbance. Unfortunately, the withdrawal of relatively large reference sample volumes may cause significant changes in haemodynamics in small ectothermic vertebrates with low cardiac outputs (Moalli *et al.*, 1980). Furthermore, an injection volume and injected sphere number must be chosen that does not significantly influence cardiac output, or its distribution within the heart, in ectothermic vertebrates with intracardiac shunts. In the rat, at least, an excessive

Fig. 5. Relationship between the percentage difference in microsphere concentration measured in two simultaneously drawn blood samples and the combined number of spheres in the samples. The curved lines represent the upper 95 per cent confidence limits for different ratios of sample volumes ranging from 1 : 10 (0.05) to 1 : 1 (0.5). The smallest expected differences occur when the total number of microspheres is large (>1000) and when the sample volumes are equal (0.5). (From Buckberg *et al.*, 1971.)

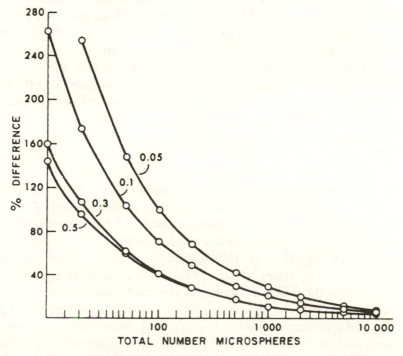

number of injected spheres ($> 10^5$) alters systemic haemodynamics. At normal resting levels of cardiac output an injection of 10^5 spheres results in numbers of spheres in excess of 200 in all organs except the adrenal glands (Ishise *et al.*, 1980).

Microspheres have been used to determine changes in the distribution of systemic cardiac output and organ blood flow occurring during 'diving' in endotherms (Jones *et al.*, 1979) and during ventilatory and apneic episodes in ectothermic vertebrates (Burggren and Moalli, 1984). An interesting further application of the microsphere technique is its use to assess the distribution of blood flow at the intracardiac shunt in amphibians and reptiles, and more peripherally at the junction between the pulmocutaneous and cutaneous arteries in anuran amphibians. Meyers *et al.* (1979) measured central shunting in conscious bullfrogs (*Rana catesbeiana*) held under conditions in which access to air was either free or denied. The technique has also been applied to quantify the extent and direction of intracardiac shunting in the lizard, *Varanus exanthematicus* and the turtle, *Chrysemys picta* (Heisler & Glass, 1985). In these experiments microspheres with different gamma labels were injected simultaneously into the left and right atria of conscious animals, moving freely within a terrarium. A different pair of labels was used in ventilatory and apneic periods, measured with a pneumotachograph, so that the influence of ventilation on shunting could be assessed. Blood was withdrawn from both atria at a known rate immediately after injection in order to determine the amount of recirculation of spheres, and a reference sample taken from the carotid artery to determine systemic cardiac output. Pulmonary cardiac output was calculated from the systemic cardiac output and the amount of L–R shunting, which was determined as the ratio of activity of microspheres injected into the left atrium to this plus the activity of this label in systemic body tissues. The intracardiac R–L shunt was calculated similarly (Berger and Heisler, 1977).

Several recent studies have utilized the microsphere technique to measure cutaneous arterial blood flow, and its distribution, in the bullfrog. The blood supply to the skin, which plays a significant role in gas exchange in anurans, is derived both from a branch of the pulmonary outflow tract, the cutaneous artery, and a diffuse systemic arterial supply. Using anaesthetized frogs, Moalli *et al.* (1980) introduced microspheres into the left pulmocutaneous artery, while spheres with a different label were then infused into the systemic circulation with the cutaneous arteries ligated in order to assess the relative contributions to skin blood flow of the pulmonary and systemic circulations. More recently, microspheres have been used to investigate the distribution of pulmocutaneous arterial flow to the lungs and skin in conscious bullfrogs (Boutilier, Glass & Heisler, 1986). The microspheres were infused *via* a non-occlusive cannula into the pulmo-cutaneous artery under control conditions, or under conditions of aquatic or gaseous hypoxia, which the authors hypothesized may influence the distribution of blood flow between the pulmocutaneous and cutaneous arteries. The relative flow distribution to the lungs and skin under each condition was measured by the ratio of the activity of each label in the respective organs, while regional skin perfusion was calculated as the ratio of specific activity (activity-tissue weight) of the region of interest to average specific activity of the whole skin.

Radionuclide-labelled blood components.

Technetium ^{99}m, a man-made element derived from molybdenum 99, has become the most commonly used radionuclide in the field of nuclear medicine. The six hour half-life of the nuclide, the absence of particulate emission, and its monoenergetic gamma photon emission at 140 kev make it ideal for visualization with current clinical imaging devices. Radiopharmaceuticals have been developed, by combining technetium ^{99}m with a variety of ligands, which make it feasible to provide an image of almost any organ. In the case of the blood pool, this has centred on the development of labelled blood components, in particular, serum albumin and red blood cells (Froelich and Callahan, 1983). Macroaggregated albumin (MAA) labelled with radionuclide may be used to obtain an image of blood flow distribution at a given time: if labelled macroaggregates with an average diameter of 40 μm are injected into the central cardiovascular area, they become lodged in capillaries, enabling regional blood flow to be visualized with the aid of a gamma camera. Thus, it has proved possible to inject macroaggreagates into the left ventricle in order to investigate the distribution of blood flow during forced submersion in Pekin ducks (Heieis & Jones, 1988). The short half-life of technetium ^{99}m, and the dispersion of the macroaggregates with a half-life of two hours, allowed serial visualizations of flow distribution to be performed in a single animal. Remotely controlled injection of MAA has enabled flow distribution to be investigated during voluntary diving in Lesser Scaup, *Aythya affinis* (Heieis & Jones, personal communication). The ability to label erythrocytes means that it is feasible to visualize changes in the distribution of blood flow dynamically in an animal that has been administered labelled red cells. The injection of labelled erythrocytes before forced submersion enabled the associated changes in central and peripheral vascular capacitance to be observed in Pekin ducks (Heieis & Jones, 1988; Fig. 6). An advantage of the technique therefore is that it is possible to obtain static (MMA) and dynamic (labelled red blood cells) views of blood flow distribution in the same individual in different physiological states. Although the use of radionuclide labelled blood products is a promising technology for the comparative physiologist, there are several associated disadvantages:

(i)   Currently, resolution of combined computer and detector systems is about ±3 mm. Therefore small scale changes in blood flow are impossible to detect.

(ii)   Flow distribution in a three-dimensional object is reduced to a two-dimensional image by gamma scanning equipment, making the quantification of flow impossible. Rotational (3-dimensional) gamma cameras have a poor effective resolution.

(iii)   Gamma imaging facilities exist almost exclusively to fulfil clinical diagnostic needs and are currently beyond the means of most individual researchers.

Fig. 6. Changes in vascular capacitance during 4 minutes of involuntary submersion in Pekin ducks, determined from the distribution of technetium ^{99}m labelled red blood cells. The arrow heads indicate the beginning and end of submersion. T, thorax; L, legs; V, viscera. Vertical axis: counts min^{-1}, horizontal axis: time in seconds. (From Heieis & Jones, 1988.)

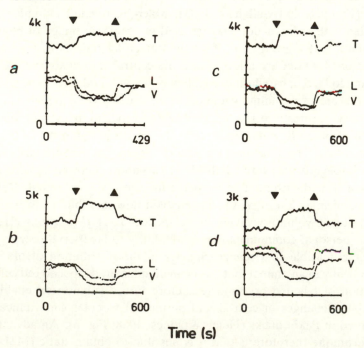

counts/min

Time (s)

Other methods

Applications of Fick's principle

The distribution of blood flow in circulations with central and/or peripheral shunts, such as those of amphibians and reptiles, may be amenable to analysis by an application of Fick's principle, using blood O_2 content and total O_2 uptake ($\dot{M}_{O_2}$) as the measured variables. The total amount of O_2 consumed by the tissues is assumed to equal oxygen consumption ($\dot{M}_{O_2}$) so that blood flows in the divisions of the arterial circulation may be calculated by a series of mass flow equations at junctions (nodes), where the algebraic sum of all inflows and outflows is assumed to be zero (Tazawa & Johansen, 1987). The technique involves heroic experimental intervention and some assumptions. For example, in experiments on pithed bullfrogs, blood O_2 content was measured at nine sampling sites in addition to measuring $\dot{M}_{O_2}$ in the intact animals (Tazawa, Mochizuki & Piiper, 1979). Ideally, if the technique were to be applied to an unanaesthetized animal, each site would have to be cannulated non-occlusively and all blood samples analyzed for O_2 content immediately, because of the measurable rate of O_2 consumption of nucleated non-mammalian red cells (Heisler & Glass, 1985).

Gas washout techniques

Cutaneous blood flow in the frog, *Rana pipiens* has recently been investigated by a gas washout technique (Malvin & Hlastala, 1986). Halothane anaesthetized frogs were equilibrated with 9.5% Freon 22 and placed in a plexiglass box. A small chamber (2.2 cm diameter) was sealed to the abdominal skin and air, hypoxic, hyperoxic or hypercapnic gas mixtures were passed through the sample chamber. The effluent gas was sampled for CO_2, Freon 22, and Halothane with a mass spectrometer. Changes in Freon or Halothane concentration in the effluent were assumed to reflect changes in skin perfusion under the chamber. An elevation in chamber O_2 increased the effluent gas concentrations, suggesting an increase in local skin perfusion, while hypoxia or hypercapnia lowered them.

Conclusions

Technological innovations, evolved for medical purposes, have historically provided comparative physiologists with new tools for the measurement of blood flow and distribution in small animals, including ectotherms. The availability of new technology was directly responsible for important advances in the appreciation of cardiovascular function in non-mammalian vertebrates some 25 years ago (Johansen, 1963). Currently, there is an increasing realization that many of the physiological phenomena that can radically alter cardiac output and its distribution (intermittent ventilation, diving, etc.) have important behavioral components that are not apparent in restrained or anaesthetized animals. It is fortunate therefore that methods which enable blood flow to be chronically measured in conscious, relatively freely moving animals are available. Each method has discrete limitations and advantages; for example radioactive microspheres provide a detailed picture of the distribution of blood flow, but not on a continuous basis. On the other hand, although miniature pulsed Doppler flow probes can provide a continuous dynamic measure of blood flow in the central arteries, they cannot provide a measure of blood flow distribution to all organs and tissues. A combination of currently available techniques may prove fruitful, and seems worth exploring. For example, in intermittently ventilating ectotherms a combination of the pulsed Doppler and microsphere techniques could provide a continuous measure of flow in the major divisions of the circulation, together with a detailed picture (or two) of the distribution of this flow during lung ventilation and apnea.

As stated by Randall (1987), it would be foolish to attempt to predict in detail, developments in the next 25 years. It may be, for example, that advances in permanent magnet flow transducer technology will diminish the current enthusiasm for the pulsed Doppler technique and microsphere techniques. Whatever the advances, I am sure that they will enable us to resolve questions such as the degree of intracardiac shunting during apnea in reptiles, or the extent of peripheral vasoconstriction during voluntary diving in birds, which stretch current methods of measurement to their limits.

Acknowledgements
I thank Dr W. W. Burggren and Dr M. D. Evered for critically reading the manuscript, which was written while the author was supported by operating grants from the MRC and NSERC (Canada).

References
Baker, D. W. & Yates, W. G. (1973). Technique for studying the sample volume of ultrasonic Doppler devices. *Medical and Biological Engineering*, **11**, 766–70.

Berger, P. J. & Heisler, N. (1977). Estimation of shunting, systemic and pulmonary output of the heart, and regional blood flow distribution in unanaesthetized lizards (*Varanus exanthematicus*) by injection of radioactively labelled microspheres. *Journal of Experimental Biology*, **71**, 111–21.

Boutilier, R. G., Glass, M. L. & Heisler, N. (1986). The relative distribution of pulmocutaneous blood flow in *Rana catesbeiana*: effects of pulmonary or cutaneous hypoxia. *Journal of Experimental Biology*, **126**, 33–9.

Brunsting, J. R. & ten Hoor, F. (1968). Factors preventing accurate *in vitro* calibration of noncannulating electromagnetic flow transducers. In *New findings in blood flometry*, ed. C. Cappelen. pp. 107–13. Oslo: Aas and Wahls Boktrykkeri.

Buckberg, G. D., Luck, J. C., Payne, D. D., Hoffman, J. I.E., Archie, J. P. & Fixler, D. E. (1971). Some sources of error in measuring regional blood flow with radioactive microspheres. *Journal of Applied Physiology*, **31**, 598–604.

Burggren, W. W. (1987). Invasive and noninvasive methodologies in ecological physiology: a plea for integration. In *New Directions in Ecological Physiology*, eds. M. Feder, A. Bennett, W. W. Burggren & R. Huey. New York: Cambridge.

Burggren, W. W. & Moalli, R. (1984). 'Active' regulation of cutaneous gas exchange by capillary recruitment in amphibians: experimental evidence and a revised model for skin respiration. *Respiration Physiology*, **55**, 379–92.

Butler, P. J. Milsom, W. K. & Woakes, A. J. (1984). Respiratory, cardiovascular and metabolic adjustments during steady state swimming in the green turtle, *Chelonia mydas. Journal of Comparative Physiology, B*, **154**, 167–74.

Cameron, J. N. (1986). *Principles of physiological measurement*. Orlanado: Academic Press.

Clark, E. B. & Hu, N. (1982). Developmental hemodynamic changes in the chick embryo from stage 18 to 27. *Circulation Research*, **51**, 810–15.

Dedichen, H. & Schenk, W. G. (1968). Influence of hematocrit changes on square-wave electromagnetic flowmeter calibration. In: *New findings in blood flowmetry*, ed. C. Cappelen. pp. 104–06. Oslo: Aas and Wahls Boktrykkeri.

Froelich, J. W. & Callahan, R. (1983). Technetium-labelled blood products for equilibrium gated studies. In *Techniques, diagnostics, and advances in nuclear cardiology*, ed. M. A. Osbakken. pp. 109–20. Springfield: Charles C. Thomas.

Gessner, U. (1961). Effects of the vessel wall on electromagnetic flow measurement. *Biophysical Journal*, **1**, 627–37.

Hartley, C. J. & Cole, J. S. (1974). An ultrasonic pulsed Doppler system for measuring blood flow in small vessels. *Journal of Applied Physiology*, **37**, 626–9.

Hartley, C. J., Lewis, R. M., Ishida, T., Chelly, J. E. & Entman, M. L. (1984). High frequency pulsed Doppler measurements of blood flow and myocardial dimensions in conscious animals. Cardiovascular instrumentation. *Proceedings of the working conference on applicability of new technology to biobehavioral research*, ed. J. A. Herd, A. M. Gotto, P. G. Kaufmann & S. M. Weiss. pp. 95–106. U.S.A.: NIH publication.

Haywood, J. R., Shaffer, R. A., Fastenow, C., Fink, G. D. & Brody, J. J. (1981). Regional blood flow measurement with pulsed Doppler flowmeter in conscious rat. *American Journal of Physiology*, **241**, H273–8.

Heieis, M. R. A. & Jones, D. R. (1988) Blood flow and volume distribution during forced submergence in Pekin ducks (*Anas platyrhynchos*). *Canadian Journal of Zoology*, in press.

Heisler, N. & Glass, M. L. (1985). Mechanisms and regulation of central vascular shunts in reptiles. In *Cardiovascular Shunts, Alfred Benzon Symposium 21*, ed. K. Johansen and W. W. Burggren, pp. 334–53. Copenhagen: Munksgaard.

Heymann, M. A., Payne, B. D., Hoffman, J. I. E. & Rudolph, A. M. (1977). Blood flow measurements with radionuclide-labelled particles. *Progress in Cardiovascular Diseases*, **20**, 55–79.

Ishise, S., Pegram, B. L., Yamamoto, J., Kitamura, Y. & Frolich, E. D. (1980). Reference sample microsphere method: cardiac output and blood flows in conscious rat. *American Journal of Physiology*, **239**, H443–9.

Johansen, K. (1963). Cardiovascular dynamics in the amphibian *Amphiuma tridactylum*. *Acta Physiologica Scandinavica*, **60**, suppl. 217, 1–82.

Jones, D. R., Bryan, R. M., West, N. H., Lord, R. H. & Clark, B.. (1979). Regional distribution of blood flow during diving in the duck (*Anas platyrhynchos*). *Canadian Journal of Zoology*, **57**, 995–1002.

Kirkebo, A. (1968). Measurements of small flows in animals with implanted probes. In *New findings in blood flowmetry*, ed. C. Cappelen. pp. 179–81. Oslo: Aas & Wahls Boktrykkeri.

Klepper, J. R. (1983). The physics of Doppler ultrasound and its measurement instrumentation. In *Cardiac Doppler diagnosis*, ed. M. P. Spencer. pp. 19–31. Boston: Martinus Nijhoff.

Kolin, A. (1936). An electromagnetic flowmeter, principle of the method and its application to blood flow measurements. *Proceedings of the Society for Experimental Biology (N.Y.)*, **35**, 53–6.

Laughlin, M. H., Armstrong, R. B., White, J. & Rouk, K. (1982). A method for using microspheres to measure muscle blood flow in exercising rats. *Journal of Applied Physiology*, **52**, 1629–35.

Malvin, G. M. & Hlastala, M. P. (1986). Regulation of cutaneous gas exchange by environmental O_2 and CO_2 in the frog. *Respiration Physiology*, **65**, 99–111.

Meyers, R. S., Moalli, R., Jackson, D. C. & Millard, R. W. (1979). Microsphere studies of bullfrog central vascular shunts during diving and breathing in air. *Journal of Experimental Zoology*, **208**, 423–30.

Meisner, H., Messmer, K. & Hagl, S. (1968). Problems concerning calibration and application of sine-wave and square-wave gated electromagnetic flowmeters in acute and chronic dog experiments. In *New findings in blood flowmetry*, ed. C. Cappelen. pp. 129–35. Oslo: Aas & Wahls Boktrykkeri.

Millard, R. W., Baig, H. & Vatner, S. F. (1977). Cardiovascular effects of radioactive microsphere suspensions and Tween 80 solutions. *American Journal of Physiology*, **232**, H331–4.

Milsom, W. K., Langille, B. L. & Jones, D. R. (1977). Vagal control of pulmonary vascular resistance in the turtle (*Chrysemys scripta*). *Canadian Journal of Zoology*, **55**, 359–67.

Moalli, R., Meyers, R. S., Jackson, D. C. & Millard, R. W. (1980). Skin circulation in the frog, *Rana catesbeiana*: distribution and dynamics. *Respiration Physiology*, **40**, 137–48.

Nakamura, T., Hayashi, K., Taenaka, Y., Umezu, M., Nakatani, T. & Takano, H. (1986). Ultrasound flowmeter with implantable miniature sensors. *Medical and Biological Engineering and Computing*, **24**, 235–42.

Noordergraaf, A. (1978). *Circulatory system dynamics*. New York: Academic Press.

Peronneau, P., Xhaard, M., Nowicki, A., Pellet, M., Delouche, P. & Hinglais, J. (1972). Pulsed Doppler ultrasonic flowmeter and flow pattern analysis. In *Blood flow measurement*, ed. C. Roberts. pp. 24–8. Baltimore: Williams & Wilkins.

Pohlman, A. G. (1909). The course of the blood through the heart of the fetal mammal, with a note on the reptilian and amphibian circulations. *Anatomical Record*, **3**, 75–109.

Randall, D. J. (1987). The next 25 years: vertebrate physiology and biochemstry. *Canadian Journal of Zoology*, **65**, 794–96.

Shelton, G. & Burggren, W. W. (1976). Cardiovascular dynamics of the Chelonia during apnoea and lung ventilation. *Journal of Experimental Biology*, **64**, 323–43.

Smits, A. W., West, N. H. & Burggren, W. W. (1986). Pulmonary fluid balance following pulmocutaneous baroreceptor denervation in the toad. *Journal of Applied Physiology*, **61**, 331–7.

Spencer, M. P. & Barefoot, C. A. (1968). Sensor design for electro-magnetic flowmeters. In *New findings in blood flowmetry*, ed. C. Cappelen. pp. 76–87. Oslo: Aas & Wahls Boktrykkeri.

Tazawa, H., Mochizuki, M. & Piiper, J. (1979). Respiratory gas transport by the incompletely separated double circulation in the bullfrog, *Rana catesbeiana. Respiration Physiology*, **36**, 77–95.

Tazawa, H. & Johansen, K. (1987). Comparative model analysis of central shunts in vertebrate cardiovascular systems. *Comparative Biochemistry and Physiology*, **86**A, 595–607.

Van Vliet, B. N. & West, N. H. (1986). Cardiovascular responses to electrical stimulation of the recurrent laryngeal nerve in conscious toads (*Bufo marinus*). *Journal of Comparative Physiology B*, **156**, 363–75.

Van Vliet, B. N. & West, N. H. (1987). Response characteristics of pulmocutaneous arterial baroreceptors in the toad, *Bufo marinus. Journal of Physiology (London)*, **388**, 55–70.

West, N. H. & Burggren, W. W. (1984). Factors influencing pulmonary and cutaneous arterial blood flow in the toad, *Bufo marinus. American Journal of Physiology*, **247**, R884–94.

Woodcock, J. P. (1975). Electromagnetic flowmeter. In *Theory and practice of blood flow measurement*. London: Butterworth & Co.

Wyatt, D. G. (1984). Blood flow and blood velocity measurement *in vivo* by electromagnetic induction. *Medical and Biological Engineering and Computing*, **22**, 193–211.

E. W. TAYLOR and C. J. H. ELLIOTT

Neurophysiological techniques

Introduction

Respiratory and cardiovascular systems typically rely on a central pattern generator (CPG) in the CNS to initiate the respiratory rhythm and a cardiac pacemaker to initiate heart beat, with heart rate controlled from the CNS. Although the comparative morphology and mechanics of both systems are well known the mechanisms for nervous control have been studied in detail in very few groups of animals. Amongst the vertebrates, only for the two ends of the evolutionary scale – fishes and mammals – is a reasonable amount of information available on the nervous circuits engaged in the control of ventilation (Ballintijn, 1988). Pacemaker function in the heart and its efferent control by the autonomic nervous system is relatively well studied in most vertebrate groups, but the central nervous interactions that control heart rate and its relation to the ventilatory rhythm, have only recently been described (Spyer, 1984; Taylor, 1985; Jordan & Spyer, 1988; Daly *et al.*, 1988). Amongst the invertebrates, the mechanics of ventilation and its central control have been extensively described in both insects (Miller, 1966) and crustaceans, where detailed analyses of the nature of the CPG and control of movements of the gill bailer or scaphognathite of crayfish, crabs and lobsters have been achieved (Bush *et al.*, 1988). However, our knowledge of the control of the ventilatory system by interneurones in the CNS lags behind that of other rhythmic systems (e.g. the crustacean stomatogastric ganglion, swimming in the leech and *Tritonia* and feeding in snails).

Experiments to locate the CPG and areas of the CNS influencing ventilation, such as the pneumotaxic centre in mammals, have classically consisted of progressive brain transection or ablation accompanied by measurement of ventilation rate, respiration related activity in the brain-stem or an indicator of central respiratory drive such as phrenic nerve activity (e.g. Shelton, 1959, 1961; Ballintijn, 1982; Feldman & Gautier, 1976; Dawes *et al.*, 1983). A serious limitation of ablation experiments, designed to isolate that part of the nervous system minimally needed to control respiration, is that the control functions of the parts removed may be taken over by the remaining elements of the control system (Ballintijn, 1982). Latterly, these studies have been combined with injection of neuropharmacological agents known to cross the blood-brain barrier (e.g. Bamford *et al.*, 1986). The influence of higher centres in the CNS on the CPG may similarly be studied by selective brain lesions or by electrical/chemical stimulation from centrally placed electrodes (e.g. Juch & Ballintijn, 1983; Spyer, 1984). As, however, this may stimulate areas exercising indirect influences over respiration it is also fraught with difficulties.

The activity of the CPG is of course influenced by inputs from peripheral receptors such as the lung receptors in mammals, gill mechanoreceptors in fishes (Ballintijn, 1988) or the oval organ on the gill bailer (scaphognathite) of crustaceans (Bush *et al.*, 1988). In each case phasic input from these receptors can entrain the CPG and this may be demonstrated by stimulation of the central cut ends of the appropriate afferent nerves (Trenchard, 1977; Iscoe *et al.*, 1979; Ballintijn *et al.*, 1983). Identification and characterization of afferent inputs influencing reflex control of ventilation or blood flow can be achieved by selective denervation: a bilateral section of cranial nerves IX and X abolished hypoxic bradycardia in the trout (Smith & Jones, 1978), whereas in the dogfish complete abolition required transection of branches of V, VII, IX and X (Butler *et al.*, 1977). Alternatively, an appropriate stimulus such as injection of a bolus of hyperoxic blood (Barrett & Taylor, 1984) may be used to identify a receptor site. Direct recordings of afferent nervous activity in nerves supplying the peripheral receptors are of course essential for complete characterization of their potential influence (e.g. de Graaf & Ballintijn, 1987; Gleeson, 1985; Pasztor, 1969).

On the efferent side of the reflex arc the influence of the CPG and cardiomotor centres on ventilation and heart rate may be studied by progressive nerve transection (e.g. Taylor *et al.*, 1977) or by stimulation (e.g. Short *et al.*, 1977). These studies revealed that vagal tone on the heart of the dogfish, as well as a reflex hypoxic bradycardia, were mediated predominantly via the pair of branchial cardiac branches of the vagus. Nervous control of the cardiorespiratory system may, however, be most effectively studied by the location of the cell bodies of specific respiratory or cardiomotor neurones, followed by detailed characterization of their spontaneous activity and the influences of afferent inputs.

This chapter describes a range of techniques for the investigation of the functional neuranatomy, and the neurophysiology of efferent motor control of ventilation and cardiorespiratory interactions in lower vertebrates and invertebrates.

Location of neurones by dye-marking

A major problem with the study of the central nervous control of respiration and circulation in vertebrates is the compact nature of CNS. The neurones comprising the CPG or the motor neurones supplying respiratory muscles have their cell bodies located deep in the brain as pools of neurones, often diffusely distributed amongst other neurones having different functions. Also the cell bodies are relatively small (10–20 μm in diameter) and difficult to locate or penetrate with the tip of a microelectrode. These comparisons are of course related to invertebrate systems. In gastropods the ganglia contain many large neuronal somata which are constant in location, size and membrane properties in all individuals. Many cells can be identified *in vivo* with a dissecting microscope and in some (e.g. *Lymnaea*) the neurones are highly coloured in a variety of shades – red, orange or white. A constant pattern of cell bodies, tracts and neuropil is equally obvious when axonal iontophoresis of dyes is applied to the peripheral nerves or connectives of arthropods.

Several techniques for marking specific neurones are available including cobalt or cobaltic lysine (e.g. Lazar *et al.*, 1983) and fluorescent markers such as lucifer yellow or fast blue (e.g. Sloniewski *et al.*, 1985; Niida & Ohono, 1984). These techniques often stain up the whole neurone, including its axons and dendrites and are ideally suited for cytoarchitectural studies. They have been very successfully applied to invertebrate neurones and are described below. First we concentrate on the use of horseradish peroxidase (HRP) histochemistry to locate neurones in the vertebrate CNS.

Retrograde intra-axonal transport of HRP

HRP is a large plant protein and an enzyme. As a protein it has the advantage that when introduced into an axon it is actively transported into the cell body where it accumulates. Here it may be visualized by enzymic reaction with a chromagen. The history of the development of this procedure and its effective use in tracing neural connections is described in detail by Warr *et al.* (1981) and M-Marcel Mesulum (1982).

The technique consists of locating, by careful dissection under general anaesthesia, the nerve supplying the organ (e.g. heart or respiratory muscle) and introducing HRP into afferent and efferent axons. Following a period of recovery from anaesthetic, during which the HRP is transported centrally, the animal is sacrificed and the brain sectioned. HRP histochemistry then reveals the central projections or cells of origin of the marked axons.

Introduction of HRP. To introduce sufficient HRP into axons to enable clear identification of their cell bodies requires that a high concentration of the protein is brought into solution at the site of mechanical damage of the appropriate nerve. This can be achieved by transection of the nerve. The central cut end is then cleared of connective tissue and placed on a small strip of Teflon, clear of accumulated blood or tissue fluid. A few crystals of lyophilized HRP (Sigma, type VI) may then be placed in contact with the cut nerve where they will dissolve to create a local pool of concentrated solution. If this can be maintained for 15–20 minutes the cut nerve will visibly be stained brown. At this point the excess HRP should be carefully removed to prevent uptake by other nerves in the vicinity and the cut end of the nerve covered with a water repellant substance such as petroleum jelly.

A technique for the introduction of HRP into axons with less danger of contamination of other nerves is to dry a concentrated solution of HRP onto the tip of a fine steel entomological pin which is then inserted through the nerve sheath (Withington-Wray *et al.*, 1986). For very large nerve trunks a further technique is to dissolve HRP as a 25% w/v solution in an appropriate saline and inject 2–10 nl into the nerve (Levings, 1987). If the nerves supplying a muscle (or organ) are too diffuse or difficult to locate, then satisfactory labelling of the cells of origin of motor nerves may be achieved by injecting relatively high concentrations of HRP solution into the muscle or organ (e.g. Gwyn *et al.*, 1985; Rokx *et al.*, 1985).

Having introduced HRP into the nerve or the innervated muscle, the animal has to be retained alive whilst intra-axonal transport carries the HRP to the cell bodies of the appropriate neurones. This transport time varies between animals. Warr *et al.* (1981) quoted rates of 70–120 mm day^{-1} for homeothermic animals, so that small mammals, such as the rat, may be supported on a ventilator under a general anaesthetic overnight for adequate central transport of HRP, whilst cats and rabbits must survive for 24–48 h. In poikilothermic animals transport rates are highly temperature dependent. Carlson *et al.* (1982) reported that transport time for retrograde HRP transport in the frog more than doubled with a reduction in temperature from 20 °C to 15 °C. In the dogfish, *Scyliorhinus canicula* retrograde transport rate was 0.5 cm day^{-1} at an acclimation temperature of 16 °C and 0.8 cm day^{-1} at 21 °C, whilst anterograde transport up sensory fibres was slower, possibly related to their smaller diameter, with a value of 0.4 cm day^{-1} at 19 °C (Barrett, 1984). In the ray, *Raia clavata* the average rate of retrograde axonal transport of HRP was 0.3 cm day^{-1} at temperatures between 15 °C and 17.5 °C (Levings, 1987).

HRP histochemistry. Once HRP is judged to have penetrated to the cell bodies the brain has to be removed for sectioning. The animal should be given a high dose of a general anaesthetic and then cannulated for perfusion by gravity feed or a peristaltic pump. First the animal *must* be exanguinated by perfusion with a suitable buffered saline; red blood cells contain peroxidase enzymes and stain darkly with the histochemical procedure for HRP. After a further injection of anaesthetic, perfusion is switched to a suitable fixative (we use 50 ml of 50% w/v glutaraldehyde, plus 30 ml of 40% formaldehyde solution made up to 1 litre with 0.1 M phosphate buffer at pH 7.4). Once fixed the brain or spinal cord is carefully dissected free of bone and other tissue and stored overnight in a bath of the fixative solution. It is then transferred to a solution of 10% w/v sucrose dissolved in 0.1 M phosphate buffer, pH 7.4 at 4 °C for 4 hours before being frozen and serially sectioned at between 30 and 80 µm (we use 60 µm). A simple sledge microtome (e.g. Leitz 1401) fitted with a Peltier cooling stage (Mectron, Frigistor Ltd.) is satisfactory for small blocks of tissue, though better results for a long series of sections on large pieces of frozen tissue, are achieved with a cryostat having a temperature controlled cabinet with an open top (e.g. Bright OTF/AS cryostat). The sections may be stored in a bath of chilled phosphate buffer (without sucrose) in identified compartments of a modified Repli dish (Sterilin) for three to four days prior to staining for HRP. Several changes of cool buffer (4 °C) at this stage will remove traces of fixative and provide longer lasting neuronal staining.

The staining procedure is started by placing the trays holding the sections sequentially in 3 × 1 minute washes of a rinse solution (0.04 M acetate buffer, pH 3.3 at 4 °C), then transferring them to a pre-reaction mixture made up from: 200 mg of sodium nitroprusside (BDH) dissolved in 200 ml of 0.02 M acetate buffer, pH 3.3; 10 mg tetramethylbenzidine (TMB) (Sigma) in 5 ml ethanol (absolute) heated on a hotplate to encourage dissolution (cool before use). These solutions are mixed immediately before use. After 20 minutes in this mixture, on a rocking platform, the reaction is initiated by addition of 7 ml of 0.3% hydrogen peroxide in glass distilled water. The reaction is then allowed to proceed for a maximum of 20 minutes; in practice you will develop an eye for the course of the reaction. If HRP staining is heavy and present in a large number of cell bodies you will see them develop, otherwise the sections take on a bluish colour as the reaction proceeds. Remove the sections from the reaction mixture as soon as your HRP has been stained – delay increases the risk of contamination and non-specific staining. When the reaction is complete the sections are washed in three changes of chilled 0.04 M acetate buffer (pH 3.3 at 4°) and stored prior to mounting on gelatin coated microscope slides.

Fig. 1. Neurones in the hindbrain and spinal cord of the ray, *Raia clavata* stained by HRP histochemistry, following retrograde intra-axonal transport of HRP from peripheral sites of labelling (Levings, 1987).

(*a*) T.S. of spinal cord 9.5 mm caudal of obex at the level of the first spinal motor root, following application of HRP to right hypobranchial nerve. The dense granular reaction product of HRP has filled the cell bodies, axons and dendrites of two motor neurones (scale bar 0.1 mm).

(*b*) T.S. of medulla 0.5 mm rostral of obex following application of HRP to the fourth branchial branch of the vagus. Preganglionic vagal neurones are located in the dorsal vagal motonucleus (DMV). Dense HRP staining has filled the cell bodies and the ventro–laterally projecting dendrites (scale bar 0.1 mm).

(*c*). T.S. of medulla 1.8 mm rostral of obex. HRP labelled branchial vagal motoneurones are located close to the ventro–lateral margin of the IVth ventricle (top left corner of the frame, labelled *c*). The labelled vagal afferent (sensory) projections terminate in the fasciculus solitarius, sited close to the dorsolateral margin of the IVth ventricle (scale bar 0.2 mm).

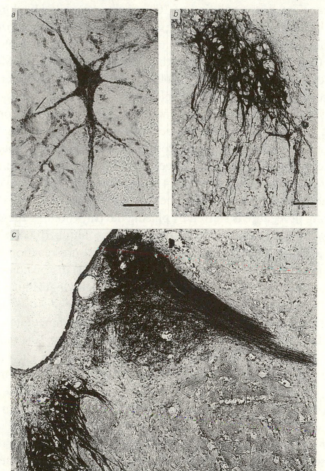

When the sections have dried onto the slide at room temperature they are counterstained by immersion in neutral red solution for one minute, rinsed in glass distilled water then left to dry at room temperature for up to 24 hours. The dry sections are then cleared in 3 × 5 minute changes of Histoclear (National Diagnostics) and mounted under coverslips in Histomount (National Diagnostics). Air-drying the sections may seem severe but it produces photographable results and avoids the problem of leaching out HRP during dehydration through alcohol series (Levings, 1987).

You now should be able to view your stained neurones under the microscope. HRP can reveal details of cytoarchitecture, including dendritic fields (Fig. 1*a*, *b*) and sensory projections (Fig. 1*c*), but is most useful for identifying the location of cells of origin of axons in selected nerves, i.e. the somatopic, viscerotopic or overall topographical organization of motor nuclei in the CNS. This may be charted by counting cells in successive sections and marking their positions on outline drawings of the sections

Fig. 2. Rostro-caudal distribution of vagal preganglionic motoneurones in sequential 60 μm thick transverse sections through the brainstem of the marine toad, *Bufo marinus*. Each spot represents one stained, neuronal body obtained after application of HRP to the whole vagus nerve on the right side of the animal. The most rostral section is at the top right and the most caudal at the bottom left of the series. The vagal motor column is continuous through 41 sections (approximately 2.5 mm) of the medulla, distributed either side of obex (located at the rostral end of the third series of sections). Approximately 20% of the motoneurones are located in a more ventro–lateral position, caudal of obex.

ROSTRO-CAUDAL DISTRIBUTION OF VAGAL PREGANGLIONIC MOTONEURONES IN THE BRAINSTEM OF *Bufo marinus*. (sequential 60 μm T.S.)

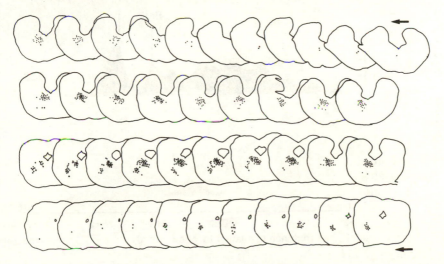

using a drawing tube (Fig. 2). Histograms of the rostro-caudal distribution of cell numbers may then be plotted and related to a fixed location in the CNS such as obex in the medulla (Fig. 3).

The techniques of HRP histochemistry have been successfully combined with the use of fluorescent markers, which by double-labelling can explore afferent and efferent projections in the CNS, and with transmitter-related histochemical procedures that allow the pharmacology of identified neurones to be described (Steward, 1981; Gwyn *et al.*, 1985).

We have used retrograde transport of HRP to trace the location within the brainstem of vagal preganglionic motoneurones in the dogfish *Scyliorhinus canicula* (Withington-Wray *et al.*, 1986). Labelled cell bodies were located in the ipselateral rhombencephalon from 2.1 mm caudal to 2.73 rostral of obex. There is a sequential topographical representation of the vagus nerve in this vagal motor column. Neurones supplying the gastrointestinal tract are located caudally; those supplying the cardiac nerves lie in the midportion of the column and the efferent supply to the intrinsic

Fig. 3 Rostro–caudal distribution of vagal preganglionic motoneurones with respect to obex in the medulla of the dogfish, *Scyliorhinus canicula*. The majority of labelled motoneurones are located medially in the DMV. A small number (8%) are located ventro–laterally and supply axons solely to the branchial cardiac branch of the vagus innervating the heart. Medial cells supplying this nerve are indicated by the unshaded portion of the upper histogram.

Motoneurones supplying axons to branchial branches 1, 2 and 3 (Br 1 – Br 111) of the vagus, innervating gill arches 2, 3 and 4 occupy the rostral part of the vagal motor column, whilst the more caudal motoneurones supply axons sequentially to the heart, oesophagus and stomach.

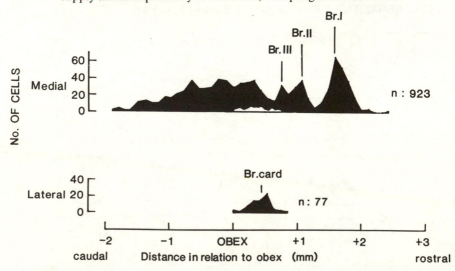

muscles in the 5 gill arches comes in sequential order from the most rostral
vagal motoneurones (Fig. 3) with the first arch supplied independently by
the glossopharyngeal nerve (Withington-Wray, *et al.*, 1986). All respiratory
and visceral vagal motoneurones are located in the dorsal motor column
(DMV), close to the fourth ventricle, but the heart is supplied with cell
bodies located both in the DMV and in a scattered ventro-lateral location
that may be homologous to the nucleus ambiguus identified in the mammalian brainstem (Barrett & Taylor, 1985*b*). These ventro–laterally located
vagal motoneurones comprise 8% of the total of stained cell bodies in the
dogfish hindbrain and supply axons solely to the heart. Further HRP studies
have revealed that this proportion rises to 12% in teleost fishes (cod and
trout) and that they may provide some axons to the gill arches, possibly with
a vasomotor function. The proportion of ventro–laterally located cells rises
in the amphibians, *Bufo marinus* (20%) and *Xenopus laevis* (32%), possibly
associated with innervation of air-breathing structures such as the lungs and
respiratory passages. In mammals 70–80% of cardiac and pulmonary vagal
motoneurones are located in ventro–lateral locations including the nucleus
ambiguus (Withington-Wray *et al.*, 1988). This interesting evolutionary
trend for progressive ventro–lateral migration of the preganglionic vagal
motoneurones controlling the heart and ventilation has been revealed by a
series of HRP studies. Its possible functional significance can only be
explored by neurophysiological investigations based on the known location
of the cell bodies in the brainstem.

The use of cobalt ions. The two main problems of early attempts to mark
cells with dyes – leakage from the cell and a lack of sensitivity – have been
overcome by using cobalt (Pitman *et al.*, 1972) with Timm's intensification
of the sulphide precipitate (Bacon & Altman, 1977). Neurones may be filled
from micropipettes containing 10% cobalt chloride solutions, but these tend
to block frequently, when depolarising current is passed. This problem is
readily solved by using a solution in which the cobalt ions are complexed
with an amino acid. Lysine and proline complexes may be made by
dropwise addition of a saturated solution of the amino acid to a 10% w/v
cobalt solution, until the volume is doubled. (To ascertain that all the cobalt
has been complexed, add a drop of dilute NaOH and check that no cobalt
hydroxide precipitates (Gallyas *et al.*, 1978).) For locust non-spiking
neurones, Siegler & Burrows (1984) found that using 0.1 M hexammineco-
balt chloride (Sigma) solution gave the most successful results. In any case,
it is usual to eject the cobalt from the electrode by passing 500 ms pulses of
3–10 nA depolarizing current, once per second.

Neurones may also be stained by axonal iontophoresis (Fig. 4). Cobalt

Fig. 4. Anatomy of ventilatory neurones.

(*a*) Central projections of ventilatory motoneurones in the thoracic ganglia of *Carcinus*, stained with cobalt by axonal iontophoresis.
(i) dorsal view of the thoracic ganglia – the depressor nerve is stained on the right side and the levator on the left. (Simmers & Bush, 1983*a*).
(ii) Lateral view of right levator and depressor neurones together.

(*b*) The morphology of the meso–thoracic spiracular closer motoneurone of the locust *Schistocerca* seen (i) dorsal and (ii) side view. This cell was injected with cobalt and subsequently intensified. (Burrows, 1982*a*).

(*c*) The morphology of a ventilatory interneurone in (i) *Schistocerca* and (ii) *Locusta*. This interneurone fires during expiration. Both stained by injection of lucifer yellow. (Pearson, 1980).

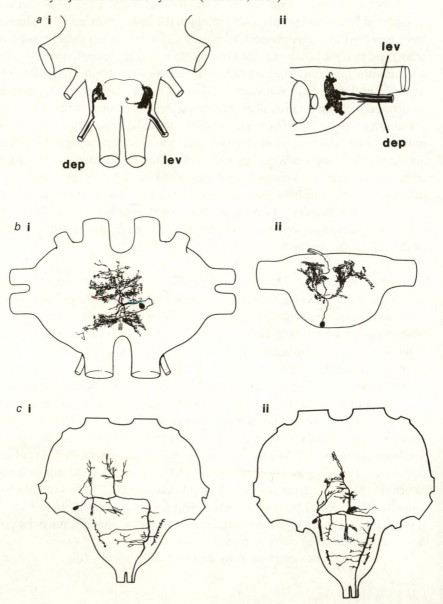

ions are readily transported in either direction. In orthopteran insects, the most reliable method is as follows: first the insect is anaesthetized by cooling to 4 °C. Then pure white petroleum jelly – 'Vaseline' (ordinary 'Vaseline' contains toxic impurities) – is used to construct a cup around the cut end of the nerve (Fig. 5). The cup is filled with a low (1–3 w/v) concentration of cobalt (or cobalt complex) and covered over with 'Vaseline'. The wound is filled with 'Vaseline' to reduce dessication. The cooling is maintained for 8–24 hours while the neurone fills. Use of metal instruments should be minimized as the intensification process will detect any metal deposits. It is important that the neurone to be filled remains alive – a moist, cool atmosphere helps; also avoid blocking the tracheae with 'Vaseline'.

When neurones have been filled, the cobalt is precipitated as cobalt sulphide. First, the 'Vaseline' cup (etc.) must be rinsed away with fresh saline. Cobalt ions can be precipitated with 0.2% ammonium sulphide in saline but use of saline saturated with hydrogen sulphide (bubble H_2S through saline for 15 minutes) is preferrable (Altman & Tyrer, 1980). The black cobalt sulphide will precipitate in less than 10 minutes. The tissue should then be washed in two changes of saline to remove any unreacted sulphide.

The tissue should then be fixed in alcoholic Bouins (see e.g. Grimstone &

Fig. 5. *In vivo* filling of nerves. (*a*) 'Vaseline' cup is constructed close to the selected nerve (*b*), filled with a drop of cobalt solution into which the cut end of the nerve is placed (*c*). The drop is then covered with 'Vaseline' (*d*) and the wound filled. (After Altman & Tyrer, 1980.)

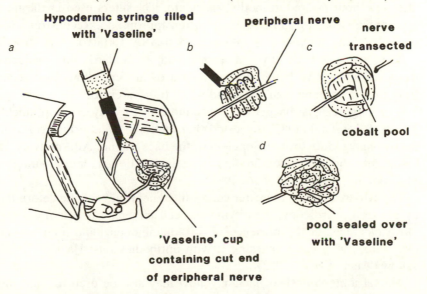

Skaer, 1972) and transferred into 70% ethanol. The picric acid may be removed by two changes (10 minutes) of 10% ammonium acetate in 70% ethanol. The nervous system is then dissected free and dehydrated in ethanol. Specimens may be observed as whole mounts in methyl salicylate. Lightly stained specimens may be intensified using a developer solution (Bacon & Altman, 1977; Altman & Tyrer, 1980). After development the ganglion may be dehydrated and cleared in methyl salicylate. If required the tissue may then be embedded in Araldite (Seyan *et al.*, 1983) and sectioned.

The use of lucifer yellow. For intracellular injection the fluorescent Naphthalimide dye, Lucifer yellow CH (Sigma) may be dissolved in distilled water (3–5%) or 1 M lithium sulphate. Micropipettes, pulled from fibre-containing or theta glass, fill readily when their blunt end is stood in either solution. The resistance of lithium sulphate electrodes is lower than similarly shaped distilled water electrodes but both fill insect and molluscan neurones well. Injections into the axons of orthopteran insects or small cell bodies of gastropods occur by diffusion and only require that the cell be held for 10 minutes, but large cells require the passage of negative current (e.g. 500 ms pulses of -2 nA, once per second).

Fixation of the tissue is done with formaldehyde as this binds the lucifer yellow to the cellular proteins. A good protocol is to use 4% formaldehyde in 0.1 M sodium phosphate buffer (pH 7.4) for 10 minutes followed by 4% formaldehyde in methanol (10 ml formalin + 90 ml methanol AR) for one hour. The tissue may then be transferred directly to absolute ethanol and after one hour cleared in methyl salicylate. The filters used in the fluorescent microscope for FITC stains are often appropriate. Reconstruction of neurones from micrographs taken on 400 ASA Ektacrome is straightforward using a front silvered mirror and a standard slide projector. Coloured slides give better discrimination of lucifer yellow fluorescence from the background than black and white transparencies.

With large ganglia background autofluorescence may develop quickly. Tauchi & Masland (1984) suggested that inclusion of ascorbic acid in the fixative may reduce the development of the background. Alternatively, the tissue may be embedded in Spurrs resin, sectioned and mounted in Fluoromount (Strausfeld *et al.*, 1983).

The cytoarchitecture of lucifer yellow filled neurones may be determined by electron microscopy, identifying the cell either by irradiating it with ultraviolet light in the presence of 3,3′-diaminobenzidine after staining (Maranto, 1982) or by marking it with antibodies raised against lucifer yellow (Taghert *et al.*, 1982).

Axonal iontophoresis of lucifer yellow may also be used to stain molluscan neurones *in vivo*, taking only 40–60 minutes (Stewart, 1981). Filling may be enhanced by applying current between the drop of lucifer yellow and the ganglia. The filled molluscan neurones may be observed *in vivo* with low levels of illumination from a mercury lamp or laser (Reichert & Krenz, 1986), and viewed as described above. In insects, axonal iontophoresis of lucifer yellow is less reliable and has not been widely used.

With higher levels of illumination, lucifer yellow can also be used for photoinactivation of neurones (Miller & Selverston, 1979, Selverston *et al.*, 1985). This allows for direct verification of the roles of particular neurones (or parts of neurones) and has been particularly useful in analysing complex rhythmic systems, especially the crustacean stomatogastric ganglion.

The use of cell specific markers in vivo. Cobalt or lucifer yellow may be introduced into any cell, but sometimes it is required to penetrate specific cell types. Some cell types may be marked by application of a false transmitter. When 5–7 diHT is added to the bathing medium of the living vertebrate retina, it is taken up solely by amacrine cells believed to use 5–HT as their transmitter (Vaney, 1986). Since the 5–7 diHT in their cell bodies fluoresces blue, it can be used to guide a micropipette to the cell body for physiological recording or lucifer yellow injection. Similarly, Tauchi & Masland (1984) used DAPI to identify cholinergic cells.

Central recordings of neuronal activity

Anatomical evidence often provides the foundation for physiological investigations including those of neural networks, where the ability to find and identify neurones regularly is essential if progress is to be made in understanding the details of synaptic connectivity and the mechanisms of central control of rhythmically active systems.

Extracellular recording from vertebrate neurones

The first step in characterizing respiratory or cardiomotor neurones in the CNS is to record their efferent activity by means of simple hook electrodes on the nerves they supply. Recordings from nerves to ventilatory muscles show regular bursts of activity which are a measure of central respiratory drive, unmodified by mechanoreceptor feedback if the animal is prevented from ventilating by injection of a paralysing agent such as curare (Barrett & Taylor, 1985*a*). Recording from the cardiac vagus in the dogfish revealed regular bursting units, synchronous with ventilatory activity, and other non-bursting units whose activity increased during hypoxia (Taylor & Butler, 1982; Barrett & Taylor, 1985*a*). The possibility that these different patterns of efferent activity may originate from the separate medial and ventro–laterally placed motoneurones located in the brainstem by HRP studies was the impetus for our progression to central recordings. The simple procedure described below is one that has successfully been applied in the medulla of the dogfish (Barrett & Taylor, 1985c).

Glass microelectrodes are pulled on a vertical electrode puller (Palmer Bioscience, microelectrode puller) using filament filled glass capillary tubing (Life Science Products, Rapidfill). Each electrode (resistance 5–15 MΩ) is filled with 4 M NaCl (the sodium salt is appropriate for extracellular recordings) containing the dye marker pontamine sky blue (Hellon, 1971) and inserted into the holder on the headstage, connected to an appropriate amplifier (Digitimer Neurolog NL 102). The electrode is then driven into the dorsal surface of the exposed medulla of a dogfish using a micromanipulator (Narashige Instruments) in a location known from the neuranatomical study to contain respiratory and cardiomotor neurones.

A relatively simple device for identifying individual respiratory moto-neurones is to record their activity with a centrally located microelectrode whilst simultaneously recording EMG from the various respiratory muscles. Synchronous firing of the neurone with an EMG recorded from a particular muscle provides good circumstantial evidence for a functional link (Ballin-tijn & Alink, 1977).

The standard method for direct identification of the cell body of a neurone supplying a particular nerve (e.g. the cardiac vagus) is to stimulate the nerve antidromically (i.e. efferent axons stimulated centrally, back into the brainstem) whilst recording from the cell body with a centrally placed microelectrode. A suitable stimulation regime from a programmer (e.g. Digitimer 4030) triggers a storage oscilloscope, provides a calibration signal then stimulates the nerve. This and more complicated experimental control may be generated by a micro-computer, that can also be pro-grammed to record resultant activity and store it on disc. Location of a neurone supplying an axon to the stimulated nerve will appear at the central electrode as an evoked action potential following the stimulus artefact. It is important to differentiate between the cell bodies of neurones supplying axons directly to the stimulated nerve and those (i.e. interneu-rones) that are in synaptic contact with these neurones. This can be determined by examination of the evoked response on the basis of the following criteria:

   (i)   Are the recorded potentials all or nothing (i.e. does each effective stimulus produce an identical response)?

   (ii)   Does the response have a constant latency (i.e. is the time delay between stimulus artefact and response constant, and does this delay accord with the measured conduction time of your nerve (this

can be measured) and the length of axon between stimulation site
and neurone)?

(iii) Is the response obtained repeatedly at high stimulation frequencies
(e.g. 100 Hz)?

(iv) Do you observe cancellation of the evoked response by collision
with a spontaneous spike?

This last criterion is the critical one but is only applicable to spontaneously active neurones. Fortunately this is a common property of respiratory
and cardiomotor neurones. If a spontaneous (and therefore orthodromic)
spike anticipates stimulation of the nerve and as a result the axon does not
conduct the antidromically evoked spike then this is spike cancellation
(Fig. 6) and it reliably indicates that spontaneous activity recorded externally from a neurone close to your electrode tip is supplied to the nerve you

Fig. 6. A series of 3 successive antidromic stimuli delivered to the
cardiac branch of the vagus of a dogfish, *Scyliorhinus canicula* whilst
recording centrally from a microelectrode placed close to the cell body of
a spontaneously active motoneurone supplying an axon to the stimulated
nerve. (The stimulus artefact is indicated by a black dot.) Stimuli 1 and 3
are successful in evoking an antidromic action potential that has a similar
latency and waveform each time. The second stimulus in the series
coincided with a spontaneous action potential that closely resembled the
evoked potentials but resulted in cancellation of the response to
electrical stimulation. (From Barrett & Taylor, 1985c.)

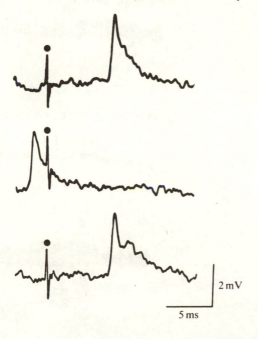

2 mV

5 ms

Fig. 7. Central recordings from antidromically identified cardiac vagal motoneurones in different locations in the medulla of the dogfish, *Scyliorhinus canicula*, together with recordings of spontaneous efferent activity in the cardiac nerve: (*a*) a rhythmically bursting unit, identified in the DMV just rostral to obex, that contributed action potentials to the regular bursts of activity in the nerve and responded to mechanical stimulation of the gill arches (prod). (*b*) a regularly firing unit identified in a ventro–lateral location, that contributed action potentials within and outside the bursts of activity recorded from the nerve. (*c*) another ventro–lateral cell, which was not spontaneously active but responded to mechanical stimulation. (Modified from Barrett & Taylor, 1985*c*.)

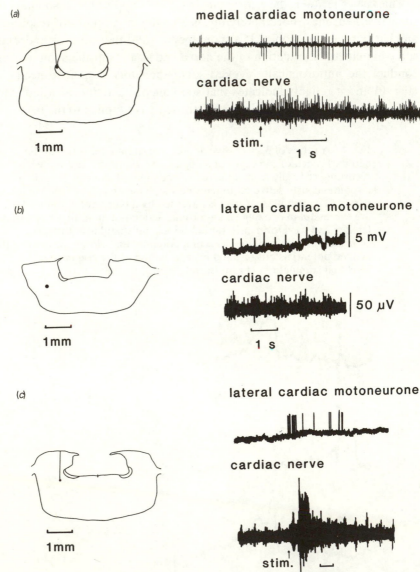

are stimulating. The stimulating regime may then be cancelled and the recording system allowed to free-run while spontaneous activity is recorded from the neurone and from the nerve it supplies (Fig. 7).

Once sufficient recordings of spontaneous activity in a neurone and, if required, its responses to appropriate stimulation of peripheral receptors or electrical stimulation of peripheral nerves or areas in the CNS, have been recorded, the location of the recorded neurone in the brainstem should be marked by iontophoretic injection of a dye (e.g. pontamine sky blue) into the recording position. This is achieved by passing a current from the electrode by applying a DC voltage (20–40 volts) for several minutes to the recording circuit. The brain is then removed and sectioned at 60 μm for location of the dye spot.

Location of efferent vagal preganglionic motoneurones antidromically stimulated from the cardiac vagus in the dogfish, revealed that their spontaneous firing characteristics related to their topographical position in the brainstem (Fig. 7). Some fired in regular bursts, contributing to the bursts of activity recorded in the whole nerve; all of these cell bodies were located medially in the dorsal vagal motor column, where the neuranatomical study revealed that they have an overlapping sequential distribution with the vagal respiratory neurones, whose firing characteristics they share. Some ventro–laterally placed cell bodies were spontaneously active, firing regularly or spasmodically but never in regular bursts, and contributing the non-bursting units to activity recorded from the cardiac nerve. Others were normally silent, although some, in common with all the spontaneously active cell bodies, responded to mechanical stimulation of the gill arches (Barrett & Taylor, 1985c; Taylor, 1985). These data lead us to the hypothesis that vagal tone on the dogfish heart is normally mediated by activity in the ventro–laterally located vagal cardiomotor neurones; whilst the medial, bursting neurones may be responsible for the generation of cardiorespiratory synchrony in resting fish, when vagal tone is low (Taylor, 1985). Further comparative studies of the lower vertebrates and mammals, may reveal the functional significance of the progressive ventro–lateral migration of vagal motoneurones to form the nucleus ambiguus (Taylor, 1985; Withington-Wray *et al.*, 1988).

Vagal motoneurones which are silent in the prepared animal, supply axons to the lungs and heart in mammals (e.g. Bennett *et al.*, 1984). Their functional properties have been studied by causing them to fire with iontophoretic injection of the amino acid DL-homocysteic acid (e.g. Gilbey *et al.*, 1985). As this is thought to specifically stimulate the cell body to fire, it represents a further check of the appropriate location of the micro-electrode tip.

This technique of extracellular recording is sufficient for location of specific neuronal cell bodies and characterization of their spontaneous and reflex activity. Insertion of multibarrelled glass micropipettes enables the neuropharmacology of neurones to be determined by the sequential iontophoretic injection of putative agonists and antagonists such as acetylcholine and atropine, or gamma aminobutyric acid and biccuculine (e.g. Gilbey *et al.*, 1985; Jordan *et al.*, 1985). Iontophoresis or injection (i.e. using hydrostatic pressure exerted by a small syringe) of HRP down a micropipette into a recording site in the CNS, will label cells that supply afferent axons to this area and by inference to the recorded neurone (Heimer & Robarts, 1981; Mesulum, 1982).

More detailed information on the control of respiratory and cardiomotor neurones may be obtained by intracellular recording, this can reveal the post-synaptic events associated with afferent input (e.g. Mifflin *et al.*, 1986). This requires a very rigid and stable electrode holder and animal clamp (e.g. Narashige Instruments), a good DC amplifier (the Neurolog NL102 is suitable) and sharp microelectrodes (20–30 MΩ). A stepping motor could be an advantage as discussed below. Once located and penetrated, the cell can be recorded from and then iontophoretically filled with HRP by passing a current (e.g. 4nA depolarizing current for one minute from an electrode filled with 4% HRP in buffered KCl; see Fig. 1 in Mifflin *et al.*, 1987). HRP histochemistry will then reveal the detailed cytoarchitecture and in particular the dendritic fields of the penetrated cell body in sufficient detail to predict local circuitry in the CNS (e.g. Juch & Ballintijn, 1983).

Impalement of invertebrate neurones

Invertebrate ganglia are robust, survive well in highly dissected preparations and contain large identifiable neurones that may readily be impaled with microelectrodes. Techniques for cell impalement vary with the individual researcher and the best advice is to start by practicing on simple preparations with large cells and little connective tissue (mouse diaphragm or frog sartorius are recommended by Purves, 1981). Successful cell impalement depends critically on the mechanical stability of both cell and electrode. If pinning the ganglia onto a wax platform or Sylgard dish via the connective tissue and/or trachea is not sufficient to immobilise your preparation then neuromuscular blocking agents may be tried (Pfluger, 1984) or the ganglion set in agar (Arshavsky *et al.*, 1985). Micropipettes should be short and stubby, but still have high resistance (20–80 MΩ). Purves (1981) and Thomas (1978) give clear directions, while Brown & Flaming (1986) describe their latest technology to pull fine tips. Other problems are often electrical – choice of a suitable amplifier, and reduction of electrical inference by screening are essential and the suggestions of Purves (1981) and Thomas (1978) should be followed.

Pipettes are best pulled from fibre-filled or theta section glass tubing. (Theta tubing may pose a problem if the septum projects beyond the tip of the barrel.) Injecting the barrel with electrolyte will cause it to migrate up and fill the tip. If the solution is expensive, standing the blunt end in the electrolyte will cause sufficient to be taken up by the capillary action of the fiber. If electrodes are hard to see, dip them in the Rotring etching ink 595617 (formerly Rotring ink K).

The micropipette should be brought up perpendicular to the ganglion surface until it is gently dimpled. Sharply but gently tapping the electrode holder or supporting micromanipulator may be sufficient to penetrate the cells (this works well in *Gryllus*). A helpful adjunct to tapping is to apply excess negative capacitance compensation until the feedback causes oscillations. This vibrates the electrode tip and 'burns' it into the cell. If the ganglion sheath is tough it may be digested with protease (e.g. 0.1% Sigma XIV for *Lymnaea* (Elliott & Benjamin, 1985) or 1% Sigma VI for *Schistocerca* (Pfluger, 1984) for two to five minutes. In other preparations (*Aplysia*, *Pleurobranchaea*) the thicker sheaths are removed by micro-dissection before cell impalement. Sometimes a piezoelectric drive (Peters & Tetzel, 1980) or stepper motor (Brown & Flaming, 1977) may be used to achieve rapid acceleration of the electrode tip.

Identification of invertebrate interneurones
In locusts two types of respiratory interneurone have been identified (Fig. 8); one cell is a co-ordinating neurone relaying the ventilatory rhythm to the spiracle motoneurones (Burrows, 1982*b*) while the other is part of the pattern generating network (Pearson, 1980). In crustaceans the generation of the rhythmic ventilatory movements of the scaphognathite depends on 'non-spiking' interneurones. An important feature of these neurones is the ability to release transmitter in a graded fashion, without the need for action potentials. In at least some of these neurones the membrane potential oscillates in phase with the ventilatory rhythm and it has been suggested that this is maintained by reciprocal inhibition between sets of antagonistic non-spiking interneurones (Bush *et al.*, 1988).

The simplest experimental technique to identify putative interneurones is to inject depolarizing or hyperpolarizing current pulses via an intracellular microelectrode and note its effect upon rhythmical discharge.

(1) If the neurone is a co-ordinating interneurone, then the rhythm will continue at the same frequency, but the firing rate of the motoneurones will be altered. For example, in Fig.8*a*, depolarizing the ventilatory inter-

Fig. 8. Physiological characterization of ventilatory interneurones in locusts.

(a) Simultaneous intracellular recordings from a co-ordinating interneurone and the mesothoracic spiracular closer motoneurone. Depolarization of the interneurone reduces the number of spikes in the motoneurone – hyperpolarization has the opposite effect. This is evidence for a synaptic connection. (Burrows, 1982b).)

(b) Intracellular recordings from the expiratory abdominal interneurone (whose morphology is shown in Fig. 7c). In each record, a depolarizing current pulse is given during the interburst interval. (i) The current pulse is just supra-threshold and has no effect on the cycle duration. (ii) A larger depolarizing current produces a burst of spikes just after the normal burst. This lengthens the cycle time. (iii) The same current given later – now the cycle is shortened. (iv) Relationship between the time at which a 300 ms, 2 nA current pulse is given and the change in cycle time.

The results of experiments (ii) and (iii), summarised in (iv), suggest that the cell is part of the pattern generating network; while experiment (i) suggests that the cell is unlikely to be an endogenous burster. (Pearson, 1980.)

(c) Dual intracellular recordings from the co-ordinating interneurone and the mesothoracic spiracular closer motoneurone shown in (a). Each interneuronal spike is used to trigger the oscilloscope. Note that the IPSP in the motoneurone has a constant latency and waveform. This would support a monosynaptic connection. (Burrows, 1982b.)

neurone reduces the number of spikes in the spiracle motoneurone and hyperpolarization has the converse effect. There is no systematic change in the ventilatory rate upon injecting current (Burrows, 1982b).

(2) If the neurone is part of the pattern generator, then current injection will alter the ventilatory rate. This can be seen for a locust interneurone (Fig. 8b) which fires during expiration (Pearson, 1980). Undisturbed cycles range from 1.2–1.4 s. In records (ii) and (iii), sufficient depolarizing current was injected to elicit a burst of spikes, which altered the duration of the next cycle (ii: 1.6 s; iii: 0.86 s). The time at which the current pulse is given has a profound effect on whether the cycle is shortened or lengthened (see Fig. 8b, iv). The subsequent cycles have the same duration as the original one. This ability to 'reset the phase of the rhythm', without altering the duration of subsequent cycles is an important part of the evidence that a neurone is part of the central pattern generator. Note that the experiment in part b(i) suggests that the interneurone is not an 'endogenous burster', as sub-threshold current pulses have no resetting ability (cycle time before = 1.28 s, after = 1.29 s). Current injection experiments may also be performed on 'non-spiking' neurones that are important in insects and crustaceans (Fig. 9). Note that no spikes are necessary for the alteration in the rhythm.

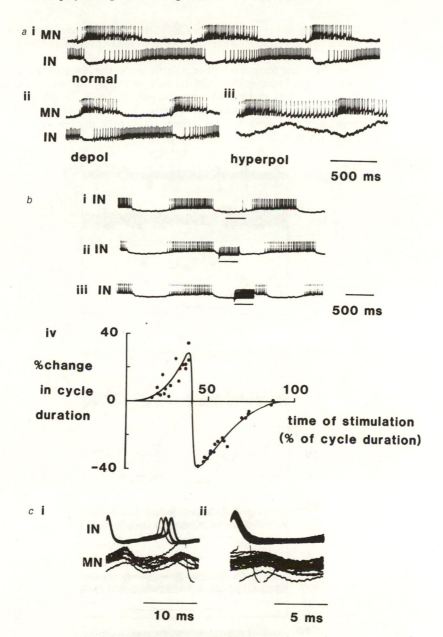

Fig. 9. Non-spiking neurones in *Carcinus*. The ventilatory rhythm is shown by the extracellular recordings from the depressor and levator nerves.

(*a*) Injecting current of either polarity resets the phase of the rhythm – the expected time of depressor bursts is shown by the bars. Note the strong oscillatory rhythm in membrane potential with no spikes.

(*b*) Records from another cell. Depolarization increases the ventilatory rate, while hyperpolarization slows the rhythm and eventually stops it. (Simmers & Bush, 1980).

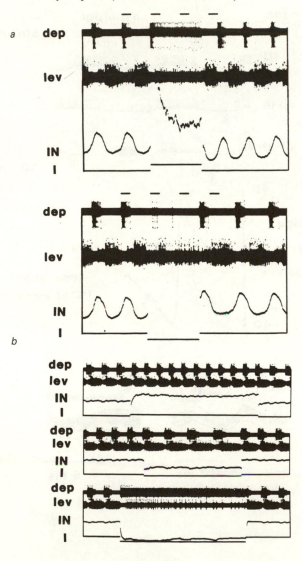

In the experiments outlined above, only a single interneurone need be penetrated by an intracellular electrode. If, however, a second cell is impaled, the range of useful information that can be obtained is enlarged. Particularly valuable is the opportunity to resolve individual synaptic potentials and the U–V oscillograph or high-speed chart recorder is sufficient to determine the sign (excitatory or inhibitory) of many PSPS. However, it is important to be wary of any depolarizing PSP that cannot give rise to an action potential, because 'reversed' IPSPS are common (they are often mediated by chloride ions). Injection of current into the post-synaptic cell should be sufficient to check if they are truly EPSPS. Often dual penetrations will provide sufficient information to show if the connection is monosynaptic, when the PSP should follow the action potential 1:1 with short and constant latency (Fig. 8c).

Conclusions

The application of HRP, cobalt and lucifer yellow dyes has produced major advances in our knowledge of the neuranatomy of both vertebrate and invertebrate central nervous systems. Researchers would be lost without these stains to guide their micropipettes and help in their analyses of the neural bases of rhythmicity. As such they are already helping those concerned with the central control of respiration and the cardio-vascular system to locate and characterize neurones functioning as central rhythm generators or motoneurones.

References

Altman, J. A. & Tyrer, N. M. (1980) Filling selected neurones with cobalt through cut axons. In *Neuroanatomical techniques: insect nervous system*, ed. N. J. Strausfeld & T. A. Miller, pp. 377–405. New York: Springer.

Arshavsky, Yu I., Beloozerova, I. N. Orlovsky, G. N., Panchin, Yu V. & Pavlova, G. A. (1985) Control of locomotion in marine mollusc *Clione limacina*. I Efferent activity during actual and fictitious swimming. *Experimental Brain Research*, **58**, 255–62.

Bacon, J. P. & Altman, J. S. (1977) A silver intensification method for cobalt-filled neurones in wholemount preparations. *Brain Research*, **138**, 359–63.

Ballintijn, C. M. (1982) Neural control of respiration in fishes and mammals. *In Exogenous and Endogenous Influences on Metabolic and Neural Control*, ed. A. D. F. Addink & N. Spronk, pp. 137–40. Oxford: Pergamon Press.

Ballintijn, C. M. (1988). Evolution of central nervous control of ventilation in vertebrates. In *Neurobiology of the Cardiorespiratory System*, ed. E. W. Taylor, pp. 3–27. Manchester: Manchester University Press.

Ballintijn, C. M. & Alink, G. M. (1977). Identification of respiratory motoneurones in the carp and determination of their firing characteristics and interconnections. *Brain Research*, **136**, 261–76.

Ballintijn, C. M., Roberts, B. L. & Luiten, P. G. M. (1983) Respiratory responses to stimulation of branchial vagus nerve ganglia of a teleost fish. *Respiration Physiology*, **51**, 241–57.

Bamford, O. S., Dawes, R. & Ward, R. A. (1986) Effects of the α adrenergic agonist clonidine and its antagonist idazoxan on the fetal lamb. *Journal of Physiology*, **381**, 29–37.

Barrett, D. J. (1984). The Xth cranial nerve (the vagus) in the elasmobranch fish *Scyliorhinus canicula*, and its role in the control of the heart. Ph.D. thesis, University of Birmingham.

Barrett, D. J. & Taylor, E. W. (1984) Changes in heart rate during progressive hyperoxia in the dogfish *Scyliorhinus canicula* L.: evidence for a venous oxygen receptor. *Comparative Biochemistry and Physiology*, **78A**, 697–703.

Barrett, D. J. & Taylor, E. W. (1985*a*). Spontaneous efferent activity in branches of the vagus nerve controlling heart rate and ventilation in the dogfish. *Journal of Experimental Biology*, **117**, 433–48.

Barrett, D. J. & Taylor, E. W. (1985*b*). The location of cardiac vagal preganglionic neurones in the brain stem of the dogfish *Scyliorhinus canicula*. *Journal of Experimental Biology*, **117**, 449–58.

Barrett, D. J. & Taylor, E. W. (1985*c*). The characteristics of cardiac vagal preganglionic motoneurones in the dogfish. *Journal of Experimental Biology*, **117**, 459–470.

Bennett, J. A., Ford, T. W., Kidd, C. & McWilliam, P. N. (1984). Characteristics of cat dorsal motor vagal motoneurones with axons in the cardiac and pulmonary branches. *Journal of Physiology*, **351**, 27 p.

Brown, K. T. & Flaming, D. G. (1977). New microelectrode techniques for intracellular work in small cells. *Neuroscience*, **2**, 813–27.

Brown, K. T. & Flaming, D. G. (1986) *Advanced micropipette techniques for cell physiology* IBRO series – Methods in the Neurosciences. vol. 9. Chichester & New York: Wiley.

Burrows, M. (1982*a*) The physiology and morphology of median nerve motor neurones in the thoracic ganglion of the locust. *Journal of Experimental Biology*, **96**, 325–41.

Burrows, M. (1982*b*) Interneurones co-ordinating the ventilatory movements of the thoracic spiracles in the locust. *Journal of Experimental Biology*, **97**, 385–400.

Bush, B. M. H., Simmers, A. J. & Pasztor, V. M. (1988). Neural control of gill ventilation in decapod crustacea. In *Neurobiology of the Cardiorespiratory System*, ed. E. W. Taylor, pp.80–112. Manchester: Manchester University Press.

Butler, P. J., Taylor, E. W. & Short, S. (1977) The effect of sectioning cranial nerves V, VII, IX and X on the cardiac response of the dogfish *Scyliorhinus canicula* to environmental hypoxia. *Journal of Experimental Biology*, **69**, 233–45.

Byrne, J. H. (1983). Identification and initial characterization of a cluster of command and pattern-generating neurons underlying respiratory pumping in *Aplysia californica. Journal of Neurophysiology*, **49**, 491–508.

Carlson, R. C., Kiff, J. & Ryugo, K. (1982). Suppression of the cell body response in axotomised frog spinal neurones does not prevent initiation of nerve regeneration. *Brain Research*, **234**, 11–25.

Daly, M. de B., Ward, J. & Wood, L. M. (1988). The peripheral chemoreceptors and cardiovascular-respiratory integration. In *Neurobiology of the Cardiorespiratory System*, ed. E. W. Taylor, pp. 342–68. Manchester: Manchester University Press.

Dawes, G. S., Gardner, W. N., Johnston, B. M. & Walker, D. W. (1983) Breathing in fetal lambs: the effects of brain stem section. *Journal of Physiology*, **335**, 535–53.

de Graaf, P. J. F. & Ballintijn, C. M. (1987). Mechanoreceptor activity in the gills of the carp. II Gill arch proprioreceptors. *Respiration Physiology*, **69**, 183–94.

Elliott, C. J. H. & Benjamin, P. R. (1985). Interactions of pattern – generating interneurons controlling feeding in *Lymnaea stagnalis. Journal of Neurophysiology*, **54**, 1396–411.

Feldman, J. L. & Gautier, H. (1976) Interaction of pulmonary afferents and pneumotaxic centre in the control of respiratory pattern in cats. *Journal of Neurophysiology*, **39**, 31–44.

Gallyas, F., Lenard, L. & Lazar, C. (1978). Improvement of cobalt transport in axons by complexing agents. *Neuroscience Letters*, **9**, 213–16.

Gilbey, M. P., Jordan, D., Spyer, K. M. & Wood, L. M. (1985). The inhibitory actions of GABA on cardiac vagal motoneurones in the cat. *Journal of Physiology*, **361**, 49P.

Gleeson, M. (1985) Changes in intrapulmonary chemoreceptor discharge in response to the adjustment of respiratory pattern during hyperventilation in domestic fowl. *Quarterly Journal of Experimental Physiology*, **70**, 503–13.

Grimstone, A. V. & Skaer, R. J. (1972). *A guidebook to microscopical methods*. Cambridge: Cambridge University Press.

Gwyn, D. J., Ritchie, T. C. & Coulter, J. D. (1985). The central distribution of vagal catecholaminergic neurons which project into the abdomen of the rat. *Brain Research*, **328**, 139–44.

Heimer, L. & Robarts, J. J. (1981). *Neuroanatomical Tract Tracing Methods*. New York: Plenum Press.

Hellon, R. F. (1971). The marking of electrode tip position in nervous tissue. *Journal of Physiology*, **241**, 12P.

Iscoe, S., Feldman, J. L. & Cohen, M. I. (1979) Properties of inspiratory termination by superlaryngeal and vagal stimulation. *Respiration Physiology*, **36**, 353–66.

Jordan, D., Gilbey, M. P., Richter, D. W., Spyer, K. M. & Wood, L. M. (1985). Respiratory-vagal interactions in the nucleus ambiguus of the cat. In *Neurogenesis of Central Respiratory Rhythm*, ed. A. L. Bianchi & M. Denavit-Saubie, pp. 370–78. Lancaster: M T P Press.

Jordan, D. & Spyer, K. M. (1988). Central neural mechanisms mediating respiratory-cardiovascular interactions. In: *Neurobiology of the Cardio-respiratory System*, ed. E. W. Taylor, pp. 323–42. Manchester: Manchester University Press.

Juch, P. J. W. & Ballintijn, C. M. (1983). Tegmental neurons controlling medullary respiratory centre neurone activity in the carp. *Respiration Physiology*, **51**, 95–107.

Kandel, E. R. (1976). *The Cellular Basis of Behaviour, an Introduction to Behavioural Neurobiology*. San Francisco: Freeman.

Lazar, G. Y., Toth, P., Csank, G. Y. & Kicliter, E. (1983). Morphology and location of tectal projection neurons in frogs: a study with HRP and cobalt filling. *Journal of Comparative Neurology*, **215**, 108–20.

Levings, J. J. (1987). Innervation of the foregut and the feeding and respiratory muscles in elasmobranch fish. M.Sc. thesis, University of Birmingham.

Maranto, A. R. (1982). Neuronal mapping: a photooxidation reaction makes lucifer yellow useful for electron microscopy. *Science*, **217**, 953–55.

Mesulum, M-M. (1982). *Tracing Neural Connections with Horseradish Peroxidase*. IBRO Handbook Series: Methods in the Neurosciences Chichester: Wiley.

Mifflin, S. W., Spyer, K. M. & Withington-Wray, D. J. (1986). Lack of respiratory modulation of baroreceptor inputs in the nucleus of the tractus solitarius of the cat. *Journal of Physiology*, **376**, 33P.

Mifflin, S. W., Spyer, K. M. & Withington-Wray, D. J. (1987). Intracellular labelling of neurones receiving carotid sinus nerve inputs in the cat. *Journal of Physiology*, **387**, 60P.

Miller, J. P. & Selverston, A. I. (1979). Rapid killing of single neurons by irradiation of intracellularly injected dye. *Science*, **206**, 702–04.

Miller, P. L. (1966). The regulation of breathing in insects. *Advances in Insect Physiology*, **3**, 279–344.

Niida, A. & Ohono, T. (1984). An extensive projection of fish dorsolateral tegmental cells to the optic tectum revealed by intra-axonal dye marking. *Neuroscience letters*, **48**, 261–66.

Pasztor, V. M. (1969). The neurophysiology of respiration in decapod crustacea. II The sensory system. *Canadian Journal of Zoology*, **47**, 435–41.

Pearson, K. G. (1980) Burst generation in coordinating interneurones of the ventilatory system of the locust. *Journal of Comparative Physiology*, **137**, 305–13.

Peters, M. & Tetzel, H. D. (1980). Piezoelectric drive for step-by-step microelectrode advancement. *Journal of Experimental Biology*, **86**, 333–6.

Pitman, R. M. Tweedle, C. D. & Cohen, M. J. (1972). Branching of central neurons: intracellular cobalt injection for light and electron microscopy. *Science*, **176**, 412–14.

Pfluger, H. J. (1984). The large fourth abdominal intersegmental interneuron: a new type of wind – sensitive ventral cord interneuron in locusts. *Journal of Comparative Neurology*, **222**, 343–57.

Purves, R. D. (1981). *Microelectrode Methods for Intracellular Recording and Iontophoresis*. London: Academic Press.

Reichert, H. & Krenz, W. D. (1986). *In vivo* visualisation of individual neurons in arthropod ganglia facilitates intracellular neuropil recording. *Journal of Comparative Physiology*, **158**, 625–37.

Rokx, J. T. M., Juch, P. J. W. & Van Willigen, J. D. (1985). On the bilateral innervation of masticatory muscles: a study with retrograde tracers. *Journal of Anatomy*, **237–43**.

Selverston, A. I., Kleindienst, H. U. & Huber, F. (1985). Synaptic connectivity between cricket auditory interneurons as studied by selective photoinactivation. *Journal of Neuroscience*, **5**, 1283–92.

Seyan, H. S., Bassemir U. K. & Strausfeld, N. J. (1983). Double marking for light and electron microscopy. In *Functional Neuroanatomy*, ed. N. J. Strausfeld, pp. 112–31. New York: Springer.

Shelton, G. (1959) The respiratory centre in the tench (*Tinca tinca* L.). I. The effects of brain transection on respiration. *Journal of Experimental Biology*, **36**, 191–202.

Shelton, G. (1961) The respiratory centre in the tench (*Tinca tinca* L.). II. Respiratory neuronal activity in the medulla oblongata. *Journal of Experimental Biology*, **38**, 79–92.

Short, S., Butler, P. J. & Taylor, E. W. (1977) The relative importance of nervous, humoral and intrinsic mechanisms in the regulation of heart rate and stroke volume in the dogfish *Scyliorhinus canicula*. *Journal of Experimental Biology*, **70**, 77–92.

Siegler, M. V. & Burrows, M. (1984). The morphology of two groups of spiking local interneurons in the metathoracic ganglion of the locust *Journal of Comparative Physiology*, **224**, 463–82.

Simmers, A. J. & Bush, B. M. H. (1980) Non-spiking neurones controlling ventilation in crabs. *Brain Research*, **197**, 247–52.

Simmers, A. J. & Bush, B. M. H. (1983*a*) Central nervous mechanisms controlling rhythmic burst generation in the ventilatory motoneurones of *Carcinus maenas*. *Journal of Comparative Physiology*, **150**, 1–21.

Simmers, A. J. & Bush, B. M. H. (1983*b*). Motor programme switching in the ventilatory system of *Carcinus maenas*: the neuronal basis of bimodal scaphognathite beating. *Journal of Experimental Biology*, **104**, 163–81.

Sloniewski, P., Usunoff, K. G. & Pilgrim, C. H. (1985). Evidence for an anterograde transport of the fluorescent tracer fast blue in the Strionigro-striated loop of the rat. *Neuroscience Letters*, **60**, 189–93.

Smith, F. M. & Jones, D. R. (1978) Localisation of receptors causing hypoxic bradycardia in trout (*Salmo gairdneri*). *Canadian Journal of Zoology*, **56**, 1260–65.

Spyer, K. M. (1984) Central control of the cardiovascular system. In *Recent Advances in Physiology*, 10, ed. P. F. Baker, pp. 163–200. Edinburgh: Churchill Livingstone.

Steward, O. (1981). Horseradish Peroxidase and Fluorescent substances and their combination with other Techniques. In *Neuroanatomical Tract Tracing Methods*, ed. L. Heimer & M. G. Robarts, New York: Plenum Press.

Stewart, W. W. (1978). Functional connections between cells as revealed by dye-coupling with a highly fluorescent naphthalimide tracer. *Cell*, **14**, 741–59.

Stewart, W. W. (1981). Lucifer dyes – highly fluorescent dyes for biological tracing. *Nature*, **292**, 17–21.

Strausfeld, N. J., Seyan, H. S., Wohlers, D. & Bacon, J. P. (1983). Lucifer yellow histology In *Functional Neuroanatomy*, ed. N. J. Strausfeld, pp. 132–55. New York: Springer.

Taghert, P. H., Bastiani, M. J., Ho, R. H. & Goodman, C. S. (1982). Guidance of pioneer growth cones: filopodial contacts and coupling revealed with an antibody to lucifer yellow *Developmental Biology*, **94**, 391–99.

Tauchi M. & Masland, R. H. (1984). The shape and arrangement of the cholinergic neurons in the rabbit retina *Proceedings of the Royal Society*, **223B**, 101–17.

Taylor, E. W. (1985). Control and coordination of gill ventilation and perfusion. *Symposium of the Society for Experimental Biology*, **39**, 123–61.

Taylor, E. W. & Butler, P. J. (1982). Nervous control of heart rate: activity in the cardiac vagus of the dogfish. *Journal of Applied Physiology*, **53**, 1330–5.

Taylor, E. W., Short, S. & Butler, P. J. (1977) The role of the cardiac vagus in response of the dogfish *Scyliorhinus canicula* to hypoxia. *Journal of Experimental Biology*, **70**, 57–75.

Thomas, R. C. (1978). *Ion sensitive intracellular microelectrodes*. London: Academic Press.

Trenchard, D. (1977) Role of pulmonary stretch receptors during breathing in rabbits, cats and dogs. *Respiration Physiology*, **29**, 231–46.

Vaney, D. I. (1986). Morphological identification of serotonin accumulating neurons in the living retina. *Science*, **233**, 444–6.

Warr, W. B., de Olmos, J. S. & Heimer, L. (1981). Horseradish Peroxidase: the Basic Procedure. In *Neuroanatomical Tract Tracing Methods*, ed. L. Heimer & M. J. Robarts, New York: Plenum Press.

Withington-Wray, D. J., Roberts, D. J. & Taylor, E. W. (1986). The topographical organisation of the vagal motor column in the elasmobranch fish *Scyliorhinus canicula* L. *Journal of Comparative Neurology*, **248**, 95–104.

Withington-Wray, D. J., Taylor, E. W. & Metcalf, J. D. (1988). The location and distribution of vagal preganglionic neurones in the hindbrain of lower vertebrates. In *Neurobiology of the Cardiorespiratory System*, ed. E. W. Taylor, pp. 304–21. Manchester: Manchester University Press.

S. F. PERRY and A. P. FARRELL

Perfused preparations in comparative respiratory physiology

Introduction

Perfusion studies permit the isolation and examination of the components of complex integrated systems. A significant advantage of perfused preparations is that the chemical, physical and hormonal properties of the perfusion fluid can be modified independently of perfusion flow, thereby eliminating many secondary interactions that may occur *in vivo*. Perfused preparations are not well-suited for descriptive studies but are used most effectively to elucidate mechanisms of a known physiological process that cannot be effectively examined in the intact animal. In these instances, the limitations of the particular preparation must be understood and care should be taken in extrapolating results from perfused preparations to the intact animal. The primary criterion by which to judge a preparation is its ability to reproduce the physiological behaviour of the tissue as it occurs *in vivo*. Unfortunately, few detailed comparisons of perfused preparations with *in vivo* function have been performed. Comparisons that have been conducted (e.g. Perry, Booth & McDonald, 1985*a*, *b*; Stagg & Shuttleworth, 1984) reveal certain preparations that can adequately simulate a given physiological process while failing to simulate another.

In the present review, we assess the suitability of the more prevalent perfusion preparations utilized in comparative respiratory physiology and recommend, where possible, the particular preparation most appropriate to study a given physiological process.

Perfusion fluids

In situ or isolated organ preparations have been perfused with a variety of fluids including variable strength seawater (Ostlund & Fange, 1962), non-colloidal or colloidal Ringer solutions (summarized in Perry *et al.*, 1984*a*), artificial haemolymph (Burnett, 1984) and blood (Saunders & Sutterlin, 1971; Kent & Pierce, 1978; Opdyke, Holcombe & Wilde, 1979; Holbert, Boland & Olson, 1979; Davie *et al.*, 1982; Perry *et al.*, 1982; Daxboeck *et al.*, 1982; Metcalfe & Butler, 1982; Olson, 1984; Perry *et al.*,

1985*a*, *b*). In the absence of plasma proteins, colloidal substitutes (Kirschner, 1969) are considered essential to minimize oedema in interstitial compartments. Polyvinylpyrrolidone (PVP; 2–4%), the most frequently used colloidal substitute, may be less effective in preventing oedema in high-pressure vascular networks than previously thought as evidenced by the severity of branchial oedema in isolated eel holobranchs perfused with Ringer containing 1–8% PVP (Ellis & Smith, 1983) and isolated trout heads perfused for longer than 45 minutes with Ringer containing 3% PVP (Bornancin, Isaia & Masoni, 1985). Conversely, Stagg & Shuttleworth (1984) did not observe lamellar oedema in isolated perfused flounder holobranchs, although the branchial circulation was perfused at a considerably lower pressure. Recently, Perry, Lauren & Booth (1984*b*) reported absence of branchial oedema in rainbow trout heads perfused with saline containing a PVP/protein mixture (4% PVP/0.2% bovine serum albumin (BSA)). Given the varied effectiveness of PVP in preventing oedema, investigators wishing to use saline should consider PVP/protein mixtures or 3–4% BSA (Jackson & Fromm, 1980, 1981; Nekvasil & Olson, 1986; Olson *et al.*, 1986) although budgetary restrictions might constrain extensive use of BSA. Other problems associated with employing saline as a perfusion fluid are its low buffering capacity, oxygen carrying capacity and viscosity. The use of homologous blood can eliminate these problems but an enhancement of performance is not always obvious (Perry *et al.*, 1985*a*, *b*). The choice of blood or saline as perfusion fluid ultimately depends on the parameter(s) being investigated (see below). Any advantages of blood must be weighed against the cost and involvement of labour in obtaining sufficient amounts for perfusion.

Regardless of the type of perfusion fluid employed, it is important to simulate the acid–base properties of the blood. Historically, gills have been perfused with hypercapnic salines (e.g. Kirschner, 1969; Rankin & Maetz, 1971; Payan & Matty, 1975) resulting in abnormally low pH or elevated bicarbonate levels if base was added to raise pH. Elevation of perfusate P_{CO_2} promotes branchial vasoconstriction independently of pH in perfused trout heads (Perry & Daxboeck, 1986) therefore the continued use of hypercapnic perfusion fluid is undesirable. Perfusate P_{CO_2}/pH may be less of a concern in perfused hearts of lower vertebrates that exhibit acid tolerance or resistance (Gesser & Poupa, 1983). Nevertheless, we recommend that perfusion fluids be equilibrated with gas mixtures to mimic the P_{CO_2} and pH of blood, *in vivo*. It is common practice to filter perfusion fluids (0.22–0.45 μm pore size) prior to use in perfused preparations.

Other important properties of perfusion fluids include oxygenation and ionic, hormonal and metabolite composition. The major ions should be and usually are very close to the plasma composition of the species under study. Excitation-contraction coupling in vertebrate hearts involves sarcolemmal Na^+/Ca^{2+} exchange. Therefore extracellular $[Na^+]$ has an important *in vitro* effect on cardiac contractility in fish, amphibians (Gesser & Jorgensen, 1983) and mammals, through its effect on intracellular $[Ca^{2+}]$. Studies with euryhaline fish or aestivating lungfish and amphibians should consider $[Na^+]$ an important variable.

Plasma $[Ca^{2+}]$ is normally well regulated *in vivo*, and the small changes in plasma $[Ca^{2+}]$ that may occur are unlikely to be a major means of modulating cardiac performance, despite claims to the contrary for lower vertebrates (Ruben & Bennett, 1981). Nonetheless, the $[Ca^{2+}]$ in perfusion fluid is critical due to its effects on cardiac or smooth muscle contraction. Approximately 50% of the total $[Ca^{2+}]$ in plasma exists as free Ca^{2+}. In salines, calcium forms insoluble salts with phosphates and may complex with colloid substitutes, such as albumin. Calcium levels must be established in relation to phosphate levels in the saline.

It is apparent that substrates other than glucose are utilized equally well or even preferred by hearts and gills *in vitro*. Cephalopod hearts can utilize amino acids. Elasmobranch hearts prefer carbohydrates to ketones, but like cyclostome hearts have little capacity to utilize fatty acids. Teleost hearts prefer carbohydrates, but can utilize fatty acids and lactate effectively. In mammalian perfused hearts, glucose, palmitate and lactate are sometimes used in combination. Lactate has been identified as an important oxidative substrate for fish gills and the potential contribution of non essential amino acids such as alanine, glycine and serine cannot be ignored (Mommsen, 1984).

Perfusion fluid oxygenation and hormone composition are discussed under separate evaluations of particular preparations.

Perfused gills

Fishes

The fish gill circulation is a complex vascular arrangement consisting of two distinct pathways, the high pressure arterio–arterial circulation and the low pressure arterio–venous circulation (see reviews by Laurent & Dunel, 1980; Laurent, 1984). The arterio–arterial circulation is a relatively low resistance circuit and in the intact animal is perfused with pulsatile high pressure flow. The importance of simulating the pulsatile nature of gill blood flow was recognized by pioneers in gill perfusion research. The heart–gill preparation (Keys, 1931*a*, *b*) allows the intact heart to pump fluid into the gill vasculature and probably most correctly mimics blood flow *in vivo*. This preparation, however, is limited in a similar manner to intact animals because of the difficulty in distinguishing between direct effects on gill function and secondary effects due to cardiovascular adjustments. The ventral aorta–gill preparation, utilizing constant pulsatile flow (Bateman & Keys, 1932; Keys & Bateman, 1932; Kirschner, 1969) was developed to overcome this problem but is no longer considered suitable for physiological studies. Currently, two principal types of perfused preparations are used in studies of fish gill physiology. These are *in situ* or isolated heads (Wood, 1974; Payan & Matty, 1975) and isolated holobranchs (Richards & Fromm, 1969; Rankin & Maetz, 1971; Shuttleworth, 1972). A variety of perfusion techniques have been used in conjunction with these preparations including constant pressure perfusion (Richards & Fromm, 1969; Rankin & Maetz, 1971; Payan & Matty, 1975), constant non-pulsatile flow perfusion (Shuttleworth, 1972; Claiborne & Evans, 1980) and constant pulsatile flow perfusion (Bergman, Olson & Fromm, 1974; Girard, 1976; Wood, 1975). With the exception of the study of Ellis & Smith (1983), the results of experiments that have compared pulsatile and non-pulsatile perfusion suggest advantages of pulsatility on preparation performance (Daxboeck & Davie, 1982; Davie & Daxboeck, 1982; Part & Svanberg, 1981).

Perfused gills have been used extensively in studies of gas transfer, haemodynamics and ionic regulation. Each of these disciplines is discussed below.

Gas transfer

Perfused preparations have contributed to our understanding of branchial gas transfer principally because convective and diffusive limitations can be assessed independently. As such, it was demonstrated that catecholamines enhance gill oxygen diffusive conductance in saline-perfused heads in the absence of ventilatory and cardiac adjustments (Pettersson & Johansen, 1982; Pettersson, 1983; Perry, Daxboeck & Dobson, 1985c). Attempts to determine whether branchial O_2 transfer is diffusion- or perfusion-limited using perfused preparations have produced conflicting results. Arterial O_2 tension was independent of $\dot{V}_b$ in spontaneously ventilating blood-perfused trout (Daxboeck *et al.*, 1982) and saline-perfused trout heads (Part *et al.*, 1984) indicating perfusion-limited gill O_2 uptake. The abnormally high levels of epinephrine (10^{-6} M) utilized in the perfused head (Part *et al.*, 1984) and the likelihood of elevated circulating catecholamines in the stressed, spontaneously ventilating trout may have obscured transit time limitations by enhancing gill O_2 diffusive conductance. In the absence of epinephrine, Pa_{O_2} was inversely proportional to $\dot{V}_b$ in saline-perfused trout heads (Perry *et al.*, 1985c) suggesting that O_2 uptake is diffusion-limited.

In Table 1, a number of respiratory parameters in rainbow trout are compared *in vivo* and in perfused preparations. Data from other species (e.g. *Gadus morhua*; Pettersson & Johansen, 1982; Pettersson, 1983; *Scyliorhinus canicula*; Metcalfe & Butler, 1982; *Anguilla australis*; Ellis & Smith, 1983) are not included due to insufficient *in vivo* data. Additionally, a number of studies have been omitted from this comparison because only normalized data (e.g. percentage changes) were reported (Part, Tuurala & Soivio, 1982a; van der Putte & Part, 1982).

A comparison with *in vivo* data reveals that isolated holobranchs (Perry *et al.*, 1982) and *in situ* heads (Wood, McMahon & McDonald, 1978; Daxboeck & Davie, 1982) are least capable of simulating gas transfer. Isolated holobranchs are poorly suited for gas transfer studies because of (i) inadequate ventilation causing formation of excessive unstirred and boundary layers, (ii) abnormal mucus production causing increased diffusion distances and (iii) reduced surface area for gas transfer and hence low absolute values of O_2 uptake ($\dot{M}_{O_2}$) and CO_2 excretion ($\dot{M}_{CO_2}$) since less than 1/8th of total gill surface area is perfused. *In situ* perfused heads allow correct simulation of dorsal aortic pressure but gas transfer is negligible probably due to lengthy surgical procedures. Blood-perfused heads (Perry *et al.*, 1985a, b) and spontaneously ventilating perfused fish (Davie *et al.*, 1982; Daxboeck *et al.*, 1982; Perry *et al.*, 1982) more closely simulate *in vivo* gas transfer in rainbow trout but significant discrepancies persist including abnormally low values of $\dot{M}_{O_2}$ and $\dot{M}_{CO_2}$. Metcalfe & Butler (1982) also reported low $\dot{M}_{O_2}$ in spontaneously ventilating blood-perfused dogfish. The reduced $\dot{M}_{O_2}$ in blood-perfused trout preparations probably is related to low oxygen carrying capacity of the blood and not impaired diffusion because values for gill O_2 extraction effectiveness appear normal (Table 1). Metcalfe & Butler (1982) attributed the low $\dot{M}_{O_2}$ in blood-perfused dogfish to reduced perfusion of gas exchange surfaces. Perry *et al.* (1982) demonstrated a direct linear relationship between haematocrit and $\dot{M}_{CO_2}$ in blood-perfused trout thus it is not surprising that $\dot{M}_{CO_2}$ is abnormally low in preparations perfused with blood of low haematocrit.

Two major problems associated with perfused head preparations are simulation of ventilatory water flow and dorsal aortic pressure. In the absence of rhythmic ventilatory pumping, it is unlikely that the complex nature of lamellar water flow can be simulated. Typically, perfused heads have been hyperventilated in an attempt to enhance lamellar ventilation and consequently, the percentage utilization of O_2 from inspired water is abnormally low (Table 1). Dorsal aortic back pressure is known to have substantial effects on branchial vascular resistance and flow patterns through gill vasculature (Wood *et al.*, 1978; Farrell, Daxboeck & Randall, 1979; Daxboeck & Davie, 1982). Minor modifications in perfused head

Table 1. *A comparison of respiratory parameters in intact rainbow trout and perfused gill preparations. Values shown are averages calculated from relevant studies. The range of values is indicated in parentheses. SVBPF = spontaneously ventilating blood-perfused fish*

Parameter	In vivo[a]	Saline-perfused head[b]	Blood-perfused head[c]	In situ head[d]	SVBPF[e]	Saline-perfused holobranchs[f]
$\dot{V}_w$ (ml kg^{-1} min^{-1})	301 (176–646)	5,015 (1712–9143)	1,356 (800–1712)	1,975 (250–3700)	169	—
$\dot{V}_b$ (ml kg^{-1} min^{-1})	19.7 (7.4–36.7)	15.9 (11.5–20.0)	12.5	18.8 (17.6–30.0)	16.2	1.4
P_{IO_2} (Torr)	150 (134–160)	155	155	155	151	152
P_{EO_2} (Torr)	99 (83–121)	—	146	150–155	120	—
Util. (%)	34 (10–46)	—	6	3	21	—
$\dot{M}_{O_2}$ (mmol kg^{-1} h^{-1})	2.2 (1.1–3.4)	0.11 (0.10–0.12)	0.60 (0.3–1.0)	ND–0.13	0.70	ND
P_{vO_2} (Torr)	29 (17–35)	47 (38–61)	28	30–34	25	15
P_{aO_2} (Torr)	111 (85–137)	117 (102–132)	125	48	103	—
O_2 ext. eff. (%)	69 (57–87)	73 (63–82)	82	3	62	—
C_{vO_2} (Vol. %)	7.1	0.26	2.0	—	2.0	—

C_{aO_2} (Vol. %)	9.0 (7.0–10.5)	0.60	3.0	—	3.5	—
Hct. (%)	25	0	11	0	11	—
P_{vCO_2} (Torr)	4.1 (2.5–5.7)	1.7	2.2	3.2	3.4	4.0
P_{aCO_2} (Torr)	2.3 (1.3–3.5)	2.8	1.6	4.8	3.7	4.8
C_{vCO_2} (mM)	8.9 (5.4–10.6)	11.0	9.2	10.6	10.3	8.7
C_{aCO_2} (mM)	6.7 (3.9–9.4)	11.0	8.1	11.6	9.0	8.9
$\dot{M}_{CO_2}$ (mmol kg^{-1} h^{-1})	2.7 (1.9–3.8)	ND	0.7–1.9	ND	1.23	ND
pH$_v$	7.76 (7.67–7.82)	8.15	8.04	7.85 (7.78–7.92)	7.76	7.73
pH$_a$	7.85 (7.70–7.99)	7.94	8.15	7.81 (7.79–7.82)	7.72	7.66

[a] Data were compiled from Cameron and Davis (1970), Davis and Cameron (1970), Eddy (1976), Eddy et al. (1977), Holeton and Randall (1967a, b), Janssen and Randall (1975), Kiceniuk and Jones (1977), Perry and Vermette (1987), Smith and Jones (1982), Stevens and Randall (1967b), Wood et al. (1982), Wright et al. (1986).
[b] Data were compiled from Part et al. (1982a), Perry et al. (1985a, c).
[c] Data were compiled from Perry et al. (1985a), Wright and Perry (unpublished data).
[d] Data were compiled from Daxboeck and Davie (1982), Wood et al. (1978).
[e] Data were compiled from Davie et al. (1982), Daxboeck et al. (1982), Perry et al. (1982).
[f] Data were compiled from Perry et al. (1982).
[g] Estimated values.

methodology (Pettersson & Johansen, 1982; Perry & Wood, 1985; Perry *et al.*, 1985*a*) now permit dorsal aortic pressure to be maintained at 10–15 cm H_2O although often at the expense of reduced dorsal aortic outflow. It is improbable that *in vivo* values of dorsal aortic pressure (about 40 cm H_2O; see Table 2) can be achieved in perfused head preparations due to inadequate control of systemic outflow vessels.

Although spontaneously ventilating blood-perfused preparations display characteristics of gas transfer most similar to intact animals, they also exhibit limitations in common with blood-perfused head preparations including low $\dot{M}_{O_2}$ and $\dot{M}_{CO_2}$ (Table 1). Moreover, the spontaneously ventilating preparations are likely to be highly stressed because experiments must be performed soon after surgery. For these reasons we believe that for short-term experiments (approximately 30 minutes) the blood-perfused head preparation is a viable alternative to the more complex and time-consuming spontaneously ventilating preparations. Future studies should probably incorporate undiluted donor blood to elevate $\dot{M}_{O_2}$ and $\dot{M}_{CO_2}$ to *in vivo* levels. Furthermore, the relationship between $\dot{V}_w$ and gas transfer should be investigated to determine if hyperventilation is indeed necessary for maintenance of adequate gas transfer. Donor blood must be obtained from chronically catheterized fish so that circulating catecholamines are not abnormally elevated but instead reflect levels in resting fish. For obvious reasons, venous blood gas tensions must be simulated.

Haemodynamics

Perfused preparations were particularly instrumental in (i) elucidating the adrenergic and cholinergic mechanisms controlling blood flow through the arterio–arterial and arterio–venous circulations (see Nilsson, 1984), (ii) establishing the compliant nature of the gill vasculature (Wood, McMahon & McDonald, 1978; Farrell *et al.*, 1979; Daxboeck & Davie, 1982; Stagg & Shuttleworth, 1984) and (iii) confirming the existence of 'plasma skimming' in blood entering the venous circulation of the gill (Olson, 1984).

Numerous types of perfused gill preparations from a variety of species have been utilized in studies of branchial haemodynamics including isolated saline- or blood-perfused holobranchs (Richards & Fromm, 1969; Rankin & Maetz, 1971; Bergman, Olson & Fromm, 1974; Smith, 1977; Bolis & Rankin, 1978, 1980; Shuttleworth, 1978; Farrell, Daxboeck & Randall, 1979; Holbert, Boland & Olson, 1979; Jackson & Fromm, 1980, 1981; Wahlqvist, 1980; Stagg, Rankin & Bolis, 1981; Stagg & Shuttleworth, 1982, 1984, 1986; Ellis & Smith, 1983; Olson, 1984; Davis & Shuttleworth, 1985; Bennet & Rankin, 1986), isolated saline- or blood-perfused heads (Keys,

Table 2. *A comparison of haemodynamic parameters in intact rainbow trout and perfused gill preparations. Values shown are avearages calculated from relevant studies. The range of values is indicated in parentheses*

Parameter	In vivo[a]	Saline-perfused head[b]	Blood-perfused head[c]	In situ head[d]	SVBPF[e]	Saline-perfused holobranchs[f]
$\dot{V}_b$ (ml min^{-1} kg^{-1})	19.7 (7.4–36.7)	18.1 (11.6–30.0)	12.5	11.8 (4.6–30.0)	16.2	*21.0 (15.2–26.7)
$P_{aff.}$ (cm H$_2$O)	59.5 (43–95)	38.8 (30–66)	46.0	63.0 (50–80)	58.8	40.5 (40–41)
$P_{eff.}$ (cm H$_2$O)	43.6 (34–68)	0–15	10–15	34.3 (17–40)	34.8	17.0 (14–20)
Rg (cm H$_2$O ml^{-1} min^{-1})	4.6 (3.4–6.0)	18.0 (13.0–21.4)	30.0	21.6 (14–26)	14.3	*23.0 (20.3–25.7)

[a] Data were compiled from Cameron & Davis (1970), Davis & Cameron (1970), Holeton & Randall (1967a, b), Kiceniuk & Jones (1977), Stevens & Randall (1967a, b), Wood & Shelton (1980).

[b] Data were compiled from Bornancin et al. (1985), Part et al. (1984), Payan & Girard (1977), Perry et al. (1985a, b), Perry & Wood (1985), Perry & Daxboeck (1986)..

[c] Data were compiled from Perry et al. (1985a, b).

[d] Data were compiled from Daxboeck & Davie (1982), Wood (1974, 1975), Wood et al. (1978).

[e] Data were compiled from Davie et al. (1982).

[f] Data were compiled from Jackson & Fromm (1980, 1981), Nekvasil & Olson (1986), Olson (1984), Smith (1977).

* Data were extrapolated to the intact animal assuming that 70% of the holobranch was perfused. Overall gill resistance was calculated according to the relationship $1/R_g = 8(1/R_{g_{holobranch}})$.

1931*a*, *b*; Girard & Payan, 1976; Payan & Girard, 1977; Pettersson & Nilsson, 1979; Claiborne & Evans, 1980; Pettersson & Johansen, 1982; Oduleye, Claiborne & Evans, 1982; Part *et al.*, 1982*a*; 1984; Part, Kiessling & Ring, 1982*b*; Evans & Claiborne, 1983; Tuurala *et al.*, 1984; Perry, Payan & Girard, 1984*c*; Perry *et al.*, 1985*a*, *c*; Perry & Daxboeck, 1986), *in situ* perfused heads (Wood, 1974, 1975; Wood *et al.*, 1978; Opdyke, Holcombe & Wilde 1979; Daxboeck & Davie, 1982) and spontaneously ventilating preparations (Saunders & Sutterlin, 1971; Kent & Pierce, 1978; Metcalfe & Butler, 1982; Davie *et al.*, 1982).

The ability of perfused preparations to simulate *in vivo* haemodynamic parameters is shown in Table 2. For clarity, only data from *Salmo gairdneri* have been included, although similar comparisons in other species in which complete data sets are available (e.g. for *Ophiodon elongatus* compare *in vivo* data of Stevens *et al.* (1972) and Farrell (1981) with perfusion data of Farrell *et al.* (1979); for *Gadus morhua* compare *in vivo* data of Johansen (1962), Jones *et al.* (1974), Wahlqvist & Nilsson (1980) and Nilsson & Holmgren (1985) with perfusion data of Pettersson & Nilsson (1979), Pettersson & Johansen (1982) and Pettersson (1983); for *Anguilla australis* compare *in vivo* data of Davie & Forster (1980) and Hipkins & Smith (1983) with perfusion data of Ellis & Smith (1983)) reveal a common anomaly. The primary abnormality in perfused gill preparations is the high branchial vascular resistance to flow (Rg). It is unlikely that isolated perfused gills will ever duplicate the low Rg in intact animals because of their inherent inability to sustain physiological efferent pressures. Attempts to raise efferent pressure to normal levels in isolated heads and holobranchs result in reductions or complete cessation of efferent perfusate outflow. Consequently, token back pressure (10–25 cm H_2O) only can be applied to isolated gills. Because gill vessels are compliant, the efferent back pressure serves to decrease Rg. Furthermore, efferent pressure is important to ensure flow into the venous circulation of the gill. Thus the partitioning of arterio–arterial and arterio–venous flows in perfused gills may not accurately reflect the situation in the intact animal. The elevated gill resistance in heads perfused with saline *in situ* at normal dorsal aortic pressure (Wood, 1974, 1975; Wood *et al.*, 1978; Daxboeck & Davie, 1982) may arise from metabolically mediated vasoconstriction or neurohumoral factors. Blockage of capillaries with damaged red blood cells may contribute to the high Rg in blood-perfused preparations.

Despite the anomolous behaviour of perfused gills, we believe that their continued use is essential to further clarify haemodynamic relationships that cannot be examined *in vivo*. Isolated heads and holobranchs have been used most frequently and each of these preparations have characteristic advantages and disadvantages. The advantages of isolated heads include the ease of preparation and the maintenance of correct gill geometry. Disadvantages include the inability to separate true branchial venous flow from cephalic drainage and the difficulty in simulating dorsal aortic pressure. Isolated trout heads often display a gradual increase in Rg (see Perry *et al.*, 1984*a*) that is overcome by the addition of epinephrine to the perfusion fluid. On the other hand, isolated heads from other species (*Opsanus beta*; Oduleye, Claiborne & Evans, 1982; *Myoxocephalus octodecimspinosus*; Claiborne & Evans, 1980) maintain stable Rg for several hours. *In situ* heads can overcome the problem of low dorsal aortic pressure by exposing the entire cut end of the head to a column of saline (Wood, 1974, 1975) but this subjects the venous circulation to unusually high pressures and venous outflow must be affected adversely. The advantages and disadvantages of isolated holobranchs in haemodynamic studies have been discussed in detail elsewhere (Olson, 1984; Perry *et al.*, 1984*a*). Briefly, the advantages are the separation of arterio–arterial and arterio–venous outflows and application of efferent back pressure. Disadvantages include the unnatural gill geometry, leaks and/or reduced arterial outflow at normal (*in vivo*) efferent pressures, inadequate ventilation and disruption of normal flow patterns if the branchial arch is ligated around the afferent cannula. Stagg & Shuttleworth (1984) observed unusually low arterial outflow in saline-perfused flounder holobranchs that may have resulted from non-restricted venous drainage and/or the relatively high efferent pressure (20–26 cm H_2O) in this preparation.

We suggest that both isolated holobranchs and heads are suitable for future studies of branchial haemodynamics. The choice of which to use may depend upon species availability and the particular function being investigated. The isolated holobranch is especially desirable for investigating the arterio–venous circulation. Blood is a superior perfusion fluid (Perry *et al.*, 1985*a*) although the collection and delivery of large quantities of homologous blood is not always possible. It is advisable to reproduce 'resting' levels of epinephrine (10^{-9}–10^{-8} M) to impart adrenergic tone (Wood, 1974) to the efferent arterio–venous anastomoses and thereby enhance arterial outflow. Spontaneously ventilating preparations most closely resemble the behaviour of intact animals (Table 2) but their advantages must be balanced against the disadvantages of labour investment and the unknown effects of post-operative stress.

Ion exchange
Although perfused preparations have been used extensively to examine ion regulation in freshwater and marine species (Table 3), it is

Table 3. *A comparison of ion transfer parameters in intact fish and perfused gill preparations. Values shown are averages calculated from relevant studies. The range of values is indicated in parentheses. Units for fluxes are umol $kg^{-1}\,h^{-1}$.*

Parameter	Species	FW/SW	In vivo	Perfused head	Perfused holobranch	References
TEP (mV)	*Salmo gairdneri*	FW	−5.1	−6.7 (−6.3–7.4)	—	21, 23
	Platichthys flesus	SW	+26.5 (+19.0–34.0)	—	+17.8	12, 24, 27
	Lagodon rhombiodes	SW	+12.9	—	+9.6	5
	Myoxocephalus octo.	SW	+7.2	+9.3	—	4
J_{inCl}	*Salmo gairdneri*	FW	182 (100–274)	116 (63–140)	—	7, 10, 11, 18, 19, 21, 22, 29, 30
J_{inNa}	*Salmo gairdneri*	FW	291 (198–470)	303 (160–478)	—	1, 6, 7, 9, 10, 11, 14, 17, 21, 22, 29, 30, 31
J_{outCl}	*Logodon rhombiodes*	SW	34,220	—	13,850	5
	Myoxocephalus octo.	SW	12,100	2,708	—	2

	Species					References
J_{outNa}	*Myoxocephalus octo.*	SW	14,720	19,996	—	2
J_{netCl}	*Salmo gairdneri*	FW	+18 (−70−+60)	−54	—	14, 15, 20, 22, 28, 29, 30
J_{netNa}	*Salmo gairdneri*	FW	+29 (−20−+50)	−0.7 (−20−+20)	—	1, 14, 15, 16, 21, 22, 28, 29, 30
	Anguilla sp.	SW	−600−1000	—	−100 (−97−100)	13, 25, 26
J_{inCa}	*Salmo gairdneri*	FW	13.9	7.9	—	20
$J_{netAmmonia}$	*Salmo gairdneri*	FW	218 (150−375)	206 (160−263)	—	11, 21, 22, 29, 30, 31
	Opsanus beta	SW	68.4	200 (160−240)	—	3, 8
	Myoxocephalus octo.	SW	230.4	321	—	8

1. Bornancin *et al.* (1985), 2. Claiborne and Evans (1981), 3. Claiborne *et al.* (1982), 4. Claiborne and Evans (1984), 5. Farmer and Evans (1981), 6. Gardaire *et al.* (1985), 7. Girard and Payan (1977), 8. Goldstein *et al.* (1982), 9. Kerstetter *et al.* (1970), 10. Kerstetter and Kirschner (1972), 11. Kerstetter and Mize (1976), 12. Macfarlane and Maetz (1975), 13. Maetz, (1971), 14. McDonald and Wood (1981), 15. McDonald (1983), 16. Part and Svanberg (1981), 17. Payan (1978), 18. Perry *et al.* (1984c), 19. Perry *et al.* (1984d), 20. Perry and Wood (1985), 21. Perry *et al.* (1985b), 22. Perry *et al.* (1987), 23. Perry (unpublished data), 24. Potts and Eddy (1973), 25. Shuttleworth (1972), 26. Shuttleworth and Freeman (1973), 27. Shuttleworth (1978), 28. Thomas *et al.* (1986), 29. Vermette and Perry (1987), 30. Wood *et al.* (1984), 31. Wright and Wood (1985).

clear that not all aspects of branchial ion transfer can be examined effectively in perfused gills (see also review by Evans *et al.*, 1982). The isolated trout head has been used exclusively to investigate ion regulation in freshwater fishes. A comparison of the data obtained from trout heads with available *in vivo* data (Table 3) is somewhat misleading because it does not reveal the progressive deterioration of many ion transfer parameters as a function of time in the absence of high levels of epinephrine in the perfusion fluid (10^{-6}–10^{-5} M). Perry *et al.* (1985b) evaluated the performance of saline- and blood-perfused trout heads with 10^{-7} M epinephrine in the perfusate. These authors reported that Na^+ and Cl^- uptake deteriorated rapidly while the transepithelial potential (TEP) was stable over the short duration of the experiment (30 min). Blood-perfusion did not substantially alter the performance. The stability of the TEP in the face of decaying rates of Na^+ and Cl^- uptake is not surprising because NaCl uptake requires active transport whereas the TEP is thought to reflect transepithelial ionic diffusion. Thus the stability of the TEP may not be a complete or valid indicator of gill viability. It is conceivable that other freshwater species might exhibit greater stability. Isolated holobranchs are the only reasonable alternative to perfused heads but these preparations are poorly suited for ion flux measurements (see Perry *et al.*, 1984a). The inherent instability of ion uptake in the perfused trout head is abolished by the addition of 10^{-6} M epinephrine (see Bornancin, Isaia & Masoni, 1985). Although we would not normally advocate using such high levels of epinephrine, there appears to be no alternative presently for studying certain mechanisms of ion transfer in freshwater fishes. As always, perfused preparations should be used only when experiments cannot be performed *in vivo* and the possible interactive effects of such high levels of catecholamines should be addressed.

In marine fish, isolated heads and holobranchs of several species have been utilized (Table 3). *Myoxocephalus* heads and *Platichthys* holobranchs yield stable TEP's that are similar to *in vivo* values. A stable electronogenic component of the TEP has been demonstrated in flounder holobranchs perfused and bathed with identical perfusion fluid (Shuttleworth, 1978; Stagg & Shuttleworth, 1984). The few measurements of ion exchanges in perfused holobranchs (Shuttleworth, 1972; Shuttleworth & Freeman, 1973; Farmer & Evans, 1981) reveal depressed flux rates. Unless the leakage of radioactive saline into the bathing medium can be prevented at reasonable perfusion flow rates, the isolated holobranch will remain poorly suited for radiotracer studies. The stability of ionic fluxes in *Myoxocephalus* heads has yet to be established. The available data from this species indicate that normal rates of Na^+ efflux can be achieved in the perfused head while Cl^-

efflux is abnormally low (Claiborne & Evans, 1981). The normal TEP and Na^+ efflux suggest that transepithelial diffusive events are unaltered but that the energetic requirements for active transport are limiting Cl^- efflux in the perfused head of *Myoxocephalus*.

Invertebrates

Isolated crustacean holobranchs have been employed primarily for investigating transepithelial ion or water transport (Croghan, Curra & Lockwood, 1965; Mantel, 1967; Berlind & Kamemoto, 1977; Pequeux & Gilles, 1978, 1981; Lucu & Siebers, 1986) and to a lesser extent, gas transfer (Burnett, 1984; Burnett & McMahon, 1985). Typically, crustacean gills have been subjected to unphysiological perfusate and water flow. Further studies would benefit by adopting the methodology of Burnett (1984) who has simulated more closely the *in vivo* conditions and assessed the metabolic viability of the perfused gill. Internal perfusion fluid (see recipe in Burnett, 1984) should be gassed appropriately to mimic haemolymph gas tensions (P_{O_2} = 10 Torr; P_{CO_2} = 3.4 Torr; pH = 7.9; Burnett & McMahon, 1985) and delivered to the gill at appropriate flows (1–3 ml min^{-1}) and pressures (2–8 cm H_2O). Efferent pressure is approximately zero. A continuous flow of water over the gill is achieved by a suction pump that creates negative pressures (−2 to −4 cm H_2O) in the gill perfusion chamber.

Perfused hearts

Studies with perfused hearts have an illustrious history. Frank (1895), using an isolated frog heart, and Patterson & Starling (1914), using an isolated dog heart and lung preparation, elucidated the effect of filling pressure on stroke volume of the heart (SV_H). Langendorff (1899) used an isolated heart to examine coronary flow for the first time. Straub (1904) examined cardiac function in *Aplysia*. Cardiac metabolism also was studied with isolated hearts. Locke & Rosenheim (1904) demonstrated that isolated hearts metabolized glucose, but after RQ values of less than 1 were found for the heart-lung preparation, fatty acids were shown to be the preferred substrate for cardiac metabolism in mammals.

Many of our present perfusion techniques therefore are modifications or improvements on earlier systems, since even landmark studies with isolated hearts were not without limitations; Starling noted his heart–lung preparation deteriorated with time and mechanical efficiency was only five per cent (Evans, 1939). The following is an overview of some of the general considerations for and limitations of perfused hearts in comparative physiology.

Anatomical considerations

Often in mammalian studies only the left side of the heart is perfused even though the whole heart is isolated. The left ventricle works isovolumetrically in the preparation of Langendorff (1899) but generates flow with the working heart preparation of Neely *et al.* (1967). In fishes, amphibians and invertebrates, the heart chambers are more discrete and more easily separated for studies with isolated atria or ventricles. This approach is valuable when studying receptors or metabolism (e.g. Ask, Stene-Larsen & Helle, 1981). However, cardiac dynamics ultimately rely upon the integrated actions of all of the cardiac chambers and ideally all chambers should be perfused when studying cardiac dynamics. For example, both atria should be perfused in the isolated octopus heart for studies of cardiac dynamics (compare Smith, 1981 with Houlihan, Duthie & Smith, 1986). Studies with perfused fish hearts provide better examples of these considerations.

Contraction of the fish atrium is probably the sole means of filling the ventricle, unlike the situation in mammals. The atrium and ventricle can therefore be considered as two in-series amplifiers, in terms of pressure generation. Clearly, studies of cardiac dynamics with isolated ventricles

Fig. 1. With the isolated working trout heart the cardiac perfusate is delivered to the atrium (A) at a constant, regulated input pressure, the heart is electrically paced and the ventricle (V) pumps the perfusate against an output pressure. Cardiac output is set by varying input pressure (to alter stroke volume) or the pacing frequency. The coronary artery is perfused with constant flow and the coronary veins drain into the atrium near the atrio-ventricular junction.

ISOLATED, WORKING TROUT HEART WITH CORONARY PERFUSION

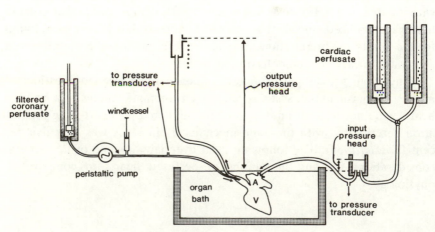

must be extrapolated to intact fish with caution. Whether the muscle of the sinus venosus also contributes significantly to cardiac filling has not been established.

It is probably impossible to isolate the intact sinus venosus from fish. However, the ventricle, atrium and ventral aorta are easily isolated and perfused by placing cannulae in the ventral aorta and the sino-atrial junction (Fig. 1) (e.g. Driedzic, Scott & Farrell, 1983; Farrell, 1987). Electrical pacing is required since the pacemaker, which is located in the sinus, is lost and the ventricle beats at a low, often irregular intrinsic rate. In addition, the loss of the sino-atrial valve means that a mechanical one-way input valve is needed in the input cannula to avoid significant backflow during atrial contraction. These problems are avoided by *in situ* heart perfusion, where the heart is left in the fish and the sinus venosus is cannulated *via* the hepatic veins (Fig.2). Other veins returning to the heart are ligated and the cardiac branch of the vagus is either crushed when the ductus is ligatured or severed below the ligature around the ductus. This type of preparation develops a regular sinus rhythm, performs physiological

Fig. 2. With the *in situ* perfused heart, the perfusate is delivered *via* a cannula placed in one of the hepatic veins (HPV) and the sinus venosus (SV) is intact. Other veins e.g. ductus Cuvier (DC), anterior jugular vein (AJV) and abdominal vein (ABV) can be ligated to prevent backflow of perfusate.

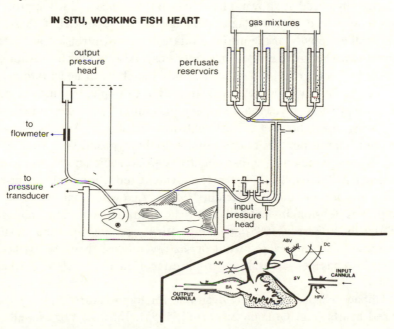

work loads (see below) for many hours, and is ideal for examining cardiac dynamics in fish (Farrell, Macleod & Driedzic, 1982; Farrell, Macleod & Chancey, 1986).

The pericardium is normally ruptured to cannulate the ventral aorta in the *in situ* perfused heart. This is unacceptable for animals in which a rigid pericardial cavity provides a *vis-a-fronte* effect on cardiac filling (e.g. tuna, sharks, some bivalves and sea squirts). A ruptured pericardium impairs $\dot{V}_b$ in sharks (Satchell, 1971). We have successfully cannulated the ventral aorta without rupturing the pericardium in the *in situ* perfused trout heart. Such a preparation would be valuable to evaluate the role of the pericardial cavity since an intact pericardium may also prevent over dilation of the atrium, which can occur in failing, isolated trout hearts (Farrell, 1987), and in failing, dog hearts (Allard *et al.*, 1983). The presence of the pericardial sac and alary ligaments would seem paramount for isolated heart studies in crustaceans.

In summary, studies on cardiac dynamics using perfused hearts should fully consider the relative importance of each chamber of the heart and the pericardial cavity. However, other types of studies (e.g. metabolic and identification of receptors) can benefit from simpler isolated heart systems.

Heart rate

Perfused hearts can develop their own intrinsic rhythm or be electrically paced at known, physiological frequencies. Each approach has advantages and disadvantages. Pacing may be essential if there is no pacemaker activity or the intrinsic rate is too low (intrinsic sinus rates are about twice ventricular rates in vertebrates). Also, it may be advantageous to control heart rate (F_H) either to study the effect of heart rate *per se* or reduce variability between preparations. For example, the rate and pattern of coronary flow, myocardial $\dot{M}_{O_2}$ and maximum $\dot{V}_b$ are dependent on F_H.

Pacing may, however, mask extrinsic effects on F_H. For example, extracellular respiratory acidosis similar to that typically seen after exhaustive exercise in teleost fish has negative chronotropic and inotropic effects (Gesser & Poupa, 1983) but the associated reduction in $\dot{V}_b$ is primarily through decreased F_H (Farrell, 1984). Likewise, extracellular Ca^{2+} improves tension in isolated ventricle strips from various teleost fish (Driedzic & Gesser, 1985) but the positive inotropy is not manifest as an increase in $\dot{V}_b$ in perfused trout or sea raven hearts because extracellular Ca^{2+} also has a negative chronotropic effect (Farrell, Macleod & Chancey, 1986).

Unpaced hearts are highly desirable for temperature studies with perfused hearts (e.g. Graham & Farrell, 1985). Intrinsic F_H, as well as the

Table 4. *A comparison of the maximum $\dot{V}_b$ in response to increased filling pressure in the in situ perfused trout heart at 5 °C and 15 °C*

	5 °C	15 °C	Ratio or Q_{10}
Fish mass (kg)	0.596	0.531	0.89
Ventricle mass (g)	0.70	0.42	0.60
Relative ventricle mass (%)	0.117	0.078*	
Maximum $\dot{V}_b$ (ml kg^{-1} min^{-1})	30.40	36.60*	1.20
F_H (beats min^{-1})	37.2	48.4*	1.30
SV_H (ml kg^{-1})	0.83	0.77	0.93
Relative myocardial power (mW g^{-1})	2.38	3.80*	1.60
Absolute myocardial power (mW)	1.67	1.60	0.95

* Denotes statistical significant difference between the two temperatures.

contractility, decrease with temperature (Table 4), but $\dot{V}_b$ need not decrease proportionally since SV_H can increase as the cardiac filling time is extended. Compensatory changes in SV_H are limited however by the normal range for SV_H.

In summary, whether the heart should or should not be paced is dictated by the nature of the study and the type of preparation. If chronotropic effects are suspected, then either a range of pacing frequencies should be examined, or, preferably, a preparation that develops an intrinsic rhythm should be used.

Cardiac work

Prior knowledge of arterial and venous pressures and $\dot{V}_b$ in the intact animal is essential for perfused heart studies. Values close to those in resting animals should be adopted for control situations since the perfused heart responds to input and output pressure in four important ways. In fact, the role of input and output pressure on cardiac dynamics should be a primary study.

Input pressure. SV_H is directly related to input (or filling) pressure of the isolated heart and characteristic curves are found for different vertebrates and invertebrates (Fig. 3). These curves can have physiological significance in terms of control of SV_H, since venous pressure may lie within the functional range of the curve (e.g. trout). Alternatively, venous pressure may not change significantly in the intact animal (e.g. mammals) or may lie outside the range of the curve (e.g. *Octopus*) and so the curve has limited significance for the intact animal. Extrinsic factors can, however, modify cardiac contractility so that a family of curves exist, implying that extrinsic factors can modulate SVH while filling pressure is constant. If supra-physiological filling pressures are required to generate physiological changes in SVH, it is likely that (i) the heart is failing, as shown for trout (Farrell *et al.*, 1986), or (ii) an extrinsic stimulatory factor is absent. A nanomolar concentration of epinephrine, similar to that in the blood of resting trout, is required for tonic stimulation of perfused trout hearts at

Fig. 3. Effects of filling pressure on stroke work in perfused hearts of selected fish and molluscs and on heart rate in selected molluscs. The effects of various cardioactive agents (e.g. epinephrine (EP), 5-hydroxytryptamine (5-HT) and FMRF-amide are also indicated. References for data are included in the text.

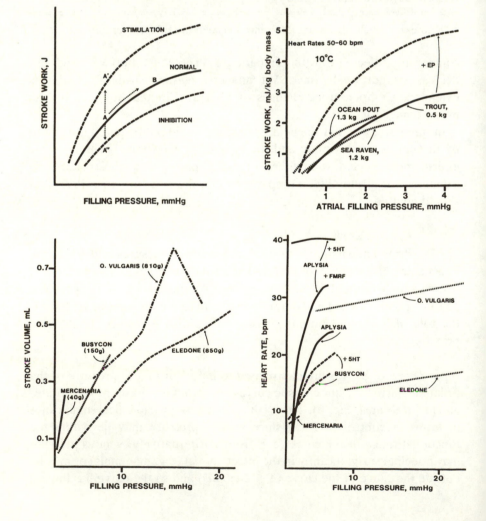

5 °C, but not necessarily at 15 °C. Perfused invertebrate hearts also may require tonic stimulation by agents such as 5-HT and FMRFamide. FMR-Famide is found in invertebrate blood in nanomolar concentrations (Smith & Hill, 1987).

Output pressure. Isolated invertebrate and vertebrate hearts are intrinsically capable of maintaining $\dot{V}_b$ independent of vascular resistance over a physiological range for arterial pressure (homeometric regulation). Outside of this pressure range there is a compromise between pressure and flow development (Farrell, 1984). Therefore, $\dot{V}_b$ may be enhanced at sub-physiological pressures or reduced at supra-physiological pressures. Pressure measurements must be corrected for the pressure drop associated with flow through the narrow cannulae used with perfused hearts.

Workload. The mechanical efficiency of the heart is directly related to its power output in isolated hearts from snails and octopus to fish and mammals (Herold, 1975; Gibbs & Chapman, 1979; Farrell *et al.*, 1985; Houlihan *et al.*, 1987). A low $\dot{V}_b$, low output pressure, high input pressure, or a combination will result in a sub-physiological workload and an inefficient heart. Most hearts appear to work with a mechanical efficiency in the 9–26% range. Anything outside this range is suspect. Low values suggest cardiac failure; higher values suggest an anaerobic contribution.

Heart rate. Increased filling pressure in invertebrate and vertebrate hearts can increase F_H. The physiological significance of this response varies. In invertebrates, where F_H may be low, F_H increases by 10–100% with filling pressure and the perfused heart may not even beat unless stretched (Fig. 3). In mammals the increase in F_H is 10–15%. More frequently with *in situ* perfused hearts from Atlantic cod, trout, sea raven and ocean pout, there is either no change or a decrease of 1–3 beats·min^{-1} at the maximum SV_H, and any increase in F_H is rarely 10%. Exceptions were found when the trout heart was failing (i.e. when intrinsic F_H was unusually low, 20 beats·min^{-1}) or when $\dot{V}_b$ was below resting levels and the heart became arrhythmic. Tonic adrenergic stimulation remedied both situations. Nevertheless, isolated hearts from trout, elasmobranchs and hagfish show up to a 75% increase in F_H with filling pressure (Jensen, 1961; 1969; 1970). These data may be significant for intact elasmobranchs and hagfish, but not for trout based on the *in situ* heart studies. This is another example of how the type of perfused preparation has limitations in terms of the questions being addressed.

Oxygenation

Invertebrate hearts generally pump oxygenated blood with a thin-walled ventricle that receives O_2 from its lumen by diffusion. The thick-walled ventricle of *Octopus* has coronary vessels that originate at the lumen and extend through the ventricular muscle to the pericardial cavity. Vertebrate hearts are generally larger than invertebrate hearts and, except for those that are fully divided, pump venous or mixed venous blood. A coronary circulation is found in mammals, birds, reptiles and the faster swimming fish (e.g. trout, tuna, shark and ratfish). Amphibians and slower moving fish use blood in the heart chamber as their sole myocardial O_2 source.

Calculations based on myocardial $\dot{M}_{O_2}$ of perfused hearts indicate that less than 4% of the O_2 contained in venous blood is extracted in sea raven (a fish without coronaries); a three-fold increase in cardiac work (probably the maximum for this species) increases extraction of 12% as myocardial $\dot{M}_{O_2}$ is directly related to cardiac work (Fig. 4) (Farrell *et al.*, 1985). Thus venous P_{O_2} rather than C_{O_2} probably limits O_2 delivery in hearts lacking a coronary circulation, although the highly trabeculated nature of the heart and facilitated O_2 diffusion by haemoglobin and myoglobin are helpful in this regard. If this supposition is true, then our concern with perfused hearts is that the P_{O_2} gradient is adequate.

O_2 content of air-equilibrated salines is about 0.4 mM and approximately five times higher when O_2-equilibrated. Thus, with the low C_{O_2} of salines it is possible that O_2 extraction reduces the PO_2 of the saline sufficiently to create diffusion limitations. However, in *Octopus*, the P_{O_2} of air-equilibrated saline is reduced by less than 10% as it passes through the lumen of the heart and by 50–60% through the coronary vessels (Houlihan *et al.*, 1987). Therefore, air-equilibrated saline seems adequate for studies with invertebrate hearts because of their low relative $\dot{M}_{O_2}$ (about one per cent of standard metabolic rate in *Octopus*).

Air- and O_2-equilibrated salines have been used with success in fish with and without a coronary circulation. Air-equilibrated saline is reduced by 10–20%, which leaves the P_{O_2} of the saline well above the normal venous P_{O_2} (Farrell *et al.*, 1985, 1986). O_2 extraction increases with arterial pressure, but it may decrease as O_2 delivery is increased with increased $\dot{V}_b$. With *in situ* hearts from 4–600 g trout, the two- to five-fold higher P_{O_2} gradient associated with air- or O_2-equilibrated saline (compared to venous blood) apparently compensates for the absence of coronary perfusion. However, in larger fish (greater than 1 kg), the coronary circulation must be perfused because the diffusion distances are greater and changes in coronary flow rate of perfused hearts alter maximum cardiac performance

(Farrell, 1987). Air- or O_2-equilibrated salines are adequate for studies with fish hearts.

The coronary circulation in vertebrates should be perfused with O_2-equilibrated saline, but even this may be inadequate as O_2 extraction from coronary blood is usually 75%. Metabolically-mediated coronary vasodilation may partially compensate, but perfusates with a higher C_{O_2} are clearly desirable. Perfluorinated carbon compounds, haemoglobin solutions

Fig. 4. Myocardial oxygen uptake and extraction as a function of changes in cardiac output and output pressure in the perfused sea raven heart.

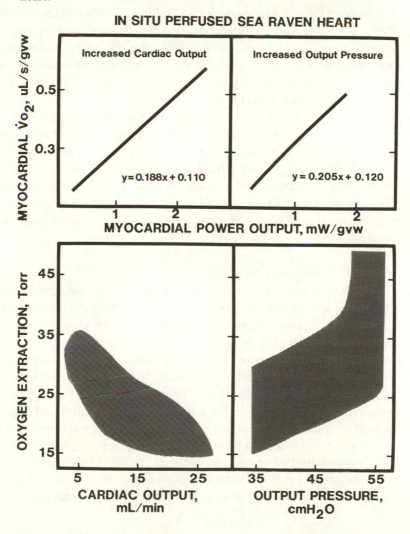

and blood-perfusion have been tried (e.g. Segel & Rendig, 1982) but they may produce unwanted coronary vasoactivity. As myocardial performance is closely related to coronary flow in mammals, adequate O_2 delivery to perfused hearts is of continuing concern.

Other perfused preparations
Fish tails and trunks

Isolated perfused fish trunks or tails have been employed to investigate physiological control of systemic resistance (Wood & Shelton, 1975, Wood, 1976, 1977; Forster, 1976; Wahlqvist & Nilsson, 1981; Davie, 1981), extravascular volume regulation (Davie, 1982), muscle substrate utilization (Moen & Klungsoyr, 1981) and lactate efflux (Turner & Wood, 1983). Perfused tails or trunks are easily prepared and display stable haemodynamic parameters (Davie, 1981). The systemic vasculature can be perfused at constant pressure to simulate dorsal aortic pressure *in vivo* (e.g. Davie, 1981) but is usually perfused at constant flow. Pulsatility is not as crucial as in perfused gills although a pulse pressure of about 2–3 cm H_2O is desirable. Arterial gas tensions should be reproduced; Turner & Wood (1983) added bovine haemoglobin (55 g/L^{-1}) to perfusion fluid to increase O_2 content but the possible deleterious osmotic consequences were not assessed. Perfused tails and trunks might benefit from the addition of perfluorinated hydrocarbons to saline to increase C_{O_2}, or blood-perfusion. Colloidal substitutes (preferably albumin) and nanomolar concentrations of epinephrine should be added to saline to minimize oedema and to impart adrenergic tone, respectively.

Extracorporeal circulations

Strictly speaking, extracorporeal circulations and external circulatory shunts are not perfused preparations. However, the partial externalization of the circulation does offer several significant advantages in physiological studies. Hughes *et al.* (1982) created an external ventral aorta circulation in the eel to permit measurement of cardiac output using a Doppler velocimeter thereby overcoming the problems associated with using electromagnetic flow probes. Thomas & Le Ruz (1982) developed an extracorporeal circulation in the trout in which arterial blood is pumped through a series of blood gas and pH measuring devices and returned to the fish *via* a venous catheter. This technique, in concert with data processing to correct for electrode response times, allows rapid changes in blood respiratory and acid–base status to be monitored (e.g. Thomas & Le Ruz, 1982; Thomas, 1983; Thomas, Fievet & Motais, 1986) that otherwise might go undetected. Furthermore, this technique eliminates the progressive decrease in haematocrit caused by repetitive blood sampling.

Acknowledgements

Our own research involving perfused preparations is supported by NSERC of Canada operating grants.

References

Allard, J. R. Gertz, E. W., Verrier, E. D., Bristow, J. D. & Hoffman, J. E. (1983). Role of the pericardium in the regulation of myocardial blood flow and its distribution in the normal and acutely failing left ventricle of the dog. *Cardiovascular Research*, **17**, 595–603.

Ask, J. A., Stene-Larsen, G. & Helle, K. B. (1981). Temperature effects on the B$_2$-adrenoceptors of the trout atrium. *Journal of Comparative Physiology*, **143**B, 161–8.

Bateman, J. B. & Keys, A. (1932). Chloride and vapour-pressure relations in the secretory activity of the gills of the eel. *Journal of Physiology* (London), **75**, 226–40.

Bennet, M. B. & Rankin, J. C. (1986). The effects of neurohypophysial hormones on the vascular resistance of the isolated perfused gill of the European eel, *Anguilla anguilla* L. *General & Comparative Endocrinology*, **64**, 60–66.

Bergman, H. L., Olson, K. R. & Fromm, P. O. (1974). The effects of vasoactive agents on the functional surface area of isolated-perfused gills of rainbow trout. *Journal of Comparative Physiology*, **94**B, 267–86.

Berlind, A. & Kamemoto, F. I. (1977). Rapid water permeability changes in eyestalkless euryhaline crabs and in isolated, perfused gills. *Comparative Biochemistry & Physiology*, **58**A, 383–85.

Bolis, L. & Rankin, J. C. (1978). Vascular effects of acetylcholine, catecholamines and detergents on isolated perfused gills of pink salmon, *Onchorynchus gorbuscha*, coho salmon, *O. kisutch* and chum salmon, *O. keta*. *Journal of Fish Biology*, **13**, 543–7.

Bolis, L. & Rankin, J. C. (1980). Interactions between vascular actions of detergents and catecholamines in perfused gills of European eel, *Anguilla anguilla* L. and brown trout, *Salmo trutta* L. *Journal of Fish Biology*, **16**, 61–73.

Bornancin, M., Isaia, J. & Masoni, A. (1985). A re-examination of the technique of isolated, perfused trout head preparation. *Comparative Biochemistry & Physiology*, **81**A, 35–41.

Burnett, L. E. (1984). Co$_2$ excretion across isolated perfused crab gills: facilitation by carbonic anhydrase. *American Zoologist*, **24**, 253–64.

Burnett, L. E. & McMahon, B. R. (1985). Facilitation of CO$_2$ excretion by carbonic anhydrase located on the surface of the basal membrane of crab gill epithelium. *Respiration Physiology*, **62**, 341–8.

Cameron, J. N. & Davis, J. C. (1970). Gas exchange in rainbow trout (*Salmo gairdneri*) with varying blood oxygen capacity. *Journal of the Fisheries Research Board of Canada*, **27**, 1069–85.

Claiborne, J. B. & Evans, D. H. (1980). The isolated, perfused head of the marine teleost fish, *Myoxocephalus octodecimspinosus*: Hemodynamic effects of epinephrine. *Journal of Comparative Physiology*, **138**, 79–85.

Claiborne, J. B. & Evans, D. H. (1981). The effect of perfusion and irrigation flow rate variations on NaCl efflux from the isolated, perfused head of the marine teleost, *Myoxocephalus octodecimspinosus*. *Marine Biology Letters*, **2**, 123–30.

Claiborne, J. B. & Evans, D. H. (1984). Transepithelial potential measurements in the isolated, perfused head of a marine teleost. *Journal of Experimental Zoology*, **230**, 321–4.

Croghan, P. C., Curra, R. A. & Lockwood, A. P. (1965). The electrical potential difference across the epithelium of isolated gills of the crayfish *Austropotamobius pallipes* (Lerreboullet). *Journal of Experimental Biology*, **42**, 463–74.

Davie, P. S. (1981). Vascular resistance responses of an eel tail preparation: alpha constriction and beta dilation. *Journal of Experimental Biology*, **90**, 65–84.

Davie, P. S. (1982). Changes in vascular and extravascular volumes of eel muscle in response to catecholamines: the function of the caudal lymphatic heart. *Journal of Experimental Biology*, **96**, 195–206.

Davie, P. S. & Daxboeck, C. (1982). Effect of pulse pressure on fluid exchange between blood and tissue in trout gills. *Canadian Journal of Zoology*, **60**, 1000–6.

Davie, P. S. & Forster, M. E. (1980). Cardiovascular responses to swimming in eels. *Comparative Biochemistry & Physiology*, **67**A, 367–73.

Davie, P. S., Daxboeck, C., Perry, S. F. & Randall, D. J. (1982). Gas transfer in a spontaneously ventilating, blood-perfused trout preparation. *Journal of Experimental Biology*, **101**, 17–34.

Davis, J. C. & Cameron, J. N. (1970). Water flow and gas exchange at the gills of rainbow trout, *Salmo gairdneri*. *Journal of Experimental Biology*, **54**, 1–18.

Davis, M. S. & Shuttleworth, T. J. (1985). Peptidergic and adrenergic regulation of electrogenic ion transport in isolated gills of the flounder (*Platichthys flesus* L.). *Journal of Comparative Physiology*, **155**B, 471–8.

Daxboeck, C. & Davie, P. S. (1982). Effects of pulsatile perfusion on flow distribution within an isolated saline-perfused trout head preparation. *Canadian Journal of Zoology*, **60**, 994–9.

Daxboeck, C., Davie, P. S., Perry, S. F. & Randall, D. J. (1982). Oxygen uptake in a spontaneously ventilating, blood-perfused trout preparation. *Journal of Experimental Biology*, **101**, 35–45.

Driedzic, W. R., Scott, D. L. & Farrell, A. P. (1983). Aerobic and anaerobic contributions to energy metabolism in perfused isolated sea raven (*Hemitriptrus americanus*) hearts. *Canadian Journal of Zoology*, **61**, 1880–3.

Driedzic, W. R. & Gesser, H. (1985). Ca^{2+} protection from the negative inotropic effect of contraction frequency on teleost hearts. *Journal of Comparative Physiology*, **156**B, 135–42.

Eddy, F. B. (1976). Acid–base balance in rainbow trout (*Salmo gairdneri*) subjected to acid stresses. *Journal of Experimental Biology*, 64, 159–71.

Eddy, F. B., Lomholt, J. P., Weber, R. E. Johansen, K. (1977). Blood respiratory properties of rainbow trout (*Salmo gairdneri*) kept in water of high C_{O_2} tension. *Journal of Experimental Biology*, 67, 37–47.

Ellis, A. G. & Smith, D. G. (1983). Edema formation and impaired O_2 transfer in Ringer-perfused gills of the eel, *Anguilla australis*. *Journal of Experimental Zoology*, 227, 371–80.

Evans, C. L. (1939). The metabolism of cardiac muscle. In *Recent Advances in Physiology*, vol. 6, pp. 157–215. Philadelphia: Blakistons Son & Co.

Evans, D. H. Claiborne, J. B., Farmer, L., Mallery, C. & Krasny, E. (1982). Fish gill ionic transport: methods and models. *Biological Bulletin*, 163, 108–30.

Farmer, L. L. & Evans, D. H. (1981). Chloride extrusion in the isolated perfused teleost gill. *Journal of Comparative Physiology*, 141B, 471–6.

Farrell, A. P. Daxboeck, C. & Randall, D. J. (1979). The effect of input pressure and flow on the pattern and resistance to flow in the isolated perfused gill of a teleost fish. *Journal of Comparative Physiology*, 133 B, 233–40.

Farrell, A. P. (1981). Cardiovascular change in the lingcod (*Ophiodon elongatus*) following adrenergic and cholinergic drug infusions. *Journal of Experimental Biology*, 91, 293–305.

Farrell, A. P., Macleod, K. R. & Driedzic, W. R. (1982). The effects of preload, afterload and epinephrine on cardiac performance in the sea raven, *Hemitripterus americanus*. *Canadian Journal of Zoology*, 60, 3165–71.

Farrell, A. P. (1984). A review of cardiac performance in the teleost heart: intrinsic and humoral regulation. *Canadian Journal of Zoology*, 62, 523–36.

Farrell, A. P., Wood, S., Hart, T. & Driedzic, W. R. (1985). Myocardial oxygen consumption in the sea raven, *Hemitripterus americanus*: the effects of volume loading, pressure loading, and progressive hypoxia. *Journal of Experimental Biology*, 117, 237–50.

Farrell, A. P., Macleod, K. R. & Chancey, B. (1986). Intrinsic mechanical properties of the perfused rainbow trout heart and the effects of catecholamines and extracellular calcium under control and acidotic conditions. *Journal of Experimental Biology*, 125, 319–45.

Farrell, A. P. (1987). Coronary flow in a perfused trout heart. *Journal of Experimental Biology*, 129, 107–23.

Forster, M. E. (1976). Effects of catecholamines on the heart and on branchial and peripheral resistance of the eel (*Anguilla anguilla*). *Comparative Biochemistry & Physiology*, 55, 27–32.

Frank, O. (1895). Zur dynamic des herzmuskels. *Zeitschrift Biologie*, 32, 370–437.

Gardaire, E., Avella, M., Isaia, J., Bornancin, M. & Mayer-Gostan, N. (1985). Estimation of sodium uptake through the gill of the rainbow trout *Salmo gairdneri*. *Experimental Biology*, 44, 181–9.

Gesser, H. & Poupa, O. (1983). Acidosis and cardiac muscle contractility:

comparative aspects. *Comparative Biochemistry & Physiology*, **76**A, 559–66.

Gesser, H. & Jorgensen, E. (1983). Effect of vanadate and of removal of extracellular Ca^{2+} and Na^+ on tension development and ^{45}Ca efflux in rat and frog myocardium. *Comparative Biochemstry & Physiology*, **76**C, 199–202.

Girard, J. P. (1976). Salt excretion by the perfused head of trout adapted to seawater and its inhibition by adrenaline. *Journal of Comparative Physiology*, **111**B, 77–91.

Girard, J. P. & Payan, P. (1976). Effect of epinephrine on vascular space of gills and head of rainbow trout. *American Journal of Physiology*, **230**, 1555–60.

Girard, J. P. & Payan, P. (1977). Kinetic analysis of sodium and chloride influxes across the gills of trout in freshwater. *Journal of Physiology* (London), **273**, 195–209.

Gibbs, C. L. & Chapman, J. B. (1979). *Cardiac Energetics*. In: *Handbook of Physiology*, vol. 1, section 2, ed. R. M. Berne, pp. 755–804. Bethesda: American Physiological Society.

Goldstein, L., Claiborne, J. B. & Evans, D. H. (1982). Ammonia excretion by the gills of two marine teleost fish: the importance of NH_4^+ permeance. *Journal of Experimental Zoology*, **219**, 395–7.

Graham, M. & Farrell, A. P. (1985). Seasonal intrinsic cardiac performance of a marine teleost. *Journal of Experimental Biology*, **118**, 173–83.

Herold, J. P. (1975). Myocardial efficiency in the isolated ventricle of the snail, *Helix pomatia* L. *Comparative Biochemistry & Physiology*, **52**A, 435–40.

Hipkins, S. F. & Smith, D. G. (1983). Cardiovascular events associated with spontaneous apnea in the Australian short-finned eel, *Anguilla australis*. *Journal of Experimental Zoology*, **227**, 339–48.

Holbert, P. W., Boland, E. J. & Olson, K. R. (1979). The effect of epinephrine and acetylcholine on the distribution of red cells within the gills of the channel catfish (*Ictalurus punctatus*). *Journal of Experimental Biology*, **79**, 135–46.

Holeton, G. F. & Randall, D. J. (1967a). Changes in blood pressure in the rainbow trout during hypoxia. *Journal of Experimental Biology*, **46**, 297–305.

Holeton, G. F. & Randall, D. J. (1967b). The effect of hypoxia upon the partial pressure of gases in the blood and water afferent and efferent to the gills of rainbow trout. *Journal of Experimental Biology*, **46**, 317–27.

Houlihan, D. F., Duthie, G. & Smith, P. G. (1986). Ventilation and circulation during exercise in *Octopus vulgaris*. *Journal of Comparative Physiology*, **156**B, 683–9.

Houlihan, D. F., Agnisola, C., Hamilton, N. M. & Genoino, I. T. (1987). Oxygen consumption of the isolated heart of *Octopus*: effects of power output and hypoxia. *Journal of Experimental Biology* (in press).

Hughes, G. M., Peyraud, C., Peyraud-Waitzenegger, M. & Soulier, P. (1982). Physiological evidence for the occurrence of pathways shunting blood away from the secondary lamellae of eel gills. *Journal of Experimental Biology*, **98**, 277–88.

Jackson, W. F. & Fromm, P. O. (1980). Effect of acute acid stress on isolated perfused gills of rainbow trout. *Comparative Biochemistry & Physiology*, **67**C, 141–5.

Jackson, W. F. & Fromm, P. O. (1981). Factors affecting ^3H_2O transfer capacity of isolated perfused trout gills. *American Journal of Physiology*, **240**, 235–45.

Janssen, R. G. & Randall, D. J. (1975). The effects of changes in pH and PCO_2 in blood and water on breathing in rainbow trout, *Salmo gairdneri*. *Respiration Physiology*, **25**, 235–45.

Jensen, D. (1961). Cardioregulation in an aneural heart. *Comparative Biochemistry & Physiology*, **2**, 181–201.

Jensen, D. (1969). Intrinsic cardiac rate regulation in the sea lamprey, *Petromyzon marinus*, and rainbow trout, *Salmo gairdneri*. *Comparative Biochemistry & Physiology*, **30**, 685–90.

Jensen, D. (1970). Intrinsic cardiac rate regulation in elasmobranch: the horned shark, *Heterodontus prancisci*, and thornback ray, *Platyrhinoidis triseriata*. *Comparative Biochemistry & Physiology*, **34**, 289–96.

Johansen, K. (1962). Cardiac output and pulsatile aortic flow in the teleost, *Gadus morhua*. *Comparative Biochemistry & Physiology*, **7**, 169–74.

Jones, D. R., Langille, B. L., Randall, D. J. & Shelton, G. (1974). Blood flow in the dorsal and ventral aortae of the cod (*Gadus morhua*). *American Journal of Physiology*, **266**, 90–95.

Kent, B. & Peirce, E. C. (1978). Cardiovascular responses to changes in blood gases in dogfish shark, *Squalus acanthias*. *Comparative Biochemistry & Physiology*, **60**C, 37–44.

Kerstetter, T. H., Kirschner, L. B. & Rafuse, D. D. (1970). On the mechanisms of sodium ion transport by the irrigated gills of rainbow trout (*Salmo gairdneri*). *Journal of General Physiology*, **56**, 342–59.

Kerstetter, T. H. & Kirschner, L. B. (1972). Active chloride transport by the gills of rainbow trout. *Journal of Experimental Biology*, **56**, 263–72.

Kerstetter, T. H. & Mize, R. (1976). Responses of trout gill ion transport systems to acute acidosis. *Journal of Experimental Biology*, **65**, 511–15.

Keys, A. B. (1931a). The heart-gill preparation of the eel and its perfusion for the study of a natural membrane *in situ*. *Zeitschrift fur Vergleichende Physiologie*, **15**, 352–63.

Keys, A. B. (1931b). Chloride and water secretion and absorption by the gills of the eel. *Zeitschrift fur Vergleichende Physiologie*, **15**, 364–88.

Keys, A. B. & Bateman, J. B. (1932). Branchial responses to adrenaline and to pitressin in the eel. *Biological Bulletin*, **63**, 327–36.

Kiceniuk, J. W. & Jones, D. R. (1977). The oxygen transport system in trout (*Salmo gairdneri*) during sustained exercise. Journal of Experimental Biology, **69**, 247–60.

Kirschner, L. B. (1969). Ventral aortic pressure and sodium fluxes in perfused eel gills. *American Journal of Physiology*, **217**, 596–604.

Langendorff, O. (1899). Zur Kenntniss des blutlaufs in den kranzgefassen des herzens. *Pflugers Archiv*, **78**, 423–40.

Laurent, P. & Dunel, S. (1980). Morphology of gill epithelia in fish. *American Journal of Physiology*, **238**, 147–59.

Laurent, P. (1984). Gill internal morphology. In: *Fish Physiology*, vol. XA, ed. W. S. Hoar and D. J. Randall, pp. 73–183. New York: Academic Press.

Locke, F. S. & Rosenheim, O. (1904). The disappearance of dextrose when perfused through the mammalian heart. *Journal of Physiology* (London), **31**, 14–15.

Lucu, C. & Siebers, D. (1986). Amiloride-sensitive sodium flux and potentials in perfused *Carcinus* gill preparations. *Journal of Experimental Biology*, **122**, 25–35.

Macfarlane, M. A. A. & Maetz, J. (1975). Acute response to a salt load of the NaCl excretion mechanisms of the gill of *Platichthys flesus* in seawater. *Journal of Comparative Physiology*, **102**B, 101–13.

Maetz, J. (1971). Fish gills: mechanisms of salt transfer in fresh water and sea water. *Philosophical Transactions of the Royal Society of London Series B.*, **262**, 209–51.

McDonald, D. G. & Wood, C. M. (1981). Branchial and renal acid and ion fluxes in the rainbow trout, *Salmo gairdneri*, at low environmental pH. *Journal of Experimental Biology*, **93**, 101–18.

McDonald, D. G. (1983). The interaction of environmental calcium and low pH on the physiology of the rainbow trout, *Salmo gairdneri* I. Branchial and renal net ion and H$^+$ fluxes. *Journal of Experimental Biology*, **102**, 123–40.

Mantel, L. H. (1967). Asymmetry potentials, metabolism and sodium fluxes in gills of the blue crab, *Callinectes sapidus*. *Comparative Biochemistry & Physiology*, **20**, 743–53.

Metcalfe, J. D. & Butler, P. J. (1982). Differences between directly measured and calculated values for cardiac output in the dogfish: a criticism of the Fick method. *Journal of Experimental Biology*, **99**, 255–68.

Moen, K. & Klungsoyr, L. (1981). Metabolism of exogenous substrates in perfused hind parts of a rainbow trout (*Salmo gairdneri*). *Comparative Biochemistry & Physiology*, **68**B, 461–6.

Mommsen, T. P. (1984). Metabolism of the fish gill. In: *Fish Physiology*, vol. XB, eds. W. S. Hoar and D. J. Randall, pp. 203–38. New York: Academic Press.

Neely, J. R., Liebermeister, H., Battersby, E. J. & Morgan, H. E. (1967). Effect of pressure development on oxygen consumption by isolated rat heart. *American Journal of Physiology* 212, 804–814.

Nekvasil, N. P. & Olson, K. R. (1986). Extraction and metabolism of circulating catecholamines by the trout gill. *American Journal of Physiology* 250, 526–531.

Nilsson, S. (1984). *Autonomic nerve function in the vertebrates*. Zoophysiology vol. 13. 253 pp. Berlin: Springer-Verlag.

Nilsson, S. & Holmgren, S. (1985). D- and L-isoprenaline have different effects on adrenoceptors in the systemic vasculature of the cod, *Gadus morhua*. *Comparative Biochemistry & Physiology*, **80**C, 105–7.

Oduleye, S. O., Claiborne, J. B. & Evans, D. H. (1982). The isolated, perfused head of the toadfish, *Opsanus beta* I. Vasoactive responses to cholinergic and adrenergic stimulation. *Journal of Comparative Physiology*, **149**B, 107–13.

Olson, K. R. (1984). Distribution of flow and plasma skimming in isolated perfused gills of three teleosts. *Journal of Experimental Biology*, **109**, 97–108.

Olson, K. R., Kullman, D., Narkates, A. J. & Oparil, S. (1986). Angiotensin extraction by trout tissues *in vivo* and metabolism by the perfused gill. *American Journal of Physiology*, **250**, 532–8.

Opdyke, D. F., Holcombe, F. R. & Wilde, D. W. (1979). Blood flow resistance in *Squalus acanthias*. *Comparative Biochemistry & Physiology*, **62**A, 711–17.

Ostlund, E. & Fange, R. (1962). Vasodilation by adrenaline and noradrenaline, and the effects of some other substances on perfused fish gills. *Comparative Biochemistry & Physiology*, **5**, 307–9.

Part, P. & Svanberg, O. (1981). Uptake of cadmium in perfused rainbow trout (*Salmo gairdneri*) gills. *Canadian Journal of Fisheries & Aquatic Sciences*, **38**, 917–24.

Part, P.,Tuurala, H. & Soivio, A. (1982a). Oxygen transfer, gill resistance and structural changes in rainbow trout (*Salmo gairdneri*) gills perfused with vasoactive agents. *Comparative Biochemistry & Physiology*, **71**C, 7–13.

Part, P., Kiessling, A. & Ring, O. (1982b). Adrenalin increases vascular resistance in perfused rainbow trout (*Salmo gairdneri*) gills. *Comparative Biochemistry & Physiology*, **72**C, 107–8.

Part, P., Tuurala, H., Nikinmaa, M. & Kiessling, A. (1984). Evidence for a non-respiratory intralamellar shunt in perfused rainbow trout gills. *Comparative Biochemistry & Physiology*, **79**A, 29–34.

Patterson, S. W. & Starling, E. H. (1914). On the mechanical factors which determine the output of the ventricles. *Journal of Physiology (London)*, **48**, 465.

Payan, P. & Matty, A. J. (1975). The characteristics of ammonia excretion by an isolated perfused head of trout (*Salmo gairdneri*): effect of temperature and C_{O_2}-free Ringer. *Journal of Comparative Physiology*, **96**B, 167–84.

Payan, P. & Girard, J. P. (1977). Adrenergic receptors regulating patterns of blood flow through the gills of trout. *American Journal of Physiology*, **232**, 18–23.

Payan, P. (1978). A study of the Na^+/NH_4^+ exchange across the gill of the perfused head of the trout (*Salmo gairdneri*). *Journal of Comparative Physiology*, **124**B, 181–8.

Pequeux, A. & Gilles, R. (1978). Na^+/NH_4^+ co-transport in isolated perfused gills of the Chinese crab *Eriocheir sinensis* acclimated to fresh water. *Experientia*, **34**, 1593–4.

Pequeux, A. & Gilles, R. (1981). Na^+ fluxes across isolated perfused gills of the Chinese crab *Eriocheir sinensis*. *Journal of Experimental Biology*, **92**, 173–86.

Perry, S. F., Davie, P. S., Daxboeck, C. & Randall, D. J. (1982). A comparison of CO_2 excretion in a spontaneously ventilating blood-perfused trout preparation and saline-perfused gill preparations: contribution of the branchial epithelium and red blood cell. *Journal of Experimental Biology*, **101**, 47–60.

Perry, S. F., Davie, P. S., Daxboeck, C., Ellis, A. G. & Smith, D. G. (1984*a*). Perfusion methods for the study of gill physiology. In: *Fish Physiology*, vol. XB, ed. W. S. Hoar and D. J. Randall, pp. 326–88. New York: Academic Press.

Perry, S. F., Lauren, D. J. & Booth, C. E. (1984*b*). Absence of branchial edema in perfused heads of rainbow trout (*Salmo gairdneri*). *Journal of Experimental Zoology*, **231**, 441–5.

Perry, S. F., Payan, P. & Girard, J. P. (1984*c*). Adrenergic control of branchial chloride transport in the isolated perfused head of the freshwater trout (*Salmo gairdneri*). *Journal of Comparative Physiology*, **154**B, 269–74.

Perry, S. F., Payan, P. & Girard, J. P. (1984*d*). Effects of perfusate HCO_3^- and PCO_2 on chloride uptake in perfused gills of rainbow trout (*Salmo gairdneri*). *Canadian Journal of Fisheries & Aquatic Sciences*, **41**, 1768–73.

Perry, S. F. & Wood, C.M. (1985). Kinetics of branchial calcium uptake in the rainbow trout: effects of acclimation to various external calcium levels. *Journal of Experimental Biology*, **116**, 411–33.

Perry, S. F., Booth, C. E. & McDonald, D. G. (1985*a*). Isolated perfused head of rainbow trout I. Gas transfer, acid-base balance and hemodynamics. *American Journal of Physiology*, **249**, 246–54.

Perry, S. F., Booth, C. E. & McDonald, D. G. (1985*b*). Isolated perfused head of rainbow trout II. Ionic fluxes. *American Journal of Physiology*, **249**, 255–61.

Perry, S. F., Daxboeck, C. & Dobson, G. P. (1985*c*). The effect of perfusion flow rate and adrenergic stimulation on oxygen transfer in the isolated, saline-perfused head of rainbow trout (*Salmo gairdneri*). *Journal of Experimental Biology*, **116**, 251–69.

Perry, S. F. & Daxboeck, C. (1986). The effects of saline gas composition and pH on haemodynamic stability in the perfused trout head preparation. *Canadian Journal of Zoology*, **64**, 274–7.

Perry, S. F., Malone, S. & Ewing, D. (1987). Hypercapnic acidosis in the rainbow trout (*Salmo gairdneri*) I. Branchial ionic fluxes and blood acid-base status. *Canadian Journal of Zoology*, **65**, 888–95.

Perry, S. F. & Vermette, M.G. (1987). The effects of prolonged epinephrine infusion on the physiology of the rainbow trout, *Salmo gairdneri* I. Blood respiratory, acid-base and ionic states. *Journal of Experimental Biology*, **128**, 235–53.

Pettersson, K. & Nilsson, S. (1979). Nervous control of the branchial vascular resistance of the Atlantic cod, *Gadus morhua*. *Journal of Comparative Physiology*, **129**B, 179–83.

Pettersson, K. & Johansen, K. (1982). Hypoxic vasoconstriction and the effects of adrenaline on gas exchange efficiency in fish gills. *Journal of Experimental Biology*, **97**, 263–72.

Pettersson, K. (1983). Adrenergic control of oxygen transfer in perfused gills of the cod, *Gadus morhua*. *Journal of Experimental Biology*, **102**, 327–35.

Potts, W. T. W. & Eddy, F. B. (1973). Gill potentials and sodium fluxes in the flounder *Platichthys flesus*. *Journal of Comparative Physiology*, **87B**, 29–48.

Rankin, J. C. & Maetz, J. (1971). A perfused teleostean gill preparation: vascular actions of neurohypophysial hormones and catecholamines. *Journal of Endocrinology*, **51**, 621–35.

Richards, B. D. & Fromm, P. O. (1969). Patterns of blood flow through filaments and lamellae of isolated-perfused rainbow trout (*Salmo gairdneri*) gills. *Comparative Biochemistry & Physiology*, **29**, 1063–70.

Ruben, J. A. & Bennett, A. F. (1981). Intense exercise, bone structure and blood calcium levels in vertebrates. *Nature*, **291**, 411–13.

Satchell, G. H. (1971). *Circulation in fishes*. Cambridge: Cambridge University Press.

Saunders, R. L. & Sutterlin, A. M. (1971). Cardiac and respiratory responses to hypoxia in the sea raven, *Hemitripterus americanus*, and an investigation of possible control mechanisms. *Journal of the Fisheries Research Board of Canada* 28, 491–503.

Segel, L. D. & Rendig, S. V. (1982). Isolated working rat heart perfusion with perfluorinated emulsion Fluosol–43. *American Journal of Physiology*, **242**, 485–9.

Shuttleworth, T. J. (1972). A new isolated perfused gill preparation for the study of the mechanisms of ionic regulation in teleosts. *Comparative Biochemistry & Physiology*, **43A**, 59–64.

Shuttleworth, T. J. & Freeman, R. F. H. (1973). The role of the gills in seawater adaptation in *Anguilla dieffenbachii* II. Net ion fluxes in isolated perfused gills. *Journal of Comparative Physiology*, **86B**, 315–21.

Shuttleworth, T. J. (1978). The effect of adrenaline on potentials in the isolated gills of the flounder (*Platichthys flesus* L.). *Journal of Comparative Physiology*, **124B**, 129–36.

Smith, D. G. (1977). Sites of cholinergic vasoconstriction in trout gills. *American Journal of Physiology*, **233**, 222–9.

Smith, F. M. & Jones, D. R. (1982). The effect of changes in blood oxygen-carrying capacity on ventilation volume in the rainbow trout (*Salmo gairdneri*). *Journal of Experimental Biology*, **97**, 325–34.

Smith, P. J. S. & Hill, R. B. (1987). Modulation of output from an isolated gastropod heart: effects of acetylcholine and FMRFamide. *Journal of Experimental Biology*, **127**, 105–20.

Stagg, R. M., Rankin, J. C. & Bolis, L. (1981). Effect of detergent on vascular responses to noradrenaline in isolated perfused gills of the eel, *Anguilla anguilla* L. *Environmental Pollution (Series A)*, **24**, 31–7.

Stagg, R. M. & Shuttleworth, T. J. (1982). The effects of copper on ionic regulation by the gills of the seawater-adapted flounder (*Platichthys flesus* L.). *Journal of Comparative Physiology*, **149B**, 83–90.

Stagg, R. M. & Shuttleworth, T. J. (1984). Hemodynamics and potentials in isolated flounder gills: effects of catecholamines. *American Journal of Physiology*, **246**, 211–20.

Stagg, R. M. & Shuttleworth, T. J. (1986). Surfactant effects on adrenergic responses in the gills of the flounder (*Platichthys flesus* L.). *Journal of Comparative Physiology*, **156**B, 727–33.

Stevens, E. D. & Randall, D. J. (1967*a*). Changes in blood pressure, heart rate, and breathing rate during moderate swimming activity in rainbow trout. *Journal of Experimental Biology*, **46**, 307–15.

Stevens, E. D. & Randall, D. J. (1967*b*). Changes of gas concentrations in blood and water during moderate swimming activity in rainbow trout. *Journal of Experimental Biology*, **46**, 329–37.

Stevens, E. D., Bennion, G. R., Randall, D. J. & Shelton, G. (1972). Factors affecting arterial pressures and blood flow from the heart in intact, unrestrained lingcod, *Ophiodon elongatus*. *Comparative Biochemistry & Physiology*, **43**A, 681–95.

Straub, W. (1904). Fortegesetze studien am aplysienherzen (dynamik, kreislauf und desen innervation) nebst bermerkungen zur vergleichenden muskelphysiologie. *Pflugers Archiv*, **103**, 429–99.

Thomas, S. & Le Ruz, H. (1982). A continuous study of rapid changes in blood acid-base status of trout during variations of water PCO_2. *Journal of Comparative Physiology*, **148**B, 123–30.

Thomas, S. (1983). Changes in blood acid-base balance in trout (*Salmo gairdneri* Richardson) following exposure to combined hypoxia and hypercapnia. *Journal of Comparative Physiology*, **152**, 53–7.

Thomas, S., Fievet, B. & Motais, R. (1986). Effect of deep hypoxia on acid-base balance in trout: role of ion transfer processes. *American Journal of Physiology*, **250**, 319–27.

Turner, J. D. & Wood, C. M. (1983). Factors affecting lactate and proton efflux from pre-exercised, isolated perfused rainbow trout trunks. *Journal of Experimental Biology*, **105**, 395–402.

Tuurala, H., Part, P., Nikinmaa, M. & Soivio, A. (1984). The basal channels of secondary lamellae in *Salmo gairdneri* gills – a non-respiratory shunt. *Comparative Biochemistry & Physiology*, **79**A, 35–9.

Van der Putte, I. & Part, P. (1982). Oxygen and chromium transfer in perfused gills of rainbow trout (*Salmo gairdneri*) exposed to hexavalent chromium at two different pH levels. *Aquatic Toxicology*, **2**, 31–45.

Vermette, M. G. & Perry, S. F. (1987). The effects of prolonged epinephrine infusion on the physiology of the rainbow trout, *Salmo gairdneri*. II. Branchial solute fluxes. *Journal of Experimental Biology*, **128**, 255–67.

Wahlqvist, I. & Nilsson, S. (1980). Adrenergic control of the cardiovascular system of the Atlantic cod, *Gadus morhua*, during 'stress'. *Journal of Comparative Physiology*, **137**B, 145–50.

Wahlqvist, I. (1980). Effects of catecholamines on isolated systemic and branchial vascular beds of the cod, *Gadus morhua*. *Journal of Comparative Physiology*, **137**B, 139–43.

Wahlqvist, I. & Nilsson, S. (1981). Sympathetic nervous control of the vasculature in the tail of the Atlantic cod, *Gadus morhua. Journal of Comparative Physiology*, **144**, 153–6.

Wood, C. M. (1974). A critical examination of the physical and adrenergic factors affecting blood flow through the gills of the rainbow trout. *Journal of Experimental Biology*, **60**, 241–65.

Wood, C. M. (1975). A pharmacological analysis of the adrenergic and cholinergic mechanisms regulating branchial vascular resistance in the rainbow trout (*Salmo gairdneri*). *Canadian Journal of Zoology*, **53**, 1569–77.

Wood, C. M. & Shelton, G. (1975). Physical and adrenergic factors affecting systemic vascular resistance in the rainbow trout: a comparison with branchial vascular resistance. *Journal of Experimental Biology*, **63**, 505–23.

Wood, C. M. (1976). Pharmacological properties of the adrenergic receptors regulating systemic vascular resistance in the rainbow trout. *Journal of Comparative Physiology*, **107**, 211–28.

Wood, C. M. (1977). Cholinergic mechanisms and the response to ATP in the systemic vasculature of the rainbow trout. *Journal of Comparative Physiology*, **122B**, 325–47.

Wood, C. M., McMahon, B. R. & McDonald, D. G. (1978). Oxygen exchange and vascular resistance in the totally perfused rainbow trout. *American Journal of Physiology*, **234**, 201–8.

Wood, C. M. & Shelton, G. (1980). Cardiovascular dynamics and adrenergic responses of the rainbow trout *in vivo. Journal of Experimental Biology*, **87**, 247–70.

Wood, C. M., McDonald, D. G. & McMahon, B. R. (1982). The influence of experimental anaemia on blood acid-base regulation *in vivo* and *in vitro* in the starry flounder (Platichthys stellatus) and the rainbow trout (*Salmo gairdneri*). *Journal of Experimental Biology*, **96**, 221–37.

Wood, C. M., Wheatly, M. G. & Hobe, H. (1984). The mechanisms of acid-base and ionoregulation in the freshwater rainbow trout during environmental hyperoxia and subsequent normoxia. III. Branchial exchanges. *Respiration Physiology*, **55**, 175–92.

Wright, P. A. & Wood, C. M. (1985). An analysis of branchial ammonia excretion in the freshwater rainbow trout: effects of environmental pH change and sodium uptake blockade. *Journal of Experimental Biology*, **114**, 329–53.

Wright, P. A., Heming, T. A. & Randall, D. J. (1986). Downstream pH changes in water flowing over the gills of rainbow trout. *Journal of Experimental Biology*, **126**, 499–512.

Molecular systems – *in vivo* and *in vitro*

G. GÄDE and M. K. GRIESHABER

Measurements of Anaerobic Metabolites

Outline of anaerobic metabolisim

Our recent advanced understanding of anaerobiosis is at least in part a consequence of improved physiological and biochemical methodology. Measurements of rates of oxygen consumption at low P_{O_2} in fully automated systems for both closed and flow-through respirometry (Forstner, 1983: Kaufmann, Forstner & Wieser, this volume), microcalorimetric studies of metabolic rates (Gnaiger, 1983; Shick, Gnaiger, Widdows, Bayne & de Zwaan, 1986; Gnaiger, Widdows & Shick, this volume) and the determination of steady state concentrations of various metabolites of the intermediary metabolism have resulted in a comprehensive understanding of the physiological and metabolic adaptations of animals to hypoxic conditions (Grieshaber, 1982; Gäde, 1983).

Accurate studies demonstrate that the ventilation of various animal species is increased by environmental hypoxia or decreased by hyperoxia. Some species are able to regulate their oxygen consumption and thus are 'oxygen independent' at hyperoxic or moderately hypoxic P_{O_2}-values. Below a certain P_{O_2} most animals adjust their oxygen consumption according to the ambient P_{O_2} and become 'oxygen conformers'. At this critical P_{O_2} oxygen consumption decreases abruptly and metabolism switches from an aerobic to an anaerobic mode (Dejours, 1975; Pörtner, Heisler & Grieshaber, 1985).

In many invertebrates the early phase of anaerobiosis is characterized by the utilization of aspartate, the transphosphorylation of a species-specific phosphagen and the degradation of glycogen via the Embden-Meyerhof-Parnas pathway (Kreutzer, Siegmund & Grieshaber, 1985).

At the very onset of anaerobiosis the tissue content of aspartate decreases. As demonstrated in marine invertebrates with the ^{14}C-labelled compounds, aspartate is metabolized to oxaloacetate which is reduced to malate by the cytosolic malate dehydrogenase. Malate itself is transported into the mitochondria where it undergoes dis-proportionation into fumarate and oxaloacetate. Fumarate acts as the mitochondrial electron acceptor, thereby being reduced to succinate with the concomitant gain of 1 mol ATP per mol succinate formed (Schroff & Schöttler, 1977).

Succinate is a temporary end product of anaerobiosis, since its concentration is very low in normoxic animals, but increases about ten- to twenty-fold below the critical P_{O_2}. The determination of this carbonic acid can provide an excellent probe for the onset of anaerobiosis. Aspartate is not as useful an indicator because the tissue contents of this amino acid vary considerably in different specimens of the same species and, thus, it is difficult to establish a significant decrease in this metabolite at the onset of anaerobiosis.

The second ATP-generating substrate during early anaerobiosis is a phosphagen (i.e. creatine phosphate, arginine phosphate or taurocyamine phosphate; Van Thoai & Robin, 1969). It is transphosphorylated to ATP in a single step reaction catalyzed by the corresponding phosphagen kinase. This reaction prevents a pronounced drop of the adenylate energy charge which is, therefore, not a good indicator of early anaerobiosis despite the ease of estimation.

The contribution of phosphagens and in particular of aspartate to anaerobic ATP-provision is limited due to low tissue contents. This also holds true for malate which serves as an anaerobic substrate in the larvae of the midge *Chaoborus crystallinus* (Englisch, Opalka & Zebe, 1982), the leech *Hirudo medicinalis* (Zebe, Salge, Wiemann & Wilps, 1981) and in *Lumbriculus variegatus* (Putzer, 1984). In contrast to these compounds, glycogen is much more abundant in many marine and fresh water invertebrates and, therefore, serves as the main substrate which can hardly be completely consumed during prolonged hypoxia (de Zwaan & Putzer, 1985).

During early anaerobiosis glycogen is mainly metabolized to pyruvate, thereby delivering 3 mol ATP per mol glycosyl unit. The redox ratio of the cytosol is balanced by the reduction of pyruvate to lactate, which is the main glycolytic end product in vertebrates, insects and crustaceans. Lactate and various opines (alanopine, octopine, strombine and tauropine) are mainly found in molluscs and annelids (Gäde & Grieshaber, 1986; Grieshaber & Kreutzer, 1986; Gäde, 1987a). In addition, some of the pyruvate is transaminated with glutamate to alanine (Felbeck, 1980). These compounds accumulate in the cytosol and are, therefore, together with succinate, characteristic end products of the early phase of anaerobiosis (Gäde & Grieshaber, 1986).

If the energy provision of the different pathways is compared in many marine invertebrates, it becomes obvious that in some species the Embden-Meyerhof-Parnas pathway provides approximately half the total amount of

ATP, while 20 to 30% is derived from the transphosphorylation of a phosphagen. The remaining part results from the metabolization of aspartate to succinate (Zandee, Holwerda & de Zwaan, 1980; Pörtner, Kreutzer, Siegmund, Heisler & Grieshaber, 1984; Kreutzer, Siegmund & Grieshaber, 1985).

If hypoxic or anoxic conditions prevail for more than 2 or 3 hours, within the habitat, all facultative anaerobes solely exploit glycogen to produce energy. During progressive hypoxia several species continue to use anaerobic glycolysis for ATP production (Gäde & Meinardus-Hager, 1986), but in many species, in particular in the good anaerobes, the major route of glycogen breakdown does not lead to pyruvate, instead it deviates at the phosphoenolpyruvate branchpoint to oxaloacetate (de Zwaan & Dando, 1984).

The control of the ratio between the carboxylation of phosphoenolpyruvate to oxaloacetate and the transphosphorylation to pyruvate and ATP is excerted via the regulation of the activity of pyruvate kinase. Holwerda, Kruitwagen & de Bont (1981) as well as Holwerda, Veenhof, van Heugten & Zandee (1983) have demonstrated the existence of two interconvertible forms of pyruvate kinase in the adductor muscle of *Mytilus edulis*. They found that with progressive hypoxia, one variant of the pyruvate kinase became predominant. It is characterized by a lower enzymatic activity, decreased affinity for phosphoenolpyruvate and an increased sensitivity to inhibition by L-alanine and protons. In *Busycotypus canaliculatum* the changes of the catalytic properties of pyruvate kinase could be assigned to a phosphorylation of a threonine residue in the protein. This may lead to conformational changes resulting in kinetic differences between normoxic and hypoxic forms of pyruvate kinase. As a consequence phosphoenolpyruvate may be metabolized preferentially to oxaloacetate (Plaxton & Storey, 1984).

Oxaloacetate is reduced to malate, which, after entering the mitochondria, gives rise to succinate. During prolonged anaerobiosis, succinate is activated to succinyl-CoA which is metabolized to propionate via the 'succinate-propionate' cycle. The specific reactions of this pathway, that had been demonstrated before in some endoparasites (Pietrzak & Saz, 1981), have been shown to operate in *Mytilus edulis* (Schultz & Kluytmans, 1983) as well as in *Arenicola marina* (Schöttler, 1986). Propionate is the definitive end product of prolonged anaerobiosis (Gäde, Wilps, Kluytmans & de Zwaan, 1975; Kluytmans, Veenhof & de Zwaan, 1975). It is released by the animal and must, therefore, be determined within the tissue as well as in the incubation medium.

Preparation of samples

Collection and quick-freezing of samples

In order to quantify the concentrations of various metabolites in biological material (tissue or haemolymph of invertebrates) by enzymatic analysis or other methods, one has to establish the following pre-requisites: the sample has to be representative (e.g. homogenous tissue of different origins or, pre- or post-branchial haemolymph) and, secondly, prior to analysis no decomposition of the metabolites to be measured should occur.

In practice the most satisfactory method for preventing changes in the composition of the tissue metabolites is quickly to expose the tissue under investigation in the living animal and subsequently freeze the tissue immediately between aluminum tongs pre-cooled with liquid nitrogen ('freeze-clamping'). Tight squeezing of the tongs produces a flat compressed layer of tissue that is thoroughly frozen in a few seconds. Especially when labile metabolites like high-energy phosphates are to be analysed, this method is superior to dissection of a piece of tissue and freezing of this whole piece in liquid nitrogen. Due to the thickness of the specimen it will take several minutes until the inner parts are frozen. The tissue sample will also be insulated by the gas envelope of the evaporating freezing agent and therefore will be prevented from completely freezing. In the deep-frozen state tissue samples can be stored, preferable at $-80\,^{\circ}\mathrm{C}$, without significant changes in metabolite concentrations. However, care should be taken to prevent dessiccation or hygroscopic activity by storage in tight stoppered glass ware or by storage in liquid nitrogen.

Disintegration and homogenization

As stated above when low molecular weight metabolites are to be analysed enzymatically, the fixation of the cellular state of these metabolites is of paramount interest. 'Quick-freezing' does not lead in most cases to a destruction of the physiological state, however, one has to bear in mind that during disintegration of a tissue the compartmental localization (and likely concentration gradients) in a given tissue are disturbed; as a result of disintegration and extraction, bound and free metabolite levels cannot be distinguished anymore. The possible instability of metabolites during disintegration has also to be considered. High-energy phosphates such as ATP, ADP, phosphagens and the reduced coenzymes nicotinamide adenine dinucleotide (NADH) and nicotinamide adenine dinucleotide phosphate (NADPH) are acid-labile, whereas triosephosphates and the oxidized coenzymes (NAD, NADP) are affected by alkaline pH; special care must be taken for the quantitative evaluation of such metabolites (Bergmeyer, 1983*a*).

In general, the following method is routinely used to prepare extracts for spectrophotometric assays: the frozen tissue is powdered with a porcelain pestle in a pre-cooled (with liquid nitrogen) porcelain mortar by frequently adding liquid nitrogen to prevent any thawing of the tissue or tissue powder. Subsequently, an acidic deproteinization in perchloric acid is used. An aliquot of the frozen tissue powder is added to a pre-weighed quantity of ice-cold perchloric acid (3 mol·L^{-1}), mixed, re-weighed and homogenized for about 20 s with an Ultra Turrax type homogenizer for extraction (smaller samples are best homogenized by sonication). After centrifugation for 20 min at 30 000 g, the supernatant is saved, the pellet reextracted (if necessary) with perchloric acid, centrifuged again and the supernatants combined. The excess of perchloric acid is removed by neutralization with KHCO$_3$ (2 mol·L^{-1}) and the resulting precipitate of KCLO$_4$ discarded after centrifugation. This method can also be scaled down using the procedure of Pette & Reichmann (1982).

Storage and stability of extracts

In general, most of the metabolites to be analysed are stable when the neutralised perchloric acid extract is stored frozen at temperatures below -20 °C. However, hydrolysis of the phosphate bond may occur and, therefore, we routinely determine the concentrations of ATP, ADP, AMP and the phosphagens (L-arginine phosphate, taurocyamine phosphate etc.) in the neutralized perchloric acid extract immediately after its preparation. The remaining extract is frozen and, after collecting numerous extracts, the desired metabolites are quantified by enzymatic analysis.

Enzymatic analysis of metabolites

The optical assay

With the help of optical methods the structure, reaction and quantity of compounds have been detected and assayed for many years. Determination of extremely small amounts of material dissolved in a liquid can be measured usually without degradation of the compound studied.

The well-known, so-called 'optical assay' of Warburg & Christian (1936) makes use of wavelengths in the ultraviolet region (200–400 nm). The radiation is passed through a cuvette containing a liquid with solutes where it interacts with those in such a way that the molecules in the liquid become excited (thus energy/light is absorbed) and the radiation leaves the cuvette attenuated, and the intensity is reduced at the detector unit. By comparing the quantities of the irradiating and received intensities, the concentration can be calculated.

The basic law of light absorbtion states that the radiation passing through a uniformly absorbing medium, the transmittance T, is proportional to the ratio of the intensity of the transmitted (I) and the incident radiation (I_0):

$$T = I/I_0.$$

This ratio is a measure of the absorbance and, hence, the concentration of the solute. Because the intensity of the transmitted radiation changes exponentially as the concentration of the solute increases linearily, it is much more convenient to use the absorbance A (also called extinction E or optical density OD) as the measured parameter which is defined as:

$$A = \log 1/T = \log I_0/I.$$

In contrast to the transmittance, which is expressed in the range from 0 to 100%, the absorbance has no dimensional units and varies from 0 to infinity. Furthermore, the Lambert-Beer law states that the absorbance is proportional to the concentration C_c of the absorbing compound in $mol \cdot L^{-1}$ and to the thickness of the layer d in mm (for example the length of the light path in the absorbing material) multiplied by a constant ε, the so called molar decadic absorption coefficient given in $L \cdot mol^{-1} \cdot mm^{-1}$:

$$A = \varepsilon \cdot C_c \cdot d.$$

A great number of metabolite determinations either involve or can be coupled to the oxidation of NAD(P)H or the respective reduction of NAD(P), and therefore these coenzymes are most often used as indicator substances for absorption photometry. NADH, for example, has an absorption maximum at 339 nm, but measurements at 340 nm or, when a gas-discharge lamp containing mercury vapour is used, at Hg 334 or even Hg 365 nm, give reliable values. However, the absorption coefficient of NADH is not only dependent on the applied wavelength, but on pH, ionic strength and temperature. For correct values see Haar, Netheler & Ziegenhorn (1983).

With 'ordinary' equipment, accurate measurements can be made with NADH concentrations of about 5 nmoles per cuvette (10 mm light path, 1 ml volume). An up to 100-fold increase in sensitivity is achieved when fluorimetric methods are used.

Enzymatic analysis by spectrofluorimetry

The basic principles of fluorescence are similar to those of photometry. Whereas the amount of light absorbed by a molecule is measured in photometry, in fluorescence the fate of the absorbed energy is observed. The molecule, after absorbing radiation, emits a small fraction of the absorbed light as fluorescence that is directly proportional to the intensity of the incident light. Thus, at low concentrations, the intensity of fluorescence

(F) is related to the intensity of the incident radiation (I_0) by the following simple relationship:

$$F = 2.303 \cdot \varepsilon \cdot I_0 \cdot \varphi \cdot C_c \cdot d,$$

where φ is the quantum yield or quantum efficiency (thus the number of quanta fluoresced divided by the number of quanta absorbed or the ratio of the number of emitted photons to the total number of absorbed photons).

A major problem with fluorimetry is 'quenching'. Any compound that absorbs the excitation light will reduce the intensity of the excitation light as it passes through the solution and will thereby diminish the emitted light, thus 'quenching' the fluorescence. This is one of the reasons why the upper limit of determining NADH should be in the range of 10 nmol in a 1 ml cuvette and as a consequence the assay procedure should be repeatedly standardized using authentic compounds (Urbanke, 1983).

Variations of the optical assay

In this chapter we will consider only the most widely used enzymatic reactions involving NAD(P)(H)-dependent dehydrogenases. As an example, the determination of the concentration of pyruvate, which is completely reduced to L-lactate (or D-lactate) is considered. The enzymatic reaction is catalysed by the enzyme L-lactate or D-lactate dehydrogenase (LDH) in the presence of the coenzyme NADH:

$$\text{pyruvate} + \text{NADH} + \text{H}^+ \xrightarrow{\text{LDH}} \text{D,L-lactate} + \text{NAD}^+$$

The enzymatic oxidation of NADH to NAD^+ with the concomitant reduction of pyruvate to lactate results in a change of the optical behaviour of NADH. Whereas the reduced coenzyme absorbs light at 339 and 260 nm, the oxidized form absorbs light only at 260 nm. In a spectrophotometer cuvette we thus follow directly the decrease of absorption near 339 nm, which is a direct measure of the amount of pyruvate.

In detail, the cuvette is filled with a suitable buffer, NADH and the extract, the pyruvate concentration of which should be measured. After some minutes the reaction is started with lactate dehydrogenase and the absorbance is monitored until no further change occurs. This procedure is termed the end-point method (Bergmeyer, 1983b).

In setting-up optimal conditions for such assays, the following prerequisites should be checked: optimal buffer system, a relatively low substrate (pyruvate) concentration (between 5 and 100 nmoles per 1 ml cuvette), a relatively high coenzyme (NADH) concentration (0.1–0.3 nM) compared to the Michaelis constant of the enzyme and a rather large amount (2–10 international units) of the enzyme (LDH). Under these conditions the reaction proceeds rapidly and the total conversion is completed in a relatively short period ($\pm$ 15 min). In general, this is achieved when the equilibrium of the reaction is in favour of the product (lactate).

However, how can the amount of a substance be determined by the end-point method when the reaction equilibrium is unfavourable for the synthesis of this compound? For this problem we can consider the same example, but in this case we want to determine the quantity of lactate:

$$\text{D-, L-lactate} + \text{NAD}^+ \xrightarrow{\text{LDH}} \text{pyruvate} + \text{NADH} + \text{H}^+$$

The equilibrium of the lactate dehydrogenase (LDH) reaction favours the production of lactate. Thus, we have to shift the equilibrium sufficiently in the direction of the product (pyruvate). This can be achieved by an increase in the substrate (lactate) and coenzyme (NAD^+) concentrations and, in addition, the use of an alkaline buffer. The most elegant way, however, is the use of a so-called 'trapping enzyme reaction'.

In our example the pyruvate produced is taken out of the equilibrium by the transamination of pyruvate to alanine catalyzed by aminotransferase (AlaT):

$$\text{pyruvate} + \text{L-glutamate} \xrightarrow{\text{AlaT}} \text{L-alanine} + \text{2-oxoglutarate.}$$

Thus, such a trapping reaction has an equilibrium in favour of the reaction product (pyruvate): the pyruvate produced in the primary reaction (by LDH) is immediately transaminated in the second (trapping) reaction and, despite the unfavourable equilibrium for lactate, this substrate can then be converted completely.

In some cases, the combination of a primary together with a trapping reaction has a second desirable effect. This is the case when the trapping reaction is also catalyzed by a NAD(P)(H)-dependent enzyme and thus the signal is doubled, as for example, in the two consecutive reactions catalyzed by ethanol dehydrogenase (EthDH) and acetaldehyde dehydrogenase (ADH):

$$\text{ethanol} + \text{NAD}^+ \xrightarrow{\text{EthDH}} \text{acetaldehyde} + \text{NADH} + \text{H}^+$$

$$\text{acetaldehyde} + \text{NAD}^+ \xrightarrow{\text{ADH}} \text{acetic acid} + \text{NADH} + \text{H}^+.$$

Other examples for end-point determination are those cases where the substance to be quantified has first to be converted in the so-called 'auxiliary reaction' to a compound that can be measured in the 'indicator reaction'. Therefore this method is also called the 'coupled reaction method'. A well-known example is the enzymatic determination of glucose. Its conversion by hexokinase to glucose-6-phosphate (auxiliary reaction) does not involve an NAD(P)(H)-dependent dehydrogenase. If, however, this reaction is coupled to the 'indicator reaction' catalysed by glucose-6-

phosphate dehydrogenase (GDPH), glucose can be quantitatively determined, as hexokinase (HK) quantitatively converts glucose into glucose-6-phosphate, which is subsequently catalysed quantitatively to gluconate-6-phosphate:

$$\text{D-glucose} + \text{ATP} \xrightarrow{\text{HK}} \text{ADP} + \text{G-6-P}$$

$$\text{G-6-P} + \text{NADP}^+ \xrightarrow{\text{GPDH}} \text{gluconate-6-P} + \text{NADPH} + \text{H}^+$$

In fact, in some cases several enzyme reactions can be coupled by the successive addition of enzymes into the same cuvette and thus several substrates can be determined sequentially in the same sample. A useful example is given in the literature (Bergmeyer, 1983*b*), where a mixture of sugars and sugar phosphates (G-6-P, G-1-P, glucose, fructose, F-6-P, DAP, GAP, F-1, 6-P_2, F-1-P) is determined.

Practical advice

In this section two mistakes that are often made and which lead to incorrect measurements of the quantity of a metabolite are discussed.

(i) After the enzyme reaction is initiated and although the reaction has already proceeded for a long time, no constant end-point is reached, i.e. the absorbance is still 'creeping'. This happens, when, for example, contaminating enzymes may be present in the enzyme solution used, which catalyze a slow reaction of other substances in the sample. In most cases the change of absorbance is linear with time and it can thus easily be eliminated by graphical extrapolation. Therefore, it is of the utmost importance to monitor the reaction with a pen-recorder.

(ii) In some end-point determinations addition of the enzyme suspension (mostly a solution in concentrated ammonium sulphate) results in an absorbance increase that would also lead to an overestimation. To counteract this, the same amount of enzyme used to initiate the reaction is added again at the end of the measurement when no (or a constant, see above) change of absorbance occurs and the resulting increase in absorbance is deducted.

Rare metabolites

Some metabolites are not routinely determined, especially those that are not involved in biomedical research. Therefore, the appropriate enzymes or authentic standards which are not always available commercially must be purified or synthesized by the investigator himself. This is, for

Table 1. *Commonly used methods for the determination of metabolites related to anaerobiosis*

Metabolites	Mode of analysis	References
Adenylates		
ATP, ADP, AMP	HPLC	Cresentini & Stocchi (1984)
ATP	SP	Trautschold *et al.* (1985)
ADP, AMP	SP	Jaworek & Welsch (1985)
Phosphagens		
phospho-L-arginine	SP	Grieshaber *et al.* (1978); Gäde (1985a)
phospholombricine	SP	Hoffmann (1981)
taurocyaminephosphate	SP	Pörtner *et al.* (1979)
Amino Acids		
L-alanine	SP	Williamson (1985)
D-alanine	SP	Graßl & Supp (1985)
L-arginine	SP	Grieshaber *et al.* (1978); Gäde (1985a)
L-aspartate	SP	Möllering (1985)
taurine	SP	Gäde (1987b)
Opines		
alanopine & strombine	HPLC	Siegmund & Grieshaber (1983); Fiore *et al.* (1984)
	GC & SP	Storey *et al.* (1982)
octopine	SP	Grieshaber *et al.* (1978); Gäde & Head (1979); Gäde (1985b)
tauropine	SP	Gäde (1988)
Carbohydrates		
D-glucose	SP	Kunst *et al.* (1984)
glycogen	SP	Keppler & Decker (1984)
Organic Acids		
acetate	SP	Bergmeyer & Möllering (1984)
	HPLC	Pörtner *et al.* (1984)
D-(−)-lactate	SP	Gawehn (1984)
L-(+)-lactate	SP	Noll (1984)
propionate	HPLC	Pörtner *et al.* (1984)
succinate	SP	Beutler (1985)
Enzyme Purification and Opine Synthesis		
octopine dehydrogenase		Grieshaber *et al.* (1978); Gäde & Head (1979)
alanopine/strombine dehydrogenase		Siegmund & Grieshaber (1983)
tauropine dehydrogenase		Gäde (1987b)
opines		Siegmund & Grieshaber (1983)

Abbreviations: SP = spectrophotometer; GC = gas chromatography;
HPLC = high pressure liquid chromatography

example, the case with the enzymes octopine and tauropine dehydrogenase and the opines. Table 1 gives references where to find methods of preparation for these enzymes and compounds.

Estimation of metabolites by chromatographic methods

Although most of the anaerobic metabolites can be quantified by means of an enzymatic assay, it is sometimes also feasible to use other methods. This holds especially true for the determination of opines (except for octopine), of glycine and propionate. No enzymatic assays are available for the estimation for glycine nor for propionate, and those published for the determination of opines are unspecific and time consuming. In particular, if the metabolic fate of a certain substrate is to be followed by isotope labelling, a separation of metabolites prior to quantification is essential.

There is no question that the method of choice for these purposes is high performance liquid chromatography (HPLC) combined with various methods for detection. Its application is almost unlimited, and the factors of analysis like separation efficiency, speed of analysis and sensitivity of detection, have been improved so enormously during the last decade that it can easily compete with the other chromatographic procedures previously used for the determination of low molecular compounds. As it is not within the scope of this article to give the theoretical and instrumental principles of HPLC, but to describe only a few applications in anaerobiosis research, we refer the reader to the text books by Snyder & Kirkland (1974) and Hamilton & Sewell (1982).

Determination of alanopine and strombine

The estimation of alanopine and strombine can be achieved directly from perchloric acid extracts, without sample preparation, on a cation exchange resin (Nr. 5311006/H from Biotronik, D-6457 Maintal, FRG). In this particular case no customer ready column is available, therefore, some details of resin preparation and column packing are given below (Siegmund & Grieshaber, 1983).

The resin is suspended in distilled water for 12 h and the supernatant decanted. The remaining resin is filtered using a Büchner funnel, resuspended in 4 M HCl and carefully stirred for 45 minutes at 90 °C. After cooling the resin is thoroughly rinsed with distilled water, resuspended in 0.5 M HCl and packed into a metal column (4 × 250 mm) using a metal reservoir that has been tightly screwed on top of the column and connected to a pressure pump (setup available from Merck-Hitachi, D-6100 Darmstadt, FRG). The reservoir must have approximately a three-fold higher volume than the column itself to maintain packing as evenly as possible. Throughout the packing procedure, which requires about 60 minutes, 0.5 M HCl is used as solvent and the pressure must be kept constant at 3000 kPa. After finishing the procedure the reservoir is removed, the column carefully closed and equilibrated with 2.5×10^{-5} M H_2SO_4 at a flow rate of 1.0 ml·min^{-1} for about 10 h. Then the column, which shows a back pressure of about 4500 kPa can be used for separation. Samples of 20 to 100 µl can be applied and separated isocratically using 2.5×10^{-5} M H_2SO_4 at a flow rate of 1.0 ml·min^{-1}.

Alanopine and strombine must be detected using a conductivity cell (BT 0330, Biotronik, D-6457 Maintal, FRG) because both compounds show only a very weak ninhydrin reaction and also a low UV absorption.

The calibration of this procedure like that of any other chromatographic method, requires authentic compounds as standards. These are synthesized enzymatically using the respective opine dehydrogenase. It is of utmost importance that the concentration of the sample used as a standard is determined as accurately as possible. The best procedure for standardization is again the enzymatic assay, even if the determination requires some time due to the low specific activity of the opine dehydrogenase in the reverse reaction. Comparing the retention time of authentic standards with those of the biological sample reveals the respective opine.

As further proof that a peak shown on the chromatogram really corresponds to a certain metabolite only, the compound under investigation may be removed by chemical or enzymatic degradation in order to exclude the presence of a co-separating metabolite. In the case of opines this can best be achieved by incubating the biological sample with near homogeneous opine dehydrogenase.

As opines are determined from neutralized, but otherwise untreated perchloric acid extracts, the column must be washed with distilled water, followed by sufficient equilibration with solvent to remove bound material after prolonged usage.

Another valuable method was published by Fiore, Nicchitta & Ellington (1984) where various opines were separated by HPLC and detected fluorimetrically after post-column derivatization with ophthalaldehyde.

The determination of propionate and acetate

In many species the definitive end product of prolonged hypoxia is propionate. It must be determined from animal tissues and, since it is excreted, also from the surrounding medium. Unfortunately there is still no procedure available that allows the measurement of propionate without sample preparation.

In order to extract propionate, the tissues should be homogenized in twice the amount (V/V) of ice-cold sodium bicarbonate buffer (0.01 M, pH 9.7) using a Sorvall Omnimixer (Omni Corp. Int., Waterbury, Connecticut, USA) at maximum speed for five minutes followed by a 30 second sonification. The homogenate (10 ml), to which 50 μl of 0.2 M valeric acid can be added as an internal standard, is then extracted with 150 ml of chloroform-methanol (2:1, v/v) and once with 50 ml of distilled water. The combined fractions are again shaken using a separating funnel and after the formation of two phases, the chloroform layer is discarded. The methanol-water fraction, which should have at least a pH of 8.0, is evaporated to dryness and stored. Immediately before analysis the dry powder is acidified with 1.0 ml of 5% o-phosphoric acid (Kluytmans, Veenhof & de Zwaan, 1975).

As this procedure is somewhat tedious, different extraction methods and sample preparations have been used. Tissues are extracted in a three-fold volume of ice-cold 0.6 N perchloric acid using an Ultra-Turrax homogenizer (Janke und Kunkel, Staufen, FRG) for 3 × 30 seconds at maximum speed. After removing the tissue by centrifugation, the supernatant is neutralized with 5 M KOH and the precipitating perchlorate is removed by centrifugation.

An aliquot of this extract (0.3 to 3 ml) is acidified with conc. sulfuric acid (10 to 20% of the extract volume) and subjected to steam distillation. The distillate is collected in 3 ml 0.1 N NaOH. The distillation is terminated when the volume of the distillate is about 15 to 20 times larger than that of the extract. The distillate, which should be slightly alkaline ($\sim$ pH 8), is evaporated to dryness and prior to analysis dissolved in a small volume (usually 0.2 to 0.5 ml) of 5% o-phosphoric acid.

Quantitative determination of propionate can either be achieved by HPLC or by gas chromatography. Using HPLC most short chain fatty acids can be separated by an isocratic method using Bio-Rad Resin Aminex HPX-87 (Bio-Rad, D-8000 München, F.R.G.; column size 300 mm × 7.8 mm) and 4 mM H_2SO_4 as the eluant at a flow rate of 0.6 ml/min, a column temperature of 20 °C and a pressure of 50 bar. Propionate can be detected monitoring its absorbance at 210 nm.

Gas chromatographic procedures separate propionate even better. A 3000 mm column packed with Chromosorb WAW 60 to 80 mesh, which is coated with 20% w/w polyethylenglycoladipate and 3% w/w o-phosphoric acid, has been widely used. A flow rate for the carrier gas (N_2) of 30 ml/min is used. Temperatures used are 250 °C for the injector block, 140 °C for the column and 240 °C for the flame ionization detector respectively (Surholt, 1977). However, much better chromatograms can be obtained with capillary columns, for example with an HP-FFAP column (No. 19095F-121; Hewlett-Packard, D-7517 Waldbronn, FRG) if an isothermal program at 110 °C is used. If a temperature program is available, column HP-20M (No. 199095W-121; Hewlett-Packard, D-7517 Waldbronn, FRG) may have some advantages, as most of the contaminating low molecular compounds other than volatile fatty acids can be washed off the column at a low temperature.

The same methods can be applied, if besides the concentration of propionate, that of acetate is also of interest. It should, however, also be mentioned that acetate can be estimated using an optical assay.

Acknowledgements
We thank Dr Christopher Bridges for critically reading this manuscript and the Fonds der Chemischen Industrie and the Deutsche Forschungsgemeinschaft for continuous financial support (Ga 241/4–2 and 4–3; Gr 456/9–4). Support by a Heisenberg Fellowship awarded from the Deutsche Forschungsgemeinschaft to G.G. is acknowledged (Ga 241/5-1 and 5-2).

References
Bergmeyer, H. U. (1983*a*). In *Methods of Enzymatic Analysis*, vol. 2, 3rd edn, Weinheim: Verlag Chemie.

Bergmeyer, H. U. (1983*b*). Determination of metabolite concentrations with end-point methods. In *Methods of Enzymatic Analysis*, vol. 1, 3rd edn, ed. H. U. Bergmeyer, pp. 163–81. Weinheim: Verlag Chemie.

Bergmeyer, H. U. & Möllering, H. (1984): Acetate: Determination with acetate kinase. In *Methods of Enzymatic Analysis*, vol. 6, 3rd edn, H. U. Bergmeyer, pp. 628–39. Weinheim: Verlag Chemie.

Beutler, H. O. (1985). Succinate. In *Methods of Enzymatic Analysis*, vol. 7, 3rd edn, ed. H. U. Bergmeyer, pp. 25–33. Weinheim: Verlag Chemie.

Crescentini, G. & Stocchi, V. (1984). Fast reversed-phase high-performance liquid chromatographic determination of nucleotides in red blood cells. *Journal of Chromatography*, **290**, 393–9.

Dejours, P. (1975). *Principles of Comparative Respiratory Physiology*. New York: American Elsevier Publishing Company.

De Zwaan, A. & Dando, P. R. (1984). Phosphoenolpyruvate-pyruvate metabolism in bivalve molluscs. *Molecular Physiology*, **5**, 285–312.

De Zwaan, A. & Putzer, V. (1985). Metabolic adaptations of intertidal invertebrates to environmental hypoxia (A comparison of environmental anoxia to exercise anoxia). In *Physiological Adaptations of Marine Animals*. Symposia of the Society for Experimental Biology, vol. 39, ed. M. Laverack, pp. 33–62. Cambridge: The Company of Biologists.

Englisch, H., Opalka, B. & Zebe, E. (1982). The anaerobic metabolism of the larvae of the midge *Chaoborus crystallinus*. *Insect Biochemistry*, **12**, 149–55.

Felbeck, H. (1980). Investigations on the role of the amino acids in anaerobic metabolism of the lugworm *Arenicola marina* L. *Journal of Comparative Physiology*, **137**, 183–92.

Fiore, G. B., Nicchitta, C. V. & Ellington, W. R. (1984). High-performance liquid chromatographic separation and quantification of alanopine and strombine in crude tissue extracts. *Analytical Biochemistry*, **139**, 413–17.

Forstner, H. (1983). An automated multiple-chamber intermittent-flow respirometer. In *Polarographic Oxygen Sensors*, ed. E. Gnaiger & H. Forstner, pp. 111–33. Heidelberg: Springer Verlag.

Gäde, G. (1983). Energy metabolism of arthropods and mollusks during environmental and functional anaerobiosis. *Journal of Experimental Zoology*, **228**, 415–29.

Gäde, G. (1985a). Arginine and arginine phosphate. In *Methods of Enzymatic Analysis*, vol. 8, 3rd edn, ed. H. U. Bergmeyer, pp. 425–31. Weinheim: Verlag Chemie.

Gäde, G. (1985b). Octopine. In *Methods of Enzymatic Analysis*, vol. 8, 3rd edn, ed. H. U. Bergmeyer, pp. 419–25. Weinheim: Verlag Chemie.

Gäde, G. (1987a). Leben ohne Sauerstoff: Die Rolle der anaeroben Glykolyse bei aquatischen wirbellosen Tieren. *Verhandlungen der Deutschen Zoologischen Gesellschaft*. Ulm **80**, 93–110.

Gäde, G. (1987b). A specific enzymatic method for the determination of taurine. *Hoppe-Seyler's Zeitschrift für Physiologische Chemie* **368**, 1519–23.

Gäde, G. (1988). Energy metabolism during anoxia and recovery in shell adductor and foot muscle of the gastropod mollusc *Haliotis Camellosa*: Formation of the novel anaerobic end product tauropine. *Biological Bulletin* **175**, 122–131.

Gäde, G. & Grieshaber, M. K. (1986). Pyruvate reductases catalyze the formation of lactate and opines in anaerobic invertebrates. *Comparative Biochemistry and Physiology*, **83**B, 255–72.

Gäde, G. & Head, E. J. H. (1979). A rapid method for the purification of octopine dehydrogenase for the determination of cell metabolites. *Experientia*, **35**, 304–5.

Gäde, G. & Meinardus-Hager, G. (1986): Anaerobic energy metabolism in Crustacea, Xiphosura and Mollusca: Lactate fermentation versus multiple fermentation products. *Zoologische Beiträge N.F.*, **30**, 187–203.

Gäde, G., Wilps, H., Kluytmans, J. H. & de Zwaan, A. (1975). Glycogen degradation and end products of anaerobic metabolism in the fresh water bivalve *Anodonta cygnea*. *Journal of Comparative Physiology*, **104**, 79–85.

Gawehn, K. (1984). D-(−)-Lactate. In *Methods of Enzymatic Analysis*, vol. 6, 3rd edn, ed. H. U. Bergmeyer, pp. 588–92. Weinheim: Verlag Chemie.

Gnaiger, E. (1983). Heat dissipation and energetic efficiency in animal anoxibiosis: Economy contra power. *Journal of Experimental Zoology*, **228**, 471–90.

Graßl, M. & Supp, M. (1985). D-Alanine. In *Methods of Enzymatic Analysis*, vol. 8, ed. H. U. Bergmeyer, pp. 336–40. Weinheim: Verlag Chemie.

Grieshaber, M. K. (1982). Metabolic regulation of energy metabolism. In *Exogenous and Endogenous Influences on Metabolic and Neural Control*, ed. A. D. F. Addink & N. Spronk, pp. 225–42. Oxford and New York: Pergamon Press.

Grieshaber, M. K. & Kreutzer, U. (1986). Opine formation in marine invertebrates. *Zoologische Beiträge N.F.*, **30**, 205–29.

Grieshaber, M. K., Kronig, E. & Koormann, R. (1978). A photometric estimation of phospho-L-arginine, arginine and octopine using homogenous octopine dehydrogenase isoenzyme 2 from the squid, *Loligo vulgaris* Lam. *Hoppe-Seyler's Zeitschrift für physiologische Chemie*, **359**, 133–6.

Haar, H. P., Netheler, H. & Ziegenhorn, J. (1983). Absorption photometry, nephelometry, turbidimetry. In *Methods of Enzymatic Analysis*, vol. 1, 3rd edn, ed. H. U. Bergmeyer, pp. 280–305. Weinheim: Verlag Chemie.

Hamilton, R. J. & Sewell, P. A. (1982). *Introduction to High Performance Liquid Chromatography*, 2nd edn. London: Chapman and Hall.

Hoffmann, K. H. (1981). Phosphagens and phosphokinases in *Tubifex sp.* *Journal of Comparative Physiology*, **143**, 237–43.

Holwerda, D. A., Kruitwagen, E. C. & de Bont, A. M. T. (1981). Regulation of pyruvate kinase and phosphoenolpyruvate carboxykinase activity during anaerobiosis in *Mytilus edulis* L. *Molecular Physiology*, **1**, 165–71.

Holwerda, D. A., Veenhof, P. R., van Heugten, H. A. & Zandee, D. I. (1983). Modulation of mussel pyruvate kinase during anaerobiosis and after temperature acclimation. *Molecular Physiology*, **3**, 225–34.

Jaworek, D. & Welsch, J. (1985). Adenosine-5'-diphosphate and adenosine-5'-monophosphate: UV-method. In *Methods of Enzymatic Analysis*. Vol. 7, 3rd edn, ed. H. U. Bergmeyer, pp. 365–70. Weinheim: Verlag Chemie.

Keppler, D. & Decker, K. (1984). Glycogen. In *Methods of Enzymatic Analysis*. Vol. 6, 3rd edn, ed. H. U. Bergmeyer, pp. 11–18. Weinheim: Verlag Chemie.

Kluytmans, J. H., Veenhof, P. R. & de Zwaan, A. (1975). Anaerobic production of volatile fatty acids in the sea mussel *Mytilus edulis*. L. *Journal of Comparative Physiology*, **104**, 71–8.

Kreutzer, U., Siegmund, B. & Grieshaber, M. K. (1985). Role of coupled substrates and alternative end products during hypoxia tolerance in marine invertebrates. *Molecular Physiology*, **8**, 371–92.

Kunst, A., Draeger, B. & Ziegenhorn, J. (1984). D-Glucose: UV-methods with hexokinase and glucose-6-phosphate dehydrogenase. In *Methods of Enzymatic Analysis*, vol. 6, 3rd edn, ed. H. U. Bergmeyer, pp. 163–72. Weinheim: Verlag Chemie.

Möllering, H. (1985). L-Aspartate and L-asparagine. In *Methods of Enzymatic Analysis*, vol. 8, 3rd edn, ed. H. U. Bergmeyer, pp. 350–7. Weinheim: Verlag Chemie.

Noll, F. (1984). L-(+)-Lactate. In *Methods of Enzymatic Analysis*, vol. 6, 3rd edn, ed. H. U. Bergmeyer, pp. 582–8. Weinheim: Verlag Chemie.

Pette, D. & Reichmann, H. (1982). A method for quantitative extraction of enzymes and metabolites from tissue samples in the milligram range. *Journal of Histochemistry and Cytochemistry*, **30**, 401–2.

Pietrzak, S. M. & Saz, H. J. (1981). Succinate decarboxylation to propionate and the associated phosphorylation in *Fasciola hepatica* and *Spirometra mansonoides*. *Molecular Biochemistry and Parasitology*, **3**, 61–70.

Plaxton, W. C. & Storey, K. B. (1984). Phosphorylation *in vivo* of red-muscle pyruvate kinase from the channeled whelk, *Busycotypus canaliculatum*, in response to anoxic stress. *European Journal of Biochemistry*, **143**, 267–72.

Pörtner, H. O., Heisler, N. & Grieshaber, M. K. (1985). Oxygen consumption and mode of energy production in the intertidal worm *Sipunculus nudus* L.: Definition and characterization of the critical P_{O2} for an oxyconformer. *Respiration Physiology*, **59**, 361–77.

Pörtner, H. O. Kreutzer, U., Siegmund, B., Heisler, N. & Grieshaber, M. K. (1984). Metabolic adaptation of the intertidal worm *Sipunculus nudus* to functional and environmental hypoxia. *Marine Biology*, **79**, 237–47.

Putzer, V. (1984). Energy production and glycolytic flux during functional and environmental anoxia in *Lumbriculus variegatus*. Abstract in the *First International Congress C.P.B.*, Liege, Belgique.

Schöttler, U. (1986). Weitere Untersuchungen zum anaeroben Energiestoffwechsel des Polychaeten *Arenicola marina* L. *Zoologische Beiträge N.F.*, **30**, 141–52.

Schroff, G. & Schöttler, U. (1977). Anaerobic reduction of fumarate in the body wall musculature of *Arenicola marina (Polychaeta)*. *Journal of Comparative Physiology*, **116**, 325–36.

Schultz, T. K. & Kluytmans, J. H. (1983). Pathway of propionate synthesis in the sea mussel *Mytilus edulis*. *Comparative Biochemistry and Physiology*, **75B**, 365–72.

Shick, J. M., Gnaiger, E., Widdows, J., Bayne, B. L. & de Zwaan, A. (1986). Activity and metabolism in the mussel *Mytilus edulis* L. during intertidal hypoxia and aerobic recovery. *Physiological Zoology*, **59**, 627–42.

Siegmund, B. & Grieshaber, M. K. (1983). Determination of *meso*-alanopine and D-strombine by high pressure liquid chromatography in extracts from marine invertebrates. *Hoppe-Seyler's Zeitschrift für Physiologische Chemie*, **364**, 807–12.

Snyder, L. R. & Kirkland, J. J. (1974): *Introduction to Modern Liquid Chromatography*. New York: Wiley.

Storey, K. B., Miller, D. C., Plaxton, W. C. & Storey, J. M. (1982). Gas-liquid chromatography and enzymatic determination of alanopine and strombine in tissues of marine invertebrates. *Analytical Biochemistry*, **125**, 50–8.

Surholt, B. (1977). Production of volatile fatty acids in the anaerobic carbohydrate catabolism of *Arenicola marina*. *Comparative Biochemistry and Physiology*, **58B**, 147–50.

Trautschold, I., Lamprecht, W. & Schweitzer, G. (1985). Adenosine-5'-triphosphate: UV-method with hexokinase and glucose-6-phosphate dehydrogenase. In *Methods of Enzymatic Analysis*, vol. 7, 3rd edn, ed. H. U. Bergmeyer, pp. 346–57. Weinheim: Verlag Chemie.

Van Thoai, N. & Robin, Y. (1969). Guanidine compounds and phosphagens. In *Chemical Zoology*, vol. 4, Annelida, Echiura and Sipunculida, eds. M. Florking & B. T. Scheer, pp. 163–203. New York: Academic Press.

Urbanke, C. (1983). Fluorimetry. In *Methods of Enzymatic Analysis*, vol. 1, 3rd edn, ed. H. U. Bergmeyer, pp. 326–40, Weinheim: Verlag Chemie.

Warburg, O. & Christian, W. (1936). Optischer Nachweis der Hydrierung und Dehydrierung des Pyridins im Gärung-Co-Ferment. *Biochemische Zeitschrift*, **286**, 81–2.

Williamson, D. H. (1985). L-Alanine: Determination with alanine dehydrogenase. In *Methods of Enzymatic Analysis*, vol. 8, 3rd edn, ed. H. U. Bergmeyer, pp. 341–4. Weinheim: Verlag Chemie.

Zandee, D. I., Holwerda, D. A. & de Zwaan, A. (1980). Energy metabolism in bivalves and cephalopods. In *Animals and Environmental Fitness*, vol. 1, ed. R. Gilles, pp. 185–206. Oxford: Pergamon Press.

Zebe, E., Salge, U. Wiemann, C. & Wilps, H. (1981). The energy metabolism of the leech *Hirudo medicinalis* in anoxia and muscular work. *Journal of Experimental Zoology*, **218**, 157–63.

R. M. G. WELLS and R. E. WEBER

The measurement of oxygen affinity in blood and haemoglobin solutions

Introduction

The haemoglobin–oxygen equilibrium curve (OEC) is a fundamental description of the oxygen transporting potential of blood. However, existing procedures for its determination appear so complex and varied that many laboratories have been discouraged from generating it for routine use. The purpose of this chapter is to review techniques for measuring oxygen affinity, and to describe three procedures in detail.

Essential concepts of affinity (P_{50}) and cooperativity (n-value)

The sigmoidal shape of the OEC for human blood shows that oxygen bound by haemoglobin (Hb) enhances the uptake of more oxygen. The sigmoidicity of the human OEC is almost insensitive to changes in the working environment of Hb. The degree of sigmoidicity, or cooperativity, is quantified by n in Hill's equation:

$$\log S/(100-S) = \log P_{50} + n\log P_{O_2}$$

where S = % saturation (HbO_2), n = 2.8, P_{O_2} = partial pressure of oxygen and P_{50} is the half-saturation oxygen partial pressure (Fig. 1a).

The equilibrium data are obtained experimentally, and used to determine P_{50}, and the equation is solved for n. The affinity of blood for oxygen is not fixed, and is adjusted such that blood leaves the lungs nearly saturated with oxygen, which is released to the tissues at lower P_{O_2}. For normal human blood, n_{50} equals 2.8, and P_{50} is about 27 mmHg at P_{CO_2} = 40 mmHg, and at pH 7.4 and 37 °C. Oxygen delivery is regulated *in vivo* by the influences of pH (Bohr effect), CO_2 (carbamate effect), organic phosphates like 2,3 diphosphoglycerate (DPG) and ATP, and temperature (Fig. 1b).

In animal Hbs, cooperativity varies from species to species and may be sensitive to various factors. The question to what extent the properties of the OEC are a manifestation of the intrinsic binding properties of the Hb, or arise from interactions with factors in the erythrocyte, may be answered by measuring OEC on both whole blood and in Hb solutions.

Whole blood versus Hb solution

It is now recognised that whole blood OEC are necessary for physiological interpretations of the oxygen transport system. Methods that depend on the use of buffered erythrocyte suspensions, often billed as 'whole blood', fail to copy the *in vivo* erythrocyte environment in which the Hb functions. In particular, the specific role and interactions of carbon dioxide, the buffering system, differences in concentration and the physical

Fig. 1. Oxygen equilibrium curves illustrating the concepts of (*a*) shape (Hill's co-operativity constant, *n*) with Hill plots inset for determination of *n*, and (*b*) position (the oxygen affinity constant, P_{50}), and factors known to affect affinity.

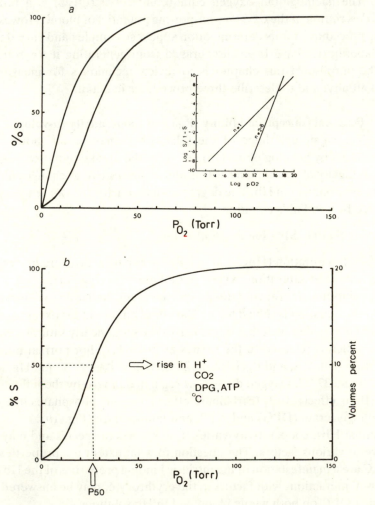

state of the Hb, and the transmembrane pH difference are discounted. The oxygen affinity of whole blood is thus lower than that of erythrocyte suspensions under comparable conditions. The difference is in part due to the Donnan distribution affected by charges on the Hb, and by the pH-organic phosphate-CO_2-inorganic ion-Hb relationships that become perturbed when buffers are substituted for plasma. Nevertheless, there are instances where the use of buffered erythrocyte suspensions provides useful information.

Hb 'stripped' from its bound phosphate and other ions shows a marked increase in oxygen affinity, and slight decrease in *n* value. In general, purification procedures tend to decrease cooperativity in all but the most stable Hb.

Review of techniques for measuring the equilibrium
The choice of method may depend on the kind of equilibrium information required and the availability of apparatus. A brief summary of various methods is given below and in Table 1. A number of other techniques are given by Oeseburg *et al.* (1972), Zander *et al.* (1978), Winslow *et al.* (1981), Imai (1982), and Asakura & Reilly (1986).

The mixing technique
The mixing method is simple and effective for a variety of applications and requires little more than an efficient tonometering system and a P_{O_2} electrode. It is based on the principle that known proportions of oxygenated and deoxygenated blood giving different percentage saturations are mixed anaerobically, and the resultant P_{O_2} is recorded with corrections for dissolved oxygen. Details are described by Edwards & Martin (1966) with modifications for a micromethod in Blunt (1974). Replication and confirmation of results with standard values are excellent with mammalian blood (Scheid & Meyer, 1978). The main sources of error lie in the tonometry system and in the slow response of the P_{O_2} electrode at low temperatures. Volumetric errors, and those arising from electrode calibration are generally not significant.

The method has recently been applied to fish blood (Nikinmaa, 1983; Cech *et al.*, 1984), though Tetens & Lykkeboe (1981) found limitations when tonometering nucleated erythrocytes. Anoxic incubation of trout blood induced a rapid increase in measured oxygen affinity through cellular oxygen consumption and decreased ATP. A possible solution to this limitation is the addition of sodium fluoride or iodoacetate to the anticoagulant to inhibit metabolism, but this might cause further complications.

Table 1. *Summary of techniques*

	Primary measurement	Simplicity of use	Precision	Dynamic or static	Blood or Hb	Recommendation for use
Van Slyke (Van Slyke & Neill, 1924)	C_{O_2}	+++	+++	S	B	+
Tucker (Tucker, 1967)	C_{O_2}	+++	+++	S	B	+++
Galvanic fuel cell (Grubb & Mills, 1981)	C_{O_2}	++	+++	S	B	++
Mixing (Edwards & Martin, 1966)	P_{O_2}	+++	++	S	B	++
Double polarographic (Duvelleroy *et al.*, 1970)	P_{O_2}	+	++	D	B	++
Electrolytic (Longmuir & Chow, 1970)	%S	++	+	D	B	+
Thin-Layer (Sick & Gersonde, 1969)	%S	+	+++	D,S	B,Hb	+++
Automated optical (Imai, 1982)	%S	+	+++	D	Hb	++
Optical tonometry (Benesch *et al.*, 1965)	%S	+++	+++	S	Hb	+++
Oximetry	%S	++	+	S	B	+

Electrolytic method

This was originally developed by Longmuir & Chow (1970) as a simple, and easy method for determining oxygen affinity. The method involves the preparation of a buffered erythrocyte suspension containing a suspension of yeast cells, mitochondria, or minced heart tissue which linearly decrease oxygen while %S is recorded spectrophotometrically. Reoxygenation could be achieved by the electrolytic generation. The principal advantage of the method is that it provides continuous curves in a short time. The major disadvantage is the presence of perturbed extra-cellular concentrations of electrolytes, often inorganic phosphate, which affects intracellular organic phosphate levels and pH.

As a result of using this technique with bird blood, unusually high affinities and polyphasic OEC were reported, and a vigorous debate ensued as to the causes (see Lutz, 1980). Although it turned out that the affinities obtained by this method were too high, the generation of continuous OEC led to the important observation, subsequently confirmed by independent methods, of heterogeneous OEC where the Hill coefficient was dependent upon saturation over a wide range of P_{O_2}. This important discovery might have been delayed had publication been suppressed on the grounds of poor methodology.

Hughes *et al.* (1976) also advocated this method for use with fish blood finding the OEC parameters '... cannot, as yet, be compared strictly with those obtained by other methods.' The results from larval lampreys using this electrolytic method are therefore questionable (Bird *et al.*, 1976). The method is not recommended for whole blood and has largely fallen into disuse.

Polarographic method

The 2-electrode polarographic method of Duvelleroy *et al.* (1970) resulted in the Radiometer Dissociation Curve Analyser, but this is no longer available. The principle of this whole-blood technique is the polarographic determination of %S from the change in P_{O_2} in a gas phase above a blood sample as equilibrium is approached. The advantage of the method is that continuous OEC are generated, and no assumptions about the validity of Hb spectra need to be made. The chief disadvantages are the length of time for an OEC (>1 h) and large blood volume required. Both of these disadvantages have been minimized in the modified applications of Teisseire *et al.* (1975, 1985).

Automated method

Imai's technique offers continuous registration of Hb OEC in solution. The principle is elegantly simple and depends on diffusion of oxygen into a solution of deoxygenated Hb contained in an optical cell fitted with an oxygen electrode. Conversely, a deoxygenating curve may be constructed if nitrogen is used. Details of the method, including construction of the cell are given in Imai (1982). The principle has lent itself to full automation. Applications to heterogeneous OEC and the extremes of saturation have been described from fish (Ikeda-Saito *et al.*, 1983), and humans (Imai, 1982). The general limitations on spectral methods apply. Some difficulty may be experienced with ensuring complete deoxygenation or oxygenation of samples in this system. The method cannot be used with whole blood though it has been adapted for use with erythrocyte suspensions (Imai *et al.*, 1980).

Thin-layer dynamic registration of OEC

The continuous curve from microlitre quantities of blood or Hb solution provides more information about saturation dependence of the equilibrium constants than can a limited number of points, especially when the curve is not of a simple hyperbolic or sigmoidal shape. Discontinuous curves may also be determined when dynamic errors are suspected. The principle of the method depends on dual or single wavelength spectrophotometry of a thin layer of blood or Hb solution respectively, and is described in detail in a later section. Excellent results have been obtained from whole blood and Hb solutions (Easton, 1979; Lapennas *et al.*, 1981). Data acquisition and on-line processing are straightforward with such a system.

Applications of the thin-layer method are found for sharks (Weber *et al.*, 1983), teleosts (Weber *et al.*, 1976; 1987), turtles (Maginniss *et al.*, 1980), frogs (Wells & Weber, 1985), and chick embryos (Lapennas & Reeves, 1983). Some of these show OEC where the Hill coefficient is markedly saturation-dependent. The technique is also suitable for studies with haemocyanin (Wells & Weber, 1982).

Optical tonometer method

This classical method for use with dilute Hb solution is described in detail later. Spectral traces are made to describe the state of oxygenation following tonometry at known P_{O_2}. The technique has been modified for use with erythrocyte suspensions (Wells, 1979) and for haemocyanin (Wells & Shumway, 1980).

Van Slyke

The manometric method described by Van Slyke & Neill (1924) is regarded as the classical physical technique for direct measurement of O_2-content and P_{50} on tonometered blood samples, and widely used to calibrate whole blood methods. A quicker micromethod based on the manometric principle has been described by Brix (1981). Determination of affinity requires point-by-point measurements of fractional oxyHb over a range of discrete P_{O_2}. Other methods are comparable in precision, quicker, and require smaller sample volumes (Easton, 1979).

Tucker's method

Tucker's (1967) direct method for measuring oxygen content on microlitre quantities of whole blood is simple, rapid, and precise. It is based on the principle that bound oxygen is released from blood injected into an oxidizing solution of ferricyanide, and the rise in P_{O_2} of the solution is proportional to oxygen content. The method is best suited for blood with relatively high O_2-carrying capacity, but has been adapted for use with low capacity invertebrate haemocyanin (Bridges *et al.*, 1979). Electrode calibration and volume determination are minor sources of error. The method may be used with a tonometering system to obtain OEC, as described later.

Simplified versions of the method include the use of two syringes instead of a special chamber (Wells, 1976), and air-equilibrated ferricyanide for electrodes responding linearly above the calibration point (Laver *et al.*, 1965). The technique has been used to describe heterogeneous Hill plots in frogs where *n* at high saturation increases to values >4 which is the theoretical maximum for tetrameric Hb (Lykkeboe & Johansen, 1978), O_2 content in tuna (Jones *et al.*, 1986), and OEC from antarctic fish at sub-zero temperatures (Tetens *et al.*, 1984). Carbon dioxide content is measured by adding a CO_2 electrode and acidifying the ferricyanide (Cameron, 1971).

Lex O₂-Con Analyzer

The Lex O_2-Con (Lexington Instruments Corp., Waltham, MA, USA) is an effective tool for measuring oxygen contents in 20-µl volumes. The principle of the operation is that O_2 is removed from blood or Hb solution by a special scrubber gas containing CO, before passing over an oxygen sensitive amperometric cell. The electrical output of the cell is proportional to O_2 content.

The principal errors are, reduced sensitivity with low O_2-capacity (e.g. when using Hb solutions), and the possibility of resistance to lysis in certain erythrocytes. One should add that CO-affinity for animal bloods may be less than that of human Hb.

A comparative study of the Lex O_2-Con and Tucker chamber using fish blood showed that both techniques gave very acceptable results, though the time for analysis was slightly less with the Tucker chamber (Hughes *et al.*, 1982). We have confirmed this in our lab using human blood. The mean content of 20-μl aliquots of air-saturated human blood was 10.4± 0.25 mol·L^{-1} (S.D., N = 5) compared with a theoretically calculated capacity of 10.3 mol·L^{-1}; for comparison, the Tucker method gave 10.4± 0.25 mol·L^{-1}. We would add that the time for analysis in the Lex O_2-Con depends on the age and condition of the fuel cell, which must eventually be replaced. The Lex O_2-Con appears to be no longer available.

Applications with fish blood (Johansen *et al.*, 1976), including examination of a Root effect (Bridges *et al.*, 1983) confirm the efficacy of the Lexington instrument. For those bold enough to build their own fuel-cell analyser, details are to be found in Grubb & Mills (1981), though as with all D.I.Y. equipment, one must caution that terms like 'simple' and 'inexpensive' are relative.

Notes on optical methods for measuring saturation

Most techniques for measuring saturation in Hb solutions are based on the spectrophotometric distinction of the oxy and deoxy states. All spectrophotometric methods are based on the assumption that Beer's Law holds:

$$[Hb:HbO_2] = k.e\ (\lambda)$$

where e is the absorbance, λ, a characteristic wavelength, and k, is a lineal constant. Though it is accepted empirically, Beer's Law does not hold strictly in all cases and slight deviations have been noted for fish and animal Hb (Burkhard & Barnikol, 1982; Ikeda-Saito *et al.*, 1983). Spectral changes may also result when Hb is bound to allosteric effectors or ions (Imaizumi *et al.*, 1978) though the extent of this error has not been examined for a range of species.

Spectrophotometric analysis of whole blood or erythrocyte suspensions poses the additional problem of light scattering. This may be largely overcome by using a spectrophotometer that allows the operator to place the sample close to the photomultiplier, and by the selection of appropriately long wavelengths. Alternatively, dual wavelength spectrophotometry permits cancellation of scattering errors (Reeves, 1980).

An inherent difficulty with spectrophotometric methods is that 100% saturation is assumed. The error introduced by this assumption becomes most serious for Hb with low affinity and cooperativity, because of the practical difficulty of achieving saturation in an equilibrium reaction, and increases exponentially at extremely high saturations – i.e. when estimating the oxygen affinity of the oxygenated (relaxed) state of the Hb.

The availability of commercial oximeters such as the Radiometer OSM 3 Hemoximeter offers the possibility of saturation measurements on small volumes of blood at known P_{O_2}. There are however, inherent weaknesses with oximetry arising from the lysis of cells required for optical clarification. Firstly, lysis increases oxygen affinity by decreasing the phosphate–haemoglobin interactions, so that all saturation measurements on blood will be too high, with the error increasing at lower saturations. Secondly, the physiological relationship between intracellular and plasma pH is destroyed so that corrections using an assumed Bohr factor will be necessary. Finally, oximeters are intended primarily for use with adult human blood, and Hbs with different spectra, and those from nucleated erythrocytes may lead to errors that have to be accounted for.

Method 1: Measuring OEC by the Tucker method
Principle

Whole blood is placed in a suitable tonometer and equilibrated with a gas mixture of known P_{O_2} and P_{CO_2}. A micro-sample of blood is taken into a gas-tight syringe and injected into a thermostatted chamber containing a solution of ferricyanide. The resulting oxidation (metHb formation) releases bound oxygen and a rise in P_{O_2} in solution is recorded from an oxygen electrode mounted in the chamber. The process is repeated for several P_{O_2} values, and finally, for aerated blood. The change in P_{O_2} corrected for dissolved oxygen, is directly proportional to the oxygen content of the blood, and the saturation calculated from the content:capacity ratios.

Oxygen content may be calculated directly from the change in P_{O_2} and from the constants of the system as described by Tucker (1967) or Wells (1976) and reviewed by Bridges (1983). A better procedure is that developed by the Zoophysiology Department at the University of Aarhus, Denmark. A chamber constant is determined empirically and is used in subsequent calculations.

The Tucker chamber

The chamber may be constructed from Perspex using the plan drawings of Tucker (1967). A better chamber is an all-glass construction (Fig. 2) which is inexpensive, and avoids problems of leaks, cleaning, poor visualization of the reaction chamber, and errors resulting from diffusion of oxygen through plastic. Alternatively, Strathkelvin Instruments (15 Lochend Rd, Bearsden, Glasgow, U.K.) have developed a micro-Tucker cell that should meet the most exacting requirements. The volume of the inner chamber is determined by weighing empty and dry, with the stirring magnet and electrode mounted. The chamber is filled with water up to the capillary tube at the top of the chamber, reweighed, and the difference in weight in mg divided by the specific gravity of water at the appropriate temperature is the chamber volume in μl. The chamber is thermostatted to a specific temperature (e.g. 30 °C) to give optimal electrode performance. This temperature need not be the same as the equilibrium temperature.

Fig. 2. Tucker apparatus for measurement of oxygen content in microlitre samples of whole blood. The volume of the inner chamber may be 1–3 ml.

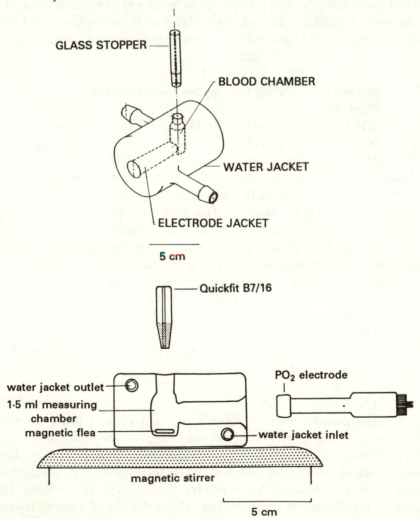

The Radiometer (Copenhagen) E-5046, or Instrumentation Laboratories 1302 P_{O_2} electrodes are most suitable when coupled to a compatible Radiometer, I.L., or Strathkelvin Instruments 781b meter. The recorder output should be coupled to a chart recorder or monitor to detect steady state. The chamber is mounted on a magnetic stirrer and a small teflon-coated rod in the chamber promotes efficient mixing.

Electrode calibration

Calibration of the electrode is achieved by filling the chamber with a 'zero solution' of ~ 10 mg Na_2SO_3 dissolved in 50 ml 0.01 mol 1^{-1} $Na_2B_4O_7$. The stirrer is started and when the electrode output is stable, the zero adjustment is made. It is important to use freshly dissolved sulphite and be sure that no air bubbles are trapped in the chamber. This adjustment is fairly stable and need be carried out only once or twice each day.

The chamber is thoroughly rinsed by filling with distilled water and emptying the chamber several times. The stirring motor is started, and the air in the empty chamber now equilibrates with the water phase in the small amount that is always left in the chamber.

The P_{O_2} of water saturated air in the empty chamber is calculated as:

$$(P_B - P_{H_2O}) \times 0.2095 \times SF$$

where P_B is barometric pressure, P_{H_2O} is water vapour pressure at the chamber temperature (31.8 Torr at 30 °C), 0.2095 is the fractional volume of oxygen in atmospheric air, and SF is the stirring factor, or the liquid-gas sensitivity ratio. It is due to the fact that the measured P_{O_2} of a gas is always a little higher than the gas tension measured in a liquid equilibrated with the same gas mixture at the same pressure, due to the diffusion barrier in the liquid and is estimated as the ratio of the measured P_{O_2} of water-saturated air to that in air-equilibrated water with the chamber three-quarters filled with distilled water, and equals approximately 1.02. The P_{O_2} adjustment should be checked several times each day.

The chamber constant

The basis for the determination of the chamber constant is to add a known amount of oxygen to the ferricyanide reagent in the Tucker chamber. Human blood is a well-studied liquid, in which the oxygen capacity in mol·L^{-1}, $(C_{HbO_2})_{max} = 0.460 \times$ Hct, where Hct is the haematocrit in per cent. This simple equation is empiric and holds for freshly drawn blood at pH ~ 7.4. The donor must be a non-smoker.

Human blood is equilibrated with humid air (which saturates to >99%) in a tonometer, at for example, 20 °C. The oxygen solubility coefficient, αb_{O_2}

of human blood at this temperature is 0.0016 mmol$\cdot$L$^{-1}\cdot$Torr^{-1} (see Christophorides & Hedley-White, 1969). The total amount of oxygen in the blood, C_{O_2tot} may then be calculated: $C_{O_2tot} = C_{HbO_2} + C_{O_2diss}$, where C_{O_2diss} is the dissolved oxygen. (For blood with a Hct of 44.5, at $P_B = 758$ Torr and at 20 °C, this gives a C_{O_2tot} of 9.2 mmol$\cdot$L^{-1} ($= 20.6$ vol%).)

The Tucker chamber is filled with degassed ferricyanide reagent. The ferricyanide reagent is made up from 0.6 g $K_3Fe(CN)_6$ and 0.3 g saponin dissolved in 100 ml distilled water. The reagent is stable for 6 months at room temperature if it is protected from the light. It is probably unnecessary to use the saponin which is added to augment lysis, but if so, the solution must be made fresh each day (Bridges, 1983). The solution is degassed in a 10 ml syringe by repeatedly withdrawing the plunger against a blocked needle, thus creating a vacuum. (Alternatively, if the saponin is omitted, a stock solution may be bubbled with nitrogen.) The stirrer is activated and care taken to exclude air bubbles. When the electrode output is stable the low P_{O_2} reading is recorded as P_{INT} (it does not have to be near zero). A known volume of fully oxygenated human blood is now injected into the chamber with the gas-tight syringe and the stirrer activated. At maximum P_{O_2}, the meter is read as P_{MAX}. The transfer of blood from tonometer to Tucker chamber must be carried out anaerobically and with the volume precisely known. The best way of doing this is to use a gas-tight, precision microsyringe such as the Unimetrics teflon tipped 5050TLC unit suitable for up to 50 µl of blood (Unimetrics Corp., 1853 Raymond, Anaheim, CA, USA).

The chamber is rinsed 3 times with distilled water and refilled with degassed reagent for a new determination. The change in P_{O_2} is calculated for each determination as follows:

$$\Delta P_{O_2} = P_{MAX} - (P_{INT} \times DF)$$

where DF is the dilution factor ($=$(volume of chamber $-$ volume of sample) /volume of chamber).

The chamber constant, CC is calculated from the mean of several determinations of ΔP_{O_2} by:

$$CC = C_{O_2tot}/\Delta P_{O_2} \text{ mmol}\cdot\text{L}^{-1}\cdot\text{Torr}^{-1}$$
$$= (0.460/2.24 \times Hct + 0.0016 \times (P_B - P_{H_2O}))/\Delta P_{O_2} \text{ at 20 °C.}$$

This procedure is repeated for different sample volumes (e.g. 10, 20, 30 µl). A plot of ΔP_{O_2} versus sample volume should give a straight line. The chamber constant for a given sample volume can now be calculated using this plot. It is convenient to calculate a series of CC and DF values, corresponding to a series of sample volumes.

The tonometering system

The 4-way, 2-gas mix microtonometry system (80–100 μl) of the Radiometer BMS2 is ideal and provides the additional facility for measurement of equilibrium pH. Unfortunately, the BMS2 is no longer available from Radiometer Denmark. Alternatively the Instrumentation Laboratory IL237 tonometer may be used to equilibrate 0.1–10 ml samples of blood. A far cheaper alternative is the Kutofix tonometry system available from L. Eschweiler & Co. (Holzkoppelweg 35, 2300 Kiel 1, FRG). An efficient microtonometer has been constructed by Grigg & Wells (1988) using the action of a DC-driven firebell alarm.

In the event that a tonometering system for equlibrating blood to different oxygen tensions is not available, then mixtures of fully oxygenation and nearly deoxygenated blood can be prepared (see Mixing Method) for precise analysis of oxygen content (S. F. Perry, pers. comm.).

The usual precautions with tonometry should be observed and errors should be assessed for each new temperature, or blood type. A range of tonometering times at say, three minute intervals for up to 18 minutes, are used to establish equilibrium times for O_2 by oxygen content measurement, and for CO_2 by pH measurement. Lysis and cell swelling may be checked by haematocrit. Metabolic changes may occur during tonometry at low P_{O_2} (Tetens & Lykkeboe, 1981) and ought to be assessed by measuring erythrocyte phosphates ATP, GTP, or DPG before and after tonometry.

Measurement of OEC by oxygen content

(i)   The chamber is filled with degassed ferricyanide, P_{INT} is read and multiplied by DF.

(ii)   A sample of blood tonometered at known P_{O_2} and P_{CO_2} is injected into the chamber as soon as P_{INT} is read. The stirrer is activated and P_{MAX} is read when the electrode output reaches a maximum.

(iii) ΔP_{O_2} is calculated as $P_{MAX} - (P_{INT} \times DF)$ and C_{O_2tot} is calculated as $\Delta P_{O_2} \times CC$. The haemoglobin bound oxygen is calculated by subtraction of dissolved oxygen, C_{O_2diss}:

$$C\,Hb_{O_2} = CC \times (P_{MAX} - (P_{INT} \times DF)) - C_{O_2diss}\ mmol \cdot L^{-1}$$

where $C_{O_2diss} = \alpha_{bO_2} \times P_{O_2}$

(iv)   The Tucker chamber is rinsed 3 times with distilled water.

(v)   A series of oxygen contents for tonometered blood of known P_{O_2} and constant P_{CO_2} are measured and O_2 capacity determined. The pH of a blood sample tonometered to a P_{O_2} close to the P_{50} is measured and the OEC plotted.

Assessment of the method

This method is well-tried and tested, requires small samples, and little specialised equipment. A reasonable complete OEC may be obtained in approx. one hour. The method is suitable for use with whole blood at any temperature but cannot be used with low capacity fluids such as dilute Hb. Provided that the tonometering system is efficient, the method is very accurate and agrees closely with theoretical predictions of content in fully saturated blood.

Method 2: Measuring OEC with an optical tonometer
Principle

A solution of Hb, or a suspension of erythrocytes is placed in a suitable tonometer and the air is removed by vacuum. The spectrum of deoxygenated Hb is measured using a spectrophotometer. Measured amounts of air are then added to the tonometer and the spectra are recorded after each additon. Finally, the fully oxygenated spectrum is recorded and the percentage oxyhaemoglobin is calculated from the serial spectra. The P_{O_2} values in the tonometer are calculated from the cumulative amounts of air added. The OEC can then be plotted.

Spectrophotometer

Almost any spectrophotometer may be used, although a scanning, dual beam instrument permits measurement of absorbance at several wavelengths without the need to re-zero the instrument. When erythrocyte suspensions are used, light scattering can be reduced by placing the tonometer as near to the photomultiplier as possible. The Pye Unicam SP 1750 has a 'second sample' position for use with turbid samples. If a dual wavelength instrument is available, then errors due to light scattering and oxygenation-dependent changes are minimized.

The measuring compartment may have to be slightly modified to accommodate the tonometer. This may be done by making a 'top hat' from black cardboard or velvet that fits over the tonometer, and prevents light from entering the compartment.

Tonometer-cuvette

This is not available commercially but can be made without complication in a glass-blowing workshop, using a design originally described by Benesch *et al.* (1965) and developed by Professor E. R. Huehns in the Clinical Haematology Dept., University College Hospital, London, UK. (Fig. 3). Alternatively, a cuvette may be glued to the tonometer. The first part is a 1 cm standard or semi-micro quartz cuvette which is joined to a round (5 × 5 cm) Pyrex glass bulb which gives a large surface area for gas exchange during equilibration. The joint must be made with a 9.5–10.5 mm graded glass seal (Jencons Lynx-seal, U.K.) because the quartz cuvette cannot be fused to Pyrex without distorting its optical surface. To the other end of the tonometer is fused a high vacuum stopcock (e.g. Quickfit TH6/2) with a small hole for admitting air. Two or three small scratches are made with a diamond pencil and extend the hole sideways for 3–4 mm to allow small amounts of air to enter the tonometer steadily. The stopcock is liberally coated with silicone grease. The weight and volume (usually about 70 ml) of the tonometer are carefully measured before use.

Method

Approx. 3 ml Hb solution or buffered cell suspension are placed in the tonometer. The concentration should be adjusted so that A_{540} is not more than 1.6. The tonometer is then evacuated via the side arm of the stopcock using a pump (e.g. Edwards High Vacuum), then disconnected and rotated for three minutes horizontally in a constant temperature waterbath. This allows equilibration between the vacuum, the buffer, and the gases in the blood. To facilitate the equilibration, a water or electrically powered spindle is fitted to a cradle holding the tonometer and rotated at approx. 78 r.p.m. This procedure is repeated three times. The cuvette is dried and placed in the spectrophotometer and the spectrum of de-oxy Hb is plotted. Deoxygenation may be judged complete when $A_{555}:A_{540} > 1.24$.

Fig. 3. Optical tonometer designed for use with Hb solutions. The tonometer has an approx. volume of 70 ml and height 200 mm.

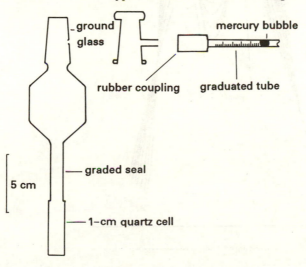

A known amount of air is allowed to enter the tonometer from a narrow calibrated glass tubing (e.g. a 3 ml burette) mounted horizontally and provided with a short length of flexible tubing for connection to the side arm of the tonometer. The distance travelled by a mercury bubble in the tube indicates the amount of air that has been added. The tonometer is then equilibrated for three minutes, or longer, and the spectrum superimposed on the deoxy Hb spectrum. The procedure is repeated until the Hb is all in the oxygenated form. A typical set of spectra is shown in Fig. 4. (Alternatively, point recording of absorbance at the wavelengths used in calculations are recorded.)

Fig. 4. Spectral changes recorded for an OEC using the optical tonometer. The absorbance at 540 and 578 represent fully oxy, and the maxima, deoxy Hb. Fractional saturations are represented by intermediate absorption at a specific wavelength.

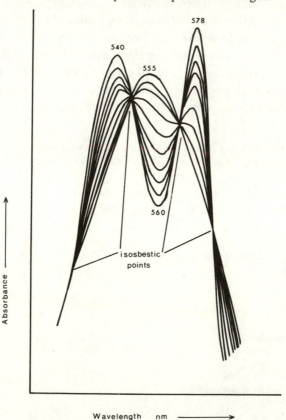

Calculation

(i)   Partial pressure of oxygen in the tonometer. This is calculated from the standard formula:

P_{O_2} (Torr) = (P_B − (H × P_{H_2O})/100) × (0.2095/(V–v)) × ((Tb/Ta) × A),

where P_B = atmospheric pressure in Torr; H = relative humidity in percent; P_{H_2O} = saturated vapour pressure at room temperature in Torr; 0.2095 = fraction of O_2 in air; V = volume of tonometer in ml; V = volume of contents in ml; and A = volume of air added in ml; Tb and Ta are the temperatures in K of the water bath and room air respectively.

(ii)   Percentage saturation of haemoglobin. The absorbance of the deoxy spectra at a particular wavelength represents 0% saturation, and the absorbance (A) of the oxy spectra at the same wavelength represents 100%. From the absorbances at 560 and 578 nm the percentage saturation for each step is calculated from the formula:

% Saturation at A nm at known P_{O_2}
= ((A deoxy − A partially oxy Hb)/(A deoxy − A oxy)) × 100

The % saturation used is the mean at 560 and 578 nm and is plotted against P_{O_2}.

Notes

(i)   any error in recording of the spectra is shown by variation of the isosbestic points. Systematic error due to a shift up or down of the spectrum is partly corrected by using the average value obtained from the two wavelengths used. A systematic shift in the isosbestic points may indicate Hb denaturation or progressive metHb formation.

(ii)   knowledge of the absolute absorbance is not necessary as only the difference in absorbances are used in calculation.

Assessment of the method

This technique is also well tried and tested, and requires little specialized equipment, provided that a spectrophotometer is available. Sufficient points for a fairly complete OEC can be obtained in approx. 45 minutes. The method is particularly suitable for use with Hb solutions and can be used to determine *n* values and the Bohr effect. Its application with suspended erythrocytes is restricted in the buffered system, which does not behave like whole blood *in vivo*. However, carbon dioxide could be used in place of the buffer.

Method 3: Measuring OEC with thin-layer techniques
Principle

Thin-layer methods provide a rapid and sensitive means of measuring OEC from whole blood, or Hb solutions. A thin film of blood or Hb solution is placed in a thermostatted chamber fitted with windows so that light of fixed wavelength passes through. Saturation is monitored spectrophotometrically as changes in extinction, while the P_{O_2} in the chamber is varied by the introduction of appropriate gas mixtures. The thin-layer rapidly comes to equilibrium and if oxygen is introduced into a deoxygenated chamber at a sufficiently slow rate, then continuous, dynamic OEC are recorded. Alternatively, if gas is added in stepwise fashion, a point-by-point, static OEC can be registered.

Whole blood techniques. Blood is a dense suspension of cells and consequently spectrophotometric measurement of saturation must be made on very thin films, and must take account of light-scattering errors. The Aminico Hemoscan produces dynamic OEC from a sample of approx. 20 μm thickness, using dual wavelength optics to correct for light scattering. The P_{O_2} of the gas phase in the chamber is continuously monitored via an oxygen electrode.

In operation, the sample is delivered via a 2 μl pipette to a cover glass and is covered with a gas-permeable membrane to prevent drying. The film is inserted into a humidified chamber and deoxygenated by purging the compartment with a CO_2/N_2 mixture to establish the 0% oxyHb and zero P_{O_2} point on an X–Y recorder chart. Then, oxygen is introduced at a controlled rate into the chamber and an OEC is registered in a few minutes. Although the OEC is registered dynamically, the gas flow is slow enough to assume full equilibrium at any point in time.

Regrettably, the Hemoscan is no longer manufactured, although the principle of operation may be put into practice using a dual wavelength spectrophotometer modified as described by Reeves (1980).

Hb solutions. The methods for dynamic and static OEC as described above are appropriate for dilute solutions. One consequence of using Hb solution is that single wavelength analysis is suitable, and a range of photometric instruments can be used. This feature has been deployed in the dynamic diffusion chamber method described by Sick & Gersonde (1969; 1972) who provide detailed drawings of a suitable chamber (manufactured by Eschweiler & Co., Kiel, FRG). Alternative designs for optical chambers are given by Dolman & Gill (1978), and by Reeves (1980). The method has been modified by Weber *et al.* (1976) to produce stepwise equilibrium using cascading gas mixes supplied from a series of Wösthoff pumps (Bochum, F.R.G.), and precise measurement of oxygen equilibrium data at extreme high and low saturations, allowing estimation of the association equilibrium constants for the oxygenated and deoxygenated states of the molecule (k_R and k_T respectively) and the constants for binding each of the four oxygen molecules (k_1, k_2, k_3 and k_4) (Weber *et al.*, 1987). Fig. 5 is an example of a typical recording from Hb solution. The P_{O_2} is calculated for each equilibration step from the %air or oxygen in the output of the cascaded gas. Lines are drawn connecting the oxy and deoxy levels on the chart. The %S is determined graphically at each step as $100(X/Y)$. Note that often a decrease in ΔA with time is observed. This reflects instability of the Hb during measurement, usually resulting from oxidation, and is automatically corrected for. Hayashi *et al.* (1973) describe an enzymatic system for prevention of metHb formation *in vitro*. Some precautions are indicated in the choice of buffer system employed for Hb solutions (see Wells, 1982).

Assessment of dynamic registration of OEC

The method is quick and gives very accurate data from normal human blood. Thin-layer equilibration is less traumatic for cells and proteins than tonometry. The system is however, rather complex, expensive to set up, and troubleshooting can be problematical. Further, a number of users believed that the dynamic method might give false P_{50} and n values when high affinity, or low co-operativity bloods were used. We have overcome these problems by reducing leakage into the measuring chamber,

Fig. 5. Typical trace for stepwise oxygenation using a thin-layer method. The numbers are percentages of air in the oxygenating mix supplied from a pump. Saturation is calculated from the ratio x/y at a given P_{O_2}.

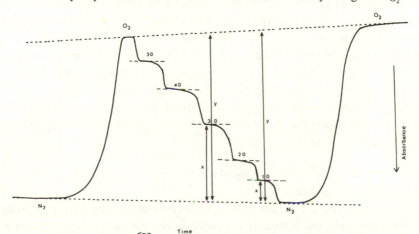

and by reducing the flow rate of oxygenating gas. A thorough analysis of errors and ways of reducing them is given by Lapennas *et al.* (1981). Alternatively, dynamic error is not a problem if OEC are determined statically as a series of stepwise changes in P_{O_2}. This method has the advantage of allowing the experimenter to be sure that equilibrium conditions are attained, and to detect and compensate for autoxidation of unstable Hb (Fig. 5). Because of the high rates of gas flow, oxygen consumption by nucleated erythrocytes does not generate errors of the magnitude that might occur with the mixing method. In the case of whole blood, sample pH must be inferred from a buffer line constructed from log P_{CO_2} vs. pH which may be conducted parallel to OEC measurement (Bridges *et al.*, 1979). For Hb solutions, pH values can be measured in separate samples equilibrated to the same gas tensions (e.g. using a Radiometer BMS-2).

Notes on thin-layer methods

(i)   Gas leakage into the Hemoscan is best reduced by making a new sample holder equipped with a sealing O-ring, and a 10 cm-length of PE-50 cannulation tubing to exit gas from the chamber via a liquid trap. The humidifier well is sealed off and an additional moisturising wick is placed inside the chamber. The wicks must always be kept moist.

(ii)   Gas flow rates must be reduced for dynamic OEC with high affinity. This may be done in the Hemoscan by reducing input gas pressure by half. If dynamic errors persist, then gas mixing pumps may have to be used to make stepwise measurements.

(iii)   Some people advocate a double-layered sandwich of teflon to slow the film from drying, but this increases dynamic error because of its resistance to oxygen diffusion. Rolls of suitable teflon film, Type C, 6 µm, '1/4 Mil' are available from the Export Division of the Dialectrix Corp., Rm. 5710 Empire State Bldg, NY 10118, USA. Silicone-polycarbonate membrane has a better oxygen conductance and is available as MEM-213 from the General Electric Co., Membrane Products Operation, Bldg. 28, Rm. 404, 1 River Rd, Schenectady, NY 12345, USA.

(iv)   The thin-layer sandwich should be scanned in a spectrophotometer to check for oxidation at the conclusion of a run, using absorbance ratios given in Benesch *et al.* (1973).

Procedures for handling blood
Anticoagulants
Heparin has the least effect on affinity measurements, erythrocyte shape, and pH, and is readily available in several salts. Between 100 and

5000 units/ml blood are used according to the clotting properties of the blood in question. The high concentration may be required for example, with blood from stressed fish. Citrate-phosphate-dextrose buffers are acidic, dilute the blood, and disturb intracellular pH. EDTA is also acidic and sequesters divalent ions that may influence Hb function. Glassware and instruments coming into contact with blood must be scrupulously clean to avoid the formation of unwanted Hb derivatives. Fluoridated tap water, for example, can be a serious contaminant. It is always advisable to check for metHb formation at the conclusion of an experiment using one of the optical techniques (Benesch *et al.*, 1973; Dacie & Lewis, 1975).

Storage

Wherever possible, blood should be analysed without delay in order to avoid organic phosphate depletion, pH reduction, metHb formation, and lysis, all of which occur during storage. These problems may also arise during prolonged tonometry. For example, trout blood rapidly depletes its ATP content when tonometered in the absence of oxygen, but the loss is insignificant if the cells are aerobic (Tetens & Lykkeboe, 1981). Nucleated erythrocytes from fish (Eddy, 1977) and birds (Scheid & Kawashiro, 1975) consume oxygen that may lead to errors in saturation estimates.

The preparation of Hb from vertebrate blood has been described in detail by Riggs (1981). Hb will slowly oxidise in domestic freezers, especially in solution at pH < 7.0. This is most marked in the case of fish and reptile Hbs. Purified solutions should always be stored at or below −80 °C, or if this is not possible, protected from oxidation by reaction with CO. Subsequently, the CO is released by photodissociation (see Imai, 1982).

Acknowledgements

The free exchange of tips and techniques between colleagues is acknowledged; in particular, thanks are due to C. R. Bridges, T. Brittain, P. S. Davie, P. S. Davies, G. C. Grigg, E. R. Huehns, G. N. Lapennas, P. L. Lutz, L. A. Maginniss, and V. Tetens.

References

Asakura, T. & Reilly, M. P. (1986). Methods for the measurement of oxygen equilibrium curves of red cell suspensions and hemoglobin solutions. In *Oxygen Transport in Red Blood Cells*, vol. 54, *Advances in the Biosciences*, ed. C. Nicolau, pp. 57–75. Oxford: Pergamon Press.

Benesch, R. E., Benesch, R. & Yung, S. (1973). Equations for the spectrophotometric analysis of hemoglobin mixtures. *Analytical Biochemistry*, **55**, 245–8.

Benesch, R., Macduff, G. & Benesch, R. E. (1965). Determination of oxygen equilibria with a versatile new tonometer. *Analytical Biochemistry*, **11**, 81–7.

Bird, D. J., Lutz, P. L. & Potter, I. C. (1976). Oxygen dissociation curves of the blood of larval and adult lampreys (*Lampetra fluviatilis*). *Journal of Experimental Biology*, **65**, 449–58.

Blunt, M. H. (1974). A micromixing method for determination of oxygen equilibria in blood. *Journal of Applied Phyisiology*, **37**, 123–5.

Bridges, C. R. (1983). pO_2 and oxygen content measurements in blood samples using polarographic oxygen sensors. In *Polarographic Oxygen Sensors*, ed. E. Gnaiger & G. Forstner, p. 219–33. Berlin: Springer-Verlag.

Bridges, C. R., Bicudo, J. E. P. W. & Lykkeboe, G. (1979). Oxygen content measurement in blood containing haemocyanin. *Comparative Biochemistry and Physiology*, **62A**, 457–62.

Bridges, C. R., Hlastala, M. P., Riepl, G. & Scheid, P. (1983). Root effect induced by CO_2 and by fixed acid in the blood of the eel, *Anguilla anguilla*. *Respiration Physiology*, **51**, 275–86.

Brix, O. (1981). A modified Van Slyke apparatus. *Journal of Applied Physiology*, **50**, 1093–7.

Burkhard, O. & Barnikol, W. K. R. (1982). Dependence of visible spectrum, $(\varepsilon(\lambda))$ of fully oxygenated hemoglobin on concentration of hemoglobin. *Journal of Applied Physiology*, **52**, 124–30.

Cameron, J. N. (1971). Rapid method for determination of total carbon dioxide in small blood samples. *Journal of Applied Physiology*, **31**, 632–4.

Cech, J. J. Laurs, R. M. & Graham, J. B. (1984). Temperature-induced changes in blood gas equilibria in the albacore, *Thunnus alalunga*, a warm-bodied tuna. *Journal of Experimental Biology*, **109**, 21–34.

Christophorides, C. & Hedley-White, J. (1969). Effect of temperature and hemoglobin concentration on solubility of O_2 in blood. *Journal of Applied Physiology*, **27**, 592–6.

Dacie, J. V. & Lewis, S. M. (1975). *Practical Haematology*, 5th ed, 629 pp. Oxford: Churchill Livingstone.

Dolman, D. & Gill, S. J. (1978). Membrane-covered thin-layer optical cell for gas-reaction studies of hemoglobin. *Analytical Biochemistry*, **87**, 127–34.

Duvelleroy, M. A., Buckles, R. G., Rosenkaimer, S., Tung, C. & Laver, M. B. (1970). An oxyhemoglobin dissociation analyzer. *Journal of Applied Physiology*, **28**, 227–33.

Easton, D. M. (1979). Oxyhemoglobin curve as expo-exponential paradigm of asymmetric sigmoid function. *Journal of Theoretical Biology*, **76**, 335–49.

Eddy, F. B. (1977). Oxygen uptake by rainbow trout blood, *Salmo gairdneri*. *Journal of Fish Biology*, **10**, 87–90.

Edwards, M. J. & Martin, R. J. (1966). Mixing technique for the oxygen-hemoglobin equilibrium and Bohr effect. *Journal of Applied Physiology*, **21**, 1898–902.

Grigg, G. C. & Wells, R. M. G. (1988). A simple and inexpensive tonometry system for use with microlitre blood samples. *Comparative Biochemistry and Physiology*, **89A**, 593–4.

Grubb, B. R. & Mills, C. D. (1981). Blood oxygen content in microlitre samples using an easy-to-build galvanic oxygen cell. *Journal of Applied Physiology*, **50**, 456–64.

Hayashi, A., Suzuki, T. & Shin, M. (1973). An enzymic reduction system for metmyoglobin and methemoglobin, and its application to functional studies of oxygen carriers. *Biochimica et Biophysica Acta*, **310**, 309–16.

Hughes, G. M., Belaud, A., Peyraud, C. & Adcock, P. J. (1982). A comparison of two methods for measurement of O_2 content of small (20 µl) samples of fish blood. *Journal of Experimental Biology*, **96**, 417–20.

Hughes, G. M., O'Neill, J. G. & van Aardt, W. J. (1976). An electrolytic method for determining oxygen dissociation curves using small blood samples: the effect of temperature on trout and human blood. *Journal of Experimental Biology*, **65**, 21–38.

Ikeda-Saito, M., Yonetani, T. & Gibson, Q. H. (1983). Oxygen equilibrium studies on hemoglobin from the bluefin tuna (*Thunnus thynnus*). *Journal of Molecular Biology*, **168**, 673–86.

Imai, K. (1982). Allosteric effects in haemoglobin. Cambridge University Press: Cambridge.

Imai, K., Hayashi, A. & Suzuki, T. (1980). Automatic oxygen equilibrium determination and its clinical application – recent topics. *Hemoglobin*, **4**, 567–72.

Imaizumi, K., Imai, K. & Tyuma, I. (1978). On the validity of the spectrophotometric determination of oxygen saturation of hemoglobin. The wavelength dependence of observed oxygen equilibrium parameter values. *Journal of Biochemistry*, **83**, 1707–13.

Johansen, K. Lykkeboe, G., Weber, R. E. & Maloiy, G. M. O. (1976). Respiratory properties of blood in awake and estivating lungfish, *Protopterus amphibius*. *Respiration Physiology*, **27**, 335–45.

Jones, D. R., Brill, R. W. & Mense, D. C. (1986). The influence of blood gas properties on gas tensions and pH of ventral and dorsal aortic blood in free-swimming tuna, *Euthynnus affinis*. *Journal of Experimental Biology*, **120**, 201–13.

Lapennas, G. N., Colacino, J. M. & Bonaventura, J. (1981). Thin-layer methods for determination of oxygen binding curves of hemoglobin solutions and blood. *Methods in Enzymology*, **76**, 449–70.

Lapennas, G. N. & Reeves, R. B. (1983). Oxygen affinity and equilibrium curve shape in blood of chicken embryos. *Respiration Physiology*, **52**, 13–26.

Laver, M. B., Murphy, A. J., Seifen, A. & Radford, E. P. (1965). Blood O_2 content measurements using the oxygen electrode. *Journal of Applied Physiology*, **20**, 1063–9.

Longmuir, I. S. & Chow, J. (1970). Rapid method for determining effect of agents on oxyhemoglobin dissociation curves. *Journal of Applied Physiology*, **28**, 343–5.

Lutz, P. L. (1980). On the oxygen affinity of bird blood. *American Zoologist*, **20**, 187–98.

Lykkeboe, G. & Johansen, K. (1978). An O_2–Hb 'paradox' in frog blood? (n-values exceeding 4.0). *Respiration Physiology*, **35**, 119–27.

Maginniss, L. A., Song, Y. K. & Reeves, R. B. (1980). Oxygen equilibria of ectotherm blood containing multiple hemoglobins. *Respiration Physiology*, **42**, 329–43.

Nikinmaa, M. (1983). Adrenergic regulation of haemoglobin oxygen affinity in rainbow trout red cells. *Journal of Comparative Physiology*, **152**, 67–72.

Oeseburg, B., Landsman, M. L. J., Mook, G. A. & Zijlstra, W. G. (1972). Direct recording of oxyhaemoglobin dissociation curve *in vivo*. *Nature*, **237**, 149–50.

Reeves, R. B. (1980). A rapid micro method for obtaining oxygen equilibrium curves on whole blood. *Respiration Physiology*, **42**, 299–315.

Riggs, A. (1981). Preparation of blood hemoglobins of vertebrates. *Methods in Enzymology*, **76**, 5–29.

Scheid, P. & Kawashiro, T. (1975). Metabolic changes in avian blood and their effects on determination of blood gases and pH. *Respiration Physiology*, **23**, 291–300.

Scheid, P. & Meyer, M. (1978). Mixing technique for study of oxygen-hemoglobin equilibrium: a critical evaluation. *Journal of Applied Physiology*, **45**, 818–22.

Sick, H. & Gersonde, K. (1969). Method for continuous registration of O_2-binding curves of hemoproteins by means of a diffusion chamber. *Analytical Biochemistry*, **32**, 362–76.

Sick, H. & Gersonde, K. (1972). Theory and application of the diffusion technique for measurement and analysis of O_2-binding properties of very autoxidizable hemoproteins. *Analytical Biochemistry*, **47**, 46–56.

Teisseire, B., Teisseire, L., Lautier, A., Herigault, R. & Laurent, D. (1975). A method of continuous recording on microsamples of the Hb–O_2 association curve. I. Technique and direct registration of standard results. *Bulletin of Physio-Pathological Respiration*, **11**, 837–51.

Teisseire, B. P., Ropars, C., Vallez, M. O., Herigault, R. A. & Nicolau, C. (1985). Physiological effects of high-P_{50} erythrocyte transfusion on piglets. *Journal of Applied Physiology*, **58**, 1810–17.

Tetens, V. & Lykkeboe, G. (1981). Blood respiratory properties of rainbow trout, *Salmo gairdneri*: responses to hypoxia acclimation and anoxic incubation of blood in vitro. *Journal of Comparative Physiology*, **145**, 117–25.

Tetens, V., Wells, R. M. G. & DeVries, A. L. (1984). Antarctic fish blood: respiratory properties and the effects of thermal acclimation. *Journal of Experimental Biology*, **109**, 265–79.

Tucker, V. A. (1967). Method for oxygen content and dissociation curves on microlitre blood samples. *Journal of Applied Physiology*, **23**, 410–14.

Van Slyke, D. D. & Neill, J. M. (1924). The determination of gases in blood and other solutions by vacuum extraction and manometric measurement. *Journal of Biological Chemistry*, **61**, 523–73.

Weber, R. E., Lykkeboe, G. & Johansen, K. (1976). Physiological properties of eel haemoglobin: hypoxic acclimation, phosphate effects and multiplicity. *Journal of Experimental Biology*, **64**, 75–88.

Weber, R. E., Jensen, F. B. & Cox, R. P. (1987). Analysis of teleost hemoglobin by Adair and Monod-Wyman-Changeux models. Effects of nucleoside triphosphates and pH on oxygenation of tench hemoglobin. *Journal of Comparative Physiology* B, **157**, 145–52.

Weber, R. E., Wells, R. M. G. & Rossetti, J. E. (1983). Allosteric interactions governing oxygen equilibria in the haemoglobin system of the spiny dogfish, *Squalus acanthias*. *Journal of Experimental Biology*, **103**, 109–20.

Wells, R. M. G. (1976). The oxygen affinity of chicken haemoglobin in whole blood and erythrocyte suspensions. *Respiration Physiology*, **27**, 21–31.

Wells, R. M. G. (1979). Haemoglobin-oxygen affinity in developing embryonic erythroid cells of the mouse. *Journal of Comparative Physiology*, **129**, 333–8.

Wells, R. M. G. (1982). Alteration of hemoglobin function by two aliphatic amine buffers. *Hemoglobin*, **6**, 523–30.

Wells, R. M. G. & Shumway, S. E. (1980). The effects of salts on haemocyanin-oxygen binding in the marine pulmonate snail *Amphibola crenata* (Martyn). *Journal of Experimental Marine Biology and Ecology*, **43**, 11–27.

Wells, R. M. G. & Weber, R. E. (1982). The Bohr effect of the hemocyanin-containing blood from the terrestrial slug *Arion ater*. *Molecular Physiology*, **2**, 149–59.

Wells, R. M. G. & Weber, R. E. (1985). Fixed acid and carbon dioxide Bohr effects as functions of hemoglobin-oxygen saturation and erythrocyte pH in the blood of the frog, *Rana temporaria*. *Pflugers Archiv*, **403**, 7–12.

Winslow, R. M., Statham, N. J. & Rossi-Bernardi, L. (1981). Continuous determination of the oxygen dissociation curve of whole blood. *Methods in Enzymology*, **76**, 511–23.

Zander, R., Lang, W. & Wolf, H. U. (1978). A new method for measuring the oxygen content in microlitre samples of gases and liquids: the oxygen cuvette. In *Oxygen Transport to Tissue*, vol. 1, *Advances in Experimental Medicine and Biology Series*, ed. I. A. Silver, M. Erencinska & H. I. Bicher, pp. 107–11. New York: Plenum Press.

N. HEISLER

Parameters and methods in Acid–Base Physiology

Introduction

During the first 50 years after the fundamentals of acid–base physiology had been laid by the work of Henderson (1909), Hasselbalch (1916), Van Slyke (1917, 1922) and other pioneers in this field, subsequent analyses have focused almost exclusively on the description of blood parameters, and, in some rare instances, on determinations of isolated tissue parameters, such as buffer titration curves and total CO_2 content. Description of the status in a fluid compartment representing a minor fraction of the body mass, however, must necessarily result in an incomplete and sometimes misleading picture of the organismic acid-base status. Particular problems are provided by analyses of circulating blood, which represents less than 8% of the body mass in vertebrates, and acts as the main convective systemic transport medium. Acid–base relevant ions produced in various intracellular body compartments are transported by the blood to the site where they are eliminated either by metabolic processes, or by permanent excretion from the body fluids. These fluxes of acid–base relevant ions, among and through body fluid compartments, have been rather discounted in the early studies of acid–base physiology, despite the fact that they most certainly represent an important factor in the overall evaluation of acid–base regulation.

Only recently has the role of these processes for the overall regulation received more attention. The transfer of acid–base relevant ions across cell membranes of systemic tissues during exogenous acid–base disturbances was directly and quantitatively determined for the first time only about 15 years ago (Heisler & Piiper, 1972). In addition, practical and accurate intracellular pH measurements on the basis of the DMO method (Waddell & Butler, 1959; see below) together with reliable closed-system determinations of intracellular buffer values (Heisler & Piiper, 1971; see below) have allowed for estimations of the amount of surplus H^+ ions buffered in tissues for only a short time as well. The development of techniques for the complete quantitative analysis of small transepithelial acid–base relevant ion transfers in aquatic species (Heisler, Weitz & Weitz, 1976; Heisler, 1978) has completed the basis for a more comprehensive evaluation of the factors involved in pH adjustment in individual tissues as well as in the whole organism.

This chapter is intended to outline approaches and methods required for such analyses. It will focus, in particular, on the estimation of transmembrane and transepithelial ion transfer processes, but will also briefly describe some peculiarities of apparently trivial methods like determination of pH and P_{CO_2} in samples of low buffer value and at low temperatures, conditions quite typical for comparative studies. The rather wide scope of the methods involved, naturally limits the space available for the description of individual procedures, and as a result this chapter often has to refer to more extensive publications.

General approaches

The acid–base status of a biological system can principally be described in two different ways. The more conventional and most often used system is based on Brønsted's (1923) definition of acids (substances capable of dissociating H^+ ions) and bases (substances capable of accepting H^+ ions and transferring them into the non-dissociated state), and on the interrelation of pH and the other two main components of the CO_2/bicarbonate buffer system, the Henderson-Hasselbalch equation (Henderson, 1909; Hasselbalch, 1916). The other system, the 'strong ion difference' concept (SID; Stewart, 1978), is in principle fully compatible, but approaches the problem on the basis of the amount of strong ions, which, for the sake of electroneutrality, balance the dissociated buffer anions or cations. The difference between the negatively and positively charged strong ions is stoichiometrically equivalent to the amount of 'buffer bases' (Singer & Hastings, 1948), and the changes in buffer bases and SID are equivalent to the term 'base excess' (Siggard-Andersen, 1963, 1974).

The strong ion difference (SID) certainly allows a complete description of the acid–base status, but suffers from practical limitations. The strong ions accompanying the buffer ions usually have to be determined on a high background of paired strong anions and cations. Accordingly the achieved accuracy is much less (Heisler, 1986a; see also below) than the direct determination of pH, P_{CO_2}, $[HCO_3^-]$ and the non-bicarbonate buffer value, which are the parameters describing the acid–base status on the basis of the conventional system. Additional uncertainty is provided by the fact that the variety of strong ions in biological, especially intracellular, compartments is often not known to its full extent and thus can hardly be completely determined. Also, the ionic activity only is relevant for determination of the

SID, rather than the overall concentration usually measured (cf. Heisler, 1986*a*; Jackson & Heisler, 1982). These problems disqualify the SID concept for biological studies, except for some marginal applications, such as an estimate of the amount of buffered H^+ ions from the amount of Ca^{2+} and Mg^{2+} ions released from bone structures. Accordingly, the typical approaches will be described in this chapter in terms of the conventional system.

Approach 1

This approach estimates the amount of bicarbonate-equivalent ions transferred across the cell membrane of an intracellular fluid compartment from the actual changes in bicarbonate concentration, and the amount of bicabornate produced or decomposed by intracellular non-bicarbonate buffers (Fig. 1; cf. Heisler, 1986*a*). When the temperature is variable the amount of bicarbonate-equivalent* ions produced by non-bicarbonate buffering ($\Delta[HCO_3^-]_{NB}$) has to be determined on the basis of the temperature-dependent buffer value matrix (Heisler & Neumann, 1980). Usually, a ternary buffer system (imidazole-like and phosphate-like non-bicarbonate buffers, and the bicarbonate buffer system) sufficiently models the *in vivo* conditions (Heisler & Neumann, 1980; see also: Heisler, 1984*a*, 1986*a*). The amount of bicarbonate-equivalent ions transferred across the cell membrane ($\Delta HCO_{3\,i\to e}^-$) can be estimated as the difference between the amount of bicarbonate produced by non-bicarbonate buffering, and the change in bicarbonate in the intracellular compartment. If these parameters are available for all or most of the intracellular fluid compartments the transepithelial ion transfer [$\Delta HCO_{3\,e\to env}^-$) can be determined as the integral of individual tissue transmembrane transfers, the amount of bicarbonate produced in the extracellular space, and the extracellular [HCO_3^-] difference (Fig. 1). This parameter, however, should preferably be determined directly, since the extent of extrapolation may often be too large to yield reliable and significant information (Heisler, 1986*a*).

Approach 2

The amount of bicarbonate-equivalent ions transferred across various epithelial sites can be measured directly as the stoichiometric difference between the changes in environmental bicarbonate and ammonia

---

* Transfer of H^+ ions, or of OH^- and HCO_3^- in the opposite direction has the same effect, but cannot be distinguished directly by the present methods. As all acid–base relevant transfer processes affect the bicarbonate concentration in the involved fluid compartments and their effect can be described on the basis of virtual transfer of bicarbonate, the analysis in this chapter utilizes the term 'bicarbonate equivalents'.

Fig. 1. General approaches for the determination of an overall pattern of the acid–base regulation. Approach 1: Various specific intracellular tissue compartments (ICS, only one indicated) are connected in parallel to the extracellular space (ECS), which in turn is serially connected to the environment. Approach 2: The environment and the extracellular space are considered as being serially connected with the overall intracellular space. 'Δ' indicates differences between two states. Indices designate certain fluid compartments, or the direction of bicarbonate equivalent ion transfer ('e' = extracellular, 'i' = intracellular, 'env' = environmental, extraepithelial). HCO_3^- = bicarbonate, NH_4^+ = ammonia, β = non-bicarbonate buffer value (indices 'Im' = imidazole-like, 'Ph' = phosphate-like, 'tot' = total, 'NB' = produced by non-bicarbonate buffering). 'V' = volume of the indicated fluid compartment.

Approach 1

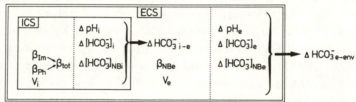

Temperature constant

$$\Delta [HCO_3^-]_{NBi} = \beta_{tot}\cdot -\Delta pH \qquad \| \qquad \Delta HCO_{3\,i-e}^- = (\beta_{tot}\cdot -\Delta pH - \Delta [HCO_3^-]_i)\,V_i$$

Temperature variable

$$\Delta [HCO_3^-]_{NBi} = \beta_{Im}(\Delta pK_{Im} - \Delta pH_i) + \beta_{Ph}(\Delta pK_{Ph} - \Delta pH_i)$$

$$\Delta HCO_{3\,i-e}^- = \{\beta_{Im}(\Delta pK_{Im} - \Delta pH_i) + \beta_{Ph}(\Delta pK_{Ph} - \Delta pH_i) - \Delta [HCO_3^-]_i\}\cdot V_i$$

(for each individual tissue)

$$\Delta HCO_{3\,e-env}^- = \sum_1^n \Delta HCO_{3\,i-e}^- - \Delta [HCO_3^-]_e \cdot V_e + \Delta HCO_{3\,NBe}^-$$

Approach 2

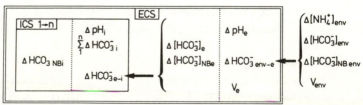

$$\Delta HCO_{3\,e-env}^- = (\Delta [HCO_3^-]_{env} - \Delta [NH_4^+]_{env} - \Delta pH_{env}\cdot \beta_{NBenv})\cdot V_{env}$$

$$\Delta HCO_{3\,e-i}^- = \Delta HCO_{3\,env-e}^- - \Delta [HCO_3^-]_e \cdot V_e + \Delta HCO_{3\,NBe}^-$$

$$\Delta HCO_{3\,NBi}^- = -\Delta HCO_{3\,e-i}^- + \sum_1^n \Delta HCO_{3\,i}^-$$

content (Fig. 1; cf. Heisler, 1984b; 1986a). When urine excretion contributes significantly to the overall transepithelial transfer, the non-bicarbonate buffering in the urine has to be taken also into account (cf. Heisler, 1986a). The overall transmembrane bicarbonate-equivalent transfer can then be estimated from the transepithelial transfer, the amount of bicarbonate produced by non-bicarbonate buffering, and the change in extracellular bicarbonate concentration (Fig. 1). Any further rough estimate as to the overall amount of bicarbonate produced by non-bicarbonate buffering may provide some general information about the buffering capabilities of the organism, but cannot be expected to yield more than qualitative results because of numerous uncertainties introduced by the impossibility to determine (HCO_3^-) and pH in all intracellular body compartments (Heisler, 1986a).

To date these approaches provide the only way to determine either the tissue-specific (approach 1) or the overall transmembrane acid–base relevant ion transfer (approach 2) in intact organisms. Furthermore, these approaches may yield an evaluation of the relative role of ionic transfer processes for the acid–base regulation during certain conditions, when the transfer terms in the above approaches are set to zero and the appropriate equations are solved for the resulting changes in pH and $[HCO_3^-]$ (cf. Heisler & Neumann, 1980; Heisler, 1984a, 1984b, 1986b, 1988b).

Although the extent of extrapolation by either of these approaches is not great, some limitations for application of these models should be kept in mind. Approach 1 is sensitive to deviations from steady state conditions, such as the transient production and accumulation of intermediary, nonvolatile metabolic end-products (e.g. lactic acid, cf. Heisler, 1986c). The amount of bicarbonate converted into molecular CO_2 by such events will appear as having been transferred to the extracellular space (Heisler, 1986a). Approach 1 is usually also limited to single point determinations due to the nature of typical methods for determination of intracellular pH in intact animals (see below), whereas approach 2 can be applied to monitor also transients of acid–base regulation. Approach 2 is quite insensitive to non-steady-state conditions, whereas analysis of such transients by application of approach 1 will certainly present considerable problems (Heisler, 1986a, c).

The model calculations outlined above have been designed for certain purposes and should not be applied without critical reconsideration and adjustment to different conditions and experimental procedures. However, as long as the essential prerequisites are fulfilled, valuable information may be gained from these indirect estimates until methods for direct measurements of the parameters in question have been developed.

Methods for determination of acid–base parameters

Because of space limitations the description of methods given below must remain incomplete, and therefore, focuses either on procedures applied most frequently, or on techniques providing significant advantages. For less widely distributed methods the reader is referred to the relevant original publications.

pH in directly accessible fluids

pH as the central parameter of acid–base physiology, is routinely measured in blood, urine, cerebrospinal fluid and other easily accessible fluids. The generally applied glass electrode technique is based on selective masking of H^+ ions in the 'gel' surface of special pH-sensitive glass, eliciting a potential across the glass membrane that is logarithmically proportional to the ratio of H^+ activity in sample and internal electrolyte solution of the electrode (Nernst equation).

Various designs of electrodes are available, differing by the characteristics of the pH-sensitive glass, the shape of the glass membrane, and the reference electrode system. At least four to five main glass qualities have to be distinguished with respect to their characteristics and their suitability for certain measurements (Table 1). The optimal reference system may be quite variable with the type of application. The electrical circuit between the inner electrolyte solution of the glass membrane and the sample is closed by a so-called internal reference, usually a chlorinated silver electrode (Ag/AgCl), the input impedance of the amplifier, and the external reference (cf. Figs. 3, 4 and 7), which, for optimal symmetry, is preferably chosen to be the same as the internal reference.

The external reference system consists of an Ag/AgCl or a calomel ($Hg/HgCl_2$) electrode, and the most important and very critical electrolyte bridge to the sample. Single electrolyte bridges may consist of a direct capillary connection from the reference electrode (1 to 3 M KCl, saturated with AgCl in Ag/AgCl references) in the form of one to three porous ceramic disks, a sleeve diaphragm, or a simple direct connection (0.5 to 1 mm diameter) between the electrolyte and the sample. The direct contact between samples and the highly concentrated and Ag/AgCl saturated KCL solution regularly results in deposition of polar salt complexes, or worse, of denatured protein. These polar deposits may lead to spurious additional potentials, which will not interfere with calibration in well buffered solutions, but may result in misestimates of pH of up to 0.2–0.3 pH units in relatively dilute samples, such as blood of lower vertebrates or even environmental water (see also Siggaard-Andersen, 1963, 1974). This problem is less severe with sleeve diaphragms, which possess a more defined

Table 1. *Characteristics and applications of different pH glass electrode*

Type[a] (Glass)	Resistance (MΩ)[b]	pH range [Na+] 0.1 M 1.0 M	Temp. range (°C)	Alkali cross-sensitivity[c]	Stability of hydration	Applications
Standard (A41)	400	0–10 0–9	15–130	++	–	General purpose, medium pH range titration of concentrated samples
LoT	80	0–9 0–8	0–70	+++	++++	Low temperature and medium range fast response and high stability applications, ΔHCO_3^- determinations
Ti	150	0–9 0–8	0–60	+	+	Capillary glass electrodes, blood measurements
U	230	0–13 0–12	10–90	+	++	High pH, high alkali, high temperature, ammonia titrations
HA	1200	0–13 0–12	10–130	–	–	Very high alkali, very high temperature and pH

[a] Denominations of glass qualities, if provided at all, vary considerably among suppliers.
[b] rough estimate for same thickness of the glass membrane and temperature
[c] increasing with rising temperature

interface, that may also be cleaned. The problem is completely avoided by a double electrolyte bridge, where a second bridge of physiological solution, or an electrolyte inert to the sample constituents, is interconnected between the reference electrolyte, and the sample, preferably with a sleeve diaphragm (cf. Fig. 4 and 7). This type of reference system yields the best results for very accurate long term measurements in fluids like the environmental water of aquatic species (cf. Heisler, 1984a, 1986b). During short-term measurements in individual samples containing high protein concentrations (e.g. blood) the best solution is an open interface, with the KCL bridge completely exchanged before every single measurement. This type of bridge is often used in commercial blood gas analyzers (e.g. BMS 1, 2 or 3, Mk 1; Radiometer, Copenhagen, Denmark).

Accurate pH measurements can be achieved only with extremely constant conditions for the pH-sensitive glass surface, and the two reference electrodes. The potentials of the internal and external reference systems are very sensitive to changes in temperature. Although the absolute potentials against standard hydrogen electrodes are similar for Ag/AgCl (+236 or +207 mV in 1 or 3 M KCL) and calomel electrodes (+244 mV), the time course to achieve a stable potential after temperature changes is rather different between the two types. Symmetry of the reference chains (e.g. Ag/AgCl for both references) can reduce this problem, but even short temperature deviations will still result in many hours of small drifts in the overall chain potential due to potential changes at the glass membrane. Similar drifts have to be expected when hydration and ionic loading of the glass surface (the 'gel') is changed by exposure to buffers and samples of different ionic strength. This problem can be avoided by the use of appropriately adjusted buffers for blood and similar fluids, urine and seawater with the buffer pH values chosen as close as possible to the expected range (< 0.1 unit). In dilute freshwater, however, calibration of the electrodes is an extremely time consuming procedure and can hardly be repeated during experimentation. Accordingly the electrode chains for measurement of ionic transepithelial transfer to have to be extremely constant (drift of less than 0.001 units/24 h, Heisler, 1984a, 1986b), and have to be selected and pretreated for this purpose. Although new electrodes can be rapidly hydrated to a certain extent by exposure to 0.1 N HCl for 4 to 6 hours or similar procedures, the slow adaptation to the sample composition (for weeks or even months) will always result in much improved stability, and also in a reduction of the resistance of the glass membrane.

Due to the high internal resistance of the glass membrane the electronic processing of the signal is rather critical. The input impedance of the preamplifier should exceed the electrode resistance by at least 3 orders of magnitude ($>10^{12}$ ohms) in order not to bias the recorded potential, and should be isolated from the ground by a similar resistance. The commercially available pH meters usually provide these features as stand-alone units. If, however, the signal output of such instruments is connected to non-floating recorders, the lack of a real differential amplifier (usually only the glass electrode input is fed into a high impedance voltage follower) will result in a variable voltage offset. The better (and in most cases less expensive) choice is to use special high impedance isolation amplifiers (e.g. Model 87, Knick, Berlin, FRG), and inexpensive low impedance recorders and voltmeters, in order to avoid ground loops especially in setups with several electrodes (cf. Heisler, Forcht, Ultsch & Anderson, 1982; Holeton, Neumann & Heisler, 1983; Claiborne & Heisler, 1984, 1986).

P_{CO_2}

P_{CO_2} can be determined in fluid samples by a number of methods. More indirect methods are those of equilibrating a fluid sample with a small bubble of gas, and then determine P_{CO_2} by gas analysis (e.g. Scholander, 1942, 1947; mass spectrometry, gas chromatography, cf. handbooks of relevant gas analysis), or the double equilibration method of Astrup (1956). The latter method is based on triplicate pH measurements before and after equilibration with gases of two different P_{CO_2} values. Interpolation on the basis of the original sample pH and the CO_2 titration curve in a log P_{CO_2}/pH plot yields the P_{CO_2} value.

Those and similar techniques are nowadays limited to marginal applications and have been replaced by the comparatively simple direct potentiometric determination of P_{CO_2} with membrane-covered pH electrodes (Stow, Baer & Randall, 1957; Gertz & Loeschke, 1958; see also: Severinghaus, 1965). Physically dissolved CO_2 diffuses through a plastic membrane, affecting pH of the inner, bicarbonate containing electrolyte solution according to the Henderson–Hasselbalch equation when equilibrium is attained. Although commercially available P_{CO_2} electrodes are generally designed for use at 37 °C and with mammalian blood, many models are also suitable for measurements at lower temperatures, as long as some basic prerequisites are fulfilled.

A critical point of P_{CO_2} electrodes is the response time. The time required to reach full equilibrium is a function of various factors, such as diffusion of CO_2 past the plastic membrane, hydration of CO_2 in the inner electrolyte which does not contain any carbonic anhydrase, and finally the response time of the pH electrode, which is mainly a function of the glass quality used (cf. Table 1). The time required to reach equilibrium is in the range of

2 min at 37 °C, but increases tremendously (Q_{10} about 2.5–3) at lower temperatures (up to 1 h when close to 0 °C, depending on various factors such as the difference in P_{CO_2} between calibration gas and sample). Since the nucleated red blood cells of lower vertebrates have a considerable metabolism, P_{CO_2} of the sample will change significantly during the response time and accordingly bias the determination. This effect can be minimized by reducing the response time by choosing the gases flushed through the electrode chamber for recalibration before and after sample measurements close to the expected sample P_{CO_2}. This will reduce the retarding effect of diffusion and hydration. The response time can also be shortened by use of special thin nylon fabric (even ordinary nylon stocking) stretched over the glass membrane (this so-called 'spacer' ensures a thin electrolyte space between glass and plastic membrane) rather than the relatively thick and dense spacer material commercially supplied. Also an appropriate choice of plastic membrane material may result in the improvement of the response time (at low temperatures the response is considerably faster with 25 μm silicone rubber membrane than with 7.5 μm Teflon). Because of the slow electrode response at temperatures close to 0°C P_{CO_2} is sometimes calculated from pH and $[HCO_3^-]$ on the basis of the Henderson–Hasselbalch equation with appropriate constants (see below: Bicarbonate concentration).

In most lower vertebrate studies, however, the temperature is too low and nucleated red cell metabolism too high to achieve any really steady readings at all. The electrode potential has then to be recorded and extrapolated to time zero (the time the sample is withdrawn from the animal), preferably with the aid of a graphic computer program. As a first approximation the extrapolation may be performed on the basis of single component correlations between P_{CO_2} (or mV, or pH) and time, although the process is certainly much more complex. The best choice is an empirically determined fit of the extrapolation curve (from sample measurements of fresh blood equilibrated with known P_{CO_2}), which takes into account the species-specific red cell metabolism, the changes in temperature during the (standardized) transit time before introduction into the electrode chamber, the electrode characteristics and other, usually unknown, factors. Extrapolation should be conducted only from the late *quasi* steady state part of the curve, which is likely to be affected mainly by the sample metabolism.

Bicarbonate concentration

The indicator substance of the approaches lined out above, bicarbonate, is almost always determined after conversion to CO_2 by acidification below pH 4, although bicarbonate can principally be measured by a double titration procedure, a method often applied to analysis of urine (cf. Heisler, 1986*a*). Further processing and detection of the physically dissolved CO_2 varies among methods. One of the first methods and very reliable ($\pm$ 0.100 mM) was the vacuum extraction with subsequent manometric determination described by Van Slyke (1917) and Van Slyke & Neill (1924). The main disadvantages of this method are the large sample volume, the high manipulatory skills of an operator, and the time needed per analysis (about 20 min). Attempts to scale this method down were not very successful because of the general loss of accuracy. Various modifications of the Van Slyke method, mainly with substitution of the manometric volume determination by mass spectrometry or gas chromatography

Fig. 2. Methods for determination of total CO_2 content in fluid samples. *Cameron Chamber*: Bicarbonate introduced by sample injection into the acidic solution in the chamber is converted into molecular CO_2 (physically dissolved CO_2), inducing a proportional change in P_{CO_2}. The apparatus is calibrated in absolute terms by injection of aqueous bicarbonate standards (for details see text).
Capnicon: Bicarbonate in the injected sample is converted into physically dissolved CO_2 which is transported by the carrier gas into the gas exchange column. The integrated change in electrical conductivity of 0.01 N NaOH induced by absorption of O_2 and production of carbonate is proportional to the amount of injected bicarbonate (see also text).

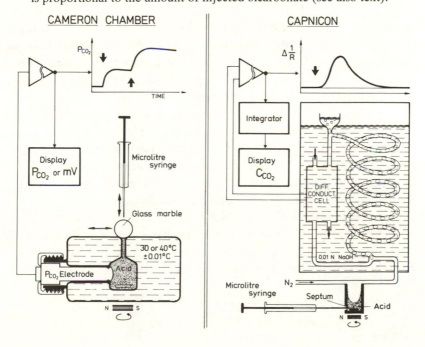

have to date not gained wide distribution, although the accuracy usually exceeds that of the Van Slyke method (e.g. mass spectrometry, Bridges & Scheid, 1982; gas chromatography, R. G. Boutilier & G. Iwama, Dalhousie University, Halifax, Nova Scotia, Canada, pers. comm.). One reason for the limited distribution of such methods may be that they require relatively extensive and expensive apparatus.

A very reliable and elegant method has been described by Cameron (1971). The apparatus is rather simple and consists of a thermostatted chamber (30–40 °C) with the sensitive surface of a P_{CO_2} electrode mounted into one side of it (Fig. 2). The reaction chamber (2 mL) is filled with an acidic solution, mixed by a magnetic stirrer. The change in P_{CO_2} upon introduction of a sample (20–50 µL) into the fixed volume of acidic fluid is directly proportional to the amount of total CO_2 in the sample. The method is sensitive to the diffusive loss of CO_2 through the chamber inlet (which should be closed down to a minimal vent directly after sample introduction), through the walls of the chamber (if constructed from plastic material instead of glass), and through the electrode itself. When these factors of error are eliminated, for instance by extrapolation to time zero (see above), the method provides a very good accuracy of about ± 2%. The time required for one analysis is in the range of 3–4 min.

The most accurate method for the determination of absolute concentrations of bicarbonate (more precisely: total CO_2) is the Capnicon III apparatus (Fig. 2; Cameron Instruments, Port Aransas, Tx, USA). The sample bicarbonate is converted into physically dissolved CO_2 by injection into stirred (for proteins containing samples) or bubbled (aqueous samples) acidic solution, and washed out by a carrier gas stream (pure N_2). The gas stream is fed through a glass gas exchange column, forming a large surface interface with about the same volume of 0.01 N NaOH drawn through the same column. The formation of carbonate from absorption of CO_2 is detected by differential electrical conductivity measurement (Maffly, 1968) between inflowing and outflowing NaOH. Integration of the differential signal yields the absolute amount of absorbed CO_2. The accuracy of this method is limited to ± 0.5% mainly by the volumetric measurements of samples and standards (20–50 µL).

Direct determination of the bicarbonate concentration by one of the above mentioned methods is most preferable, but not always possible. Accordingly, the bicarbonate concentration is often calculated from pH and P_{CO_2} on the basis of the Henderson–Hasselbalch equation:

$$\mathrm{pH} = \mathrm{p}K_1' + \log \frac{[HCO_3^-]}{\alpha_{CO_2} \cdot P_{CO_2}}$$

The required 'constants' α_{CO_2} and pK_1' depend in theory on very few physicochemical parameters, as long as only the free water phase (exclusively relevant for calculation based on the Henderson–Hasselbalch equation) is taken into account. None of these is actually species-specific. The solubility of O_2, α_{CO_2}, varies only with temperature and concentration of dissolved substances:

$$\alpha_{CO_2} = 0.1008 - 29.80 \times 10^{-3}\,[M] + (1.218 \times 10^{-3}\,[M] - 3.639 \times 10^{-3})\,t$$
$$- (19.57 \times 10^{-6}\,[M] - 69.59 \times 10^{-6})\,t^2$$
$$+ (71.71 \times 10^{-9}\,[M] - 559.6 \times 10^{-9})\,t^3 \qquad (mmol/H\,Torr))$$

t   = temperature (°C, range 0–40 °C),

M = molarity of dissolved species (mol/L) refers to the volume of solute water; for solubility per L of solution the appropriate correction is required (i.e. for the volume of protein and salts).

pK_1' is dependent on temperature and ionic strength, and to a lesser extent on pH and $[Na^+]$, if $[HCO_3^-]$ is measured as total CO_2 less physically dissolved CO_2 (cf. Heisler, 1986b):

$$pK_1''_{app} = 6.583 - 13.41 \times 10^{-3}t + 228.2 \times 10^{-6}t^2$$
$$- 1.516 \times 10^{-6}t^3 - 0.341\,I^{0.323}$$
$$- \log\{1 + 0.00039\,[Pr] + (10^{pH - 10.64 + 0.011\,t + 0.737\,I^{0.323}}$$
$$\times (1 + 10^{1.92 - 0.01\,t - 0.737\,I^{0.323}} + \log[Na^+] + (-0.494\,I + 0.651)(1 + 0.0065[Pr]))\}$$

t   = temperature (°C, range 0–40 °C)

I   = ionic strength of non-protein ions ($I = 0.5\,\Sigma\,([x]\,Z^2)$, where

[x]   = concentration in mol/L and Z = number of charges of the respective ion

$[Na^+]$ = sodium concentration (mol/L)

[Pr]   = protein concentration (g/L)

These equations have been established on the basis of theoretical considerations, and have been adjusted by empirical factors derived from several thousand measurements in biological fluids of various species and in aqueous solutions (Heisler, 1984b). Calculation of 'bicarbonate' (total CO_2 less dissolved CO_2) based on α and pK_1'' values derived from these formulas is probably the closest approximation available to date to the direct gasometric determination with Van Slyke apparatus, Cameron chamber or Capnicon III. Nevertheless the reader should always be aware of the fact that direct determination is certainly preferable and can never be replaced by indirect estimates (for more details of these formulas, theoretical background and comparison with other approaches see Heisler, 1984*b*, 1986*a*; Siggaard-Andersen, 1963, 1974; Boutilier, Heming & Iwama, 1984).

Transepithelial Ion Transfer

Any bicarbonate equivalent transfer between the extracellular space of an animal and the environment can be expressed by the general relationship (cf. Fig. 1):

$$\Delta HCO_{3\ e \to env}^{-} = (\Delta[HCO_3^-]_{env} - \Delta[NH_4^+]_{env} - \frac{\Delta pH_{env} \cdot \beta_{NBenv}) \cdot V_{env}}{A}$$

Term A represents the contribution of non-bicarbonate buffering in extra-epithelial fluids like urine or the environmental water. The contribution of urinary excretion to the acid–base regulation (usually determined by tit-ration, 'titratable acidity', cf. Heisler, 1986*a*) is rather small in aquatic species (cf. Heisler, 1984*b*, 1986*a*, 1987*b*), but represents a sizeable portion of, or is exclusively responsible for, the transepithelial ion transfer in semi-terrestrial and terrestrial animals. Non-bicarbonate buffering is insignifi-cant in normal environmental water, such that this term can often be neg-lected for marine and freshwater aquatic and semi-terrestrial animals.

The transfer of bicarbonate-equivalent ions in aquatic species can be determined in a number of ways like application of the Fick principle on the blood side, or in a water flow-through system (cf. Heisler, 1984*b*). The best resolution and accuracy, however, is obtained in a closed water recircu-lation system (Fig. 3*c*; Heisler, Weitz & Weitz, 1976). A closed system, consisting of an animal box, an oxygenator/decarbonator device and a circu-lation pump has the advantage of accumulating the transferred ions in the water, such that the experimental error is kept free of the cumulative scatter of individual measurements, and the uncertainties in integrating the trans-ferred amount by interpolation from a few spot determinations and from the average flow rate through an open system (Heisler, 1984*b*). Care should be taken, however, to avoid undue accumulation of toxic metabolic end-products, like ammonia, by carefully flushing the system with water of the same quality and temperature at appropriate intervals (cf. Heisler, 1986*b*).

The changes in the environmental bicarbonate concentration can prin-cipally be determined by the same methods as described for blood and other body fluids (see above). The sensitivity of those methods, however, is often too low to detect the small changes in water bicarbonate induced by ionic transfer processes (cf. Heisler, 1978; 1984*b*; 1986*a*). The resolution of small changes in water [HCO_3^-] on a large background (typically 2 mM) has turned out to be handled best by potentiometric mH measurements.

According to the Henderson–Hasselbalch equation the $[HCO_3^-]$ is a function of pH, P_{CO_2} and the changes of α_{CO_2} and pK_1'. The first approach in this matter was to measure pH after elimination of the influence of P_{CO_2} and temperature, and thus of α_{CO_2} and pK_1'. This was achieved by equilibration of individual water samples with constant P_{CO_2} at constant temperature, and measurement of water pH with a blood pH electrode, resulting in an accuracy of $\pm$ 1.6% (cf. Heisler, 1975). Replacing the blood capillary electrode by an indwelling large surface electrode improved the accuracy to $\pm$ 1% (method used by: Heisler, Weitz & Weitz, 1976). Further improvement was possible by the application of double electrolyte bridge references (see above), selection of electrode chains for long term stability and by avoiding any manual handling of the electrodes, by continuously pumping the equilibrated water past the sensitive electrode surface. This approach resulted in the 'Δ-bicarbonate system' (Fig. 3b) used by Heisler (1975; accuracy $\pm$ 0.2%). Absolute calibration was performed, and any uncertainties with respect to the total volume of the environmental water were eliminated, by known addition of precisely weighed amounts of solid reaction grade bicarbonate to the water in the recirculation system (Heisler, 1975). Meanwhile, this system has been further improved by better temperature control at higher temperatures (30–40 °C; $\pm$ 0.01 °C), by improved electrode and electronics design, and general electrode handling (see above) resulting in a drift of <0.001 pH units/24 h on a background concentration of 2 mM (equivalent to $\pm$ 0.1%; Boutilier & Heisler, 1988) or <0.010 units/3 months (Claiborne & Heisler, 1984). Almost the same accuracy ($\pm$ 0.001 pH units on a background of 2 mM) can be achieved with a modification of the 'Δ-bicarbonate system' for individual samples stored refrigerated during the course of an experiment (Fig. 4; N. Heisler & W. Nüsse, unpublished).

A slightly different approach to eliminate the influence of P_{CO_2} is to monitor pH and P_{CO_2} simultaneously during the course of the experiment (Fig. 5). The sensitivities of pH and P_{CO_2} electrodes are electronically standardized by equilibration of a 'standard water' with two different P_{CO_2} values and adjustment of the deflection of both electrodes to the same value (Fig. 5). Any deviations of potential changes between pH and P_{CO_2} electrodes is then proportional to the changes in the water bicarbonate concentration. Absolute calibration of this set-up is performed, similar to that for the 'Δ-bicarbonate system'. Using electrodes selected for high stability (every third Radiometer P_{CO_2} electrode, E 5037–0, is suitable for this purpose and usually lasts for years; the other two can be used for blood measurements) an accuracy of + 0.3 to 0.4% can be achieved (method applied by Heisler, 1982).

Fig. 3. Apparatus for the determination of transepithelial bicarbonate-equivalent ion transfer in aquatic and semi-aquatic species. (a) System for microprocessor-operated determination of the ammonia concentration of environmental water. Samples and standards (selected by appropriate actuation of solenoid valves (V_1 and V_2) are mixed with a small fraction of 10 N NaOH and introduced into the chamber of an ammonia electrode by roller pumps (P_1 and P_2). Stirring of the chamber fluid is continued when the pumps are stopped after appropriate chamber flush. The reading is printed in terms of absolute concentration after stabilization of the electrode potential. Each measurement is preceded by an automatic two-point calibration (see text).

(b) System for determination of bicarbonate concentration changes in the environmental water. Water from the animal system (C) is pumped through three successively connected columns with fritted glass bottoms and equilibrated with a gas of constant P_{CO_2}. The water thus normalized for temperature and P_{CO_2} is fed over special large surface glass electrodes with double electrolyte bridge references before being returned to the animal system. Any changes in water pH are an expression of bicarbonate concentration changes. The apparatus is calibrated in absolute terms by known addition of accurately weighed amounts of solid reagent grade bicarbonate (see text).

(c) Closed water recirculation system. The water from the animal box is thermostatted and conditioned with respect to respiratory gases by recirculation through an oxygenator/decarbonator device and a particle filter unit. The water/animal volume ratio is typically in the range of 6 to 15.

Ammonia, as the other main acid-base relevant ion released to the environment, can be determined in water with commercially available, hydrophobic-membrane-covered pH-sensitive electrodes (HNU Systems, Orion, Ingold). Although less suitable for manual operation (the electrodes are rather sensitive to pressure changes, static electricity and changes in the mode of stirring), these electrodes can be used for semi- or fully-automated high precision microprocessor-controlled measurements (Fig. 3a). Samples and standards are mixed with 10 M NaOH to ensure alkalinization to pH 12 and accordingly complete transfer of the ammonium into diffusable ammonia (pK' of ammonia about 9.6, Cameron & Heisler, 1983). The changes of pH in the internal filling solution induced by the entry of ammonia, are exponentially proportional to the original sample ammonium concentration (similar to the CO_2 electrode). Automated assemblies allow full two-point calibrations before and after each sample measurement without attendance, and achieve routine accuracies of less than $\pm$ 0.005 mM, and of $\pm$ 0.002 mM during optimized conditions (method applied for example by Heisler, 1975, 1982; Holeton & Heisler, 1983; Holeton et al., 1983; Cameron & Heisler, 1983; Claiborne & Heisler, 1984, 1986; Boutilier

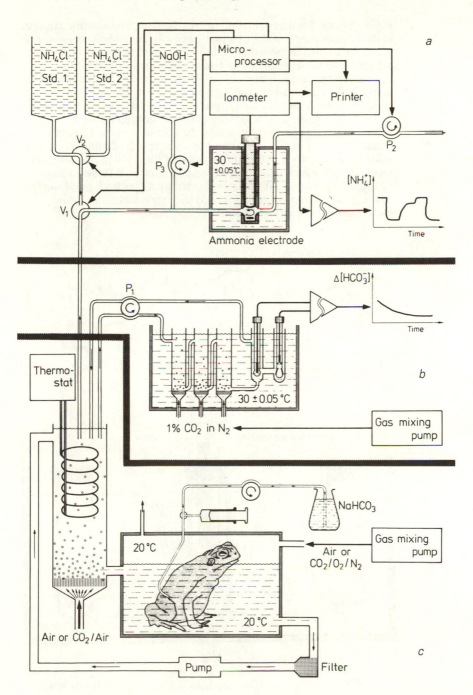

Ammonia electrode

$[NH_4^+]$

Time

$\Delta[HCO_3^-]$

Time

Micro-processor

Ionmeter

Printer

NH_4Cl Std. 1

NH_4Cl Std. 2

NaOH

V_2

V_1

P_3

P_2

P_1

$30 \pm 0.05°C$

$30 \pm 0.05°C$

$1\% CO_2$ in N_2

Gas mixing pump

Thermo-stat

NaHCO$_3$

Gas mixing pump

Air or $CO_2/O_2/N_2$

20°C

20°C

Air or CO_2/Air

Pump

Filter

a

b

c

Fig. 4. System for determination of bicarbonate concentration changes in the environmental water of aquatic animals. Individual samples, taken according to the time schedule of the experiment and stored refrigerated, are simultaneously equilibrated with a constant P_{CO_2}, at constant temperature in twelve equilibration vessels (two shown in the graph, right and left) surrounding a central electrode set-up. After complete equilibration (temperature and P_{CO_2}) samples are transferred by gas pressure one at a time through the connecting tubing into the central electrode chamber by closing the vent hole of the respective equilibration vessel. pH is recorded by a single unit, double electrolyte bridge glass electrode. After stabilization of the reading, the sample is removed by vacuum, and the next sample introduced (schematized recording, arrows). Calibration (indicated by 'C') is performed similar to the Δ-bicarbonate system (Fig. 3*b*) by known addition of solid bicarbonate to the animal system. See also text.

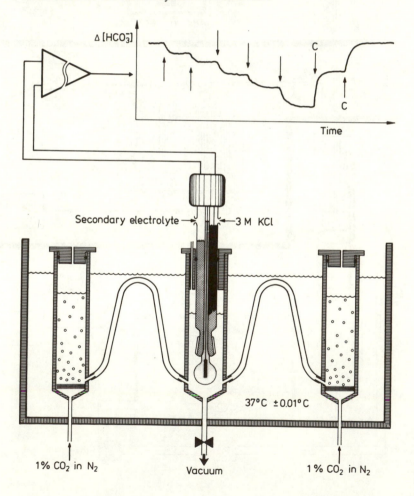

Fig. 5. Determination of water bicarbonate changes by simultaneous monitoring of pH and P_{CO_2}. Indwelling electrodes (upper panel) are calibrated by equilibration of experimental water with two P_{CO_2} values (Gas I, Gas II). The sensitivity of the electrodes is normalized after stabilization of the respective reading for Gas I and Gas II for the same deflection by electronic adjustment (middle panel, (1) position adjustment, and (2) gain adjustment). Any deviations from parallel tracing is due to changes in water bicarbonate. Absolute calibration is performed as above.

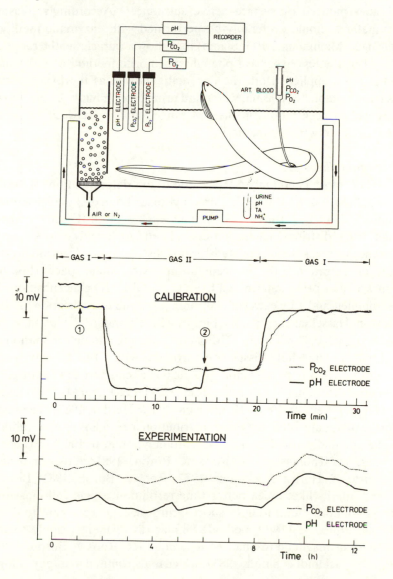

& Heisler, 1988). Problems arising in automated systems operated with seawater, caused by deposits of carbonate and borate in the sample chamber and solenoid valves, have to be handled by flushing cycles with strong acid.

Although ammonia electrodes function perfectly well with aqueous samples, they are not very reliable during operation with samples containing protein (e.g. plasma). Problems occur due to deposits, and to leaky membranes induced by surface active substances. Accordingly, plasma determinations should preferably be performed with enzymatic methods (Boehringer, Mannheim, FRG; Sigma) that allow accuracies in the range of 2–5%. Also, the less expensive phenylhypochlorite method of Solorzano (1969) is often applied. Both the enzymatic as well as the hypochlorite method are suitable for plasma, water and urine determinations, and allow, at least with concentrations in the µM range, similar accuracies as non-automated potentiometric methods.

Intracellular pH

The most commonly used method for determination of intracellular pH (pH_i) is the DMO (5,5-dimethyl-oxazolidinedione) distribution technique (Waddell & Butler, 1959). The weak acid DMO moves freely through, and is distributed in its non-dissociated form evenly across the cell membrane, whereas the dissociated form is distributed reciprocally to the H^+ ion concentration in the intracellular and extracellular spaces (Fig. 6). The intracellular pH is determined from the total DMO concentrations in tissue samples and the extracellular fluid by double application of the Henderson–Hasselbalch equation (Fig. 6) with a standard deviation in the range of $\pm$ 0.02–0.03 pH units. Radioactive labeling and determination of the volume of extracellular tissue compartments with ^{3}H-inulin have made the application of this method more practicable as compared to the original photometric determination (for details cf. Heisler & Piiper, 1972; Heisler, 1975; Heisler, Weitz & Weitz, 1976*b*; Boutilier & Heisler, 1987). The main advantage of this and other indicator distribution methods is its applicability to awake and undisturbed animals. Accordingly, other techniques (microelectrodes, cf. Thomas, 1978; Roos & Boron, 1981; direct indicator photometry, cf. Herbst & Piontek, 1972; Roos & Boron, 1981; Boron, 1983) with merits like: much better time resolution, single cell measurements, continuous monitoring of short term transients etc., can, to date, hardly substitute the DMO method. Similar disadvantages apply to the NMR (nuclear magnetic resonance) technique (cf. Roos & Boron, 1981; Boron, 1983; Schmidt & Smith, 1983), which is also limited to unphysiologically restrained or anaesthetized animals, or to isolated preparations. Future development of this technique, however, may allow also measurement in small animals under physiological conditions.

Buffering
The buffering capability is a key parameter for the evaluation of the amount of surplus acid–base relevant ions bound in intracellular, extracellular and extraepithelial fluid compartments. Determination by strong acids and base titration is the most direct approach, but does not always provide correct results. While aqueous fluids like urine and environmental water

Fig. 6. Principle of the indicator distribution technique for the determination of intracellular pH. Free membrane penetration of the non-ionized form is an indispensible prerequisite for this method, whereas any permeation of the ionized form must be negligible. Duplicate application of the Henderson–Hasselbalch equation and equal concentration of the non-ionized form as a prerequisite allows determination of the intracellular pH from total intracellular and extracellular concentrations of the indicator. The intracellular concentration is usually determined from the fractional extracellular space, and tissue- and extracellular indicator concentrations (cf. Heisler & Piiper, 1972; Heisler, 1975). See also text.

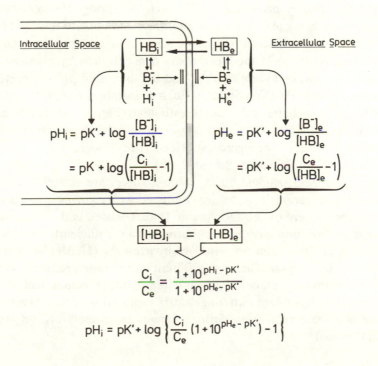

can be successfully titrated, all protein- and nucleotide-containing compartments may present considerable problems for this approach. Uncatalyzed removal of CO_2 produced by bicarbonate buffering may be a limiting factor and lead accordingly to spurious pH values. Furthermore, extremely low and high pH values close to the site of titrant addition may lead to hydrolysis of non-buffering substances, producing physiological buffer molecules, a phenomenon well known from creatine phosphate (Meyerhof & Lohmann, 1928). The involvement of this mechanism is indicated by reports of buffer values much too high (e.g. Castellini & Somero, 1981) as compared to the theoretical limit of the tissue osmolarity (Burton, 1978). Although direct titration does not necessarily exclude determination of reasonable data (for review: Heisler, 1986*a*), the more indirect CO_2 equilibration method is apparently more reliable and actually the only method checked against a non-homogenized tissue preparation (cf. Heisler & Piiper, 1971, 1972). A further advantage provided by this method is the possibility to correct for metabolic activity, and the option to establish temperature dependent buffer value matrices (cf. Heisler & Piiper, 1971; Heisler & Neumann, 1980).

The principle of this method is titration of non-bicarbonate buffers by changes in P_{CO_2}. The sample is equilibrated in an intermittently rotating cuvette with gases of different P_{CO_2} (Fig. 7). The resulting changes of pH can be recorded by means of a double electrolyte bridge electrode (with the advantage of being able to directly monitor completion of the equilibration process), or determined in individual samples taken from the tonometer. After determination of the bicarbonate concentration by calculation (see above) from the equilibration P_{CO_2} and the measured pH, or, preferably, direct measurement (Van Slyke, Cameron chamber, Capnicon), the total non-bicarbonate buffer value can be estimated (Fig. 8), taking into account any eventual dilution of the original sample, e.g. by the addition of Ringer solution to tissue homogenates (Heisler & Piiper, 1971; Heisler & Neumann, 1980; Bridges & Scheid, 1982).

Temperature-dependent buffer value matrices are established by equilibration with various P_{CO_2} values combined with changes in temperature. In general a ternary buffer system (bicarbonate, and imidazole- and phosphate-like non-bicarbonate buffers) can sufficiently describe the physiological behaviour of tissue compartments (Heisler & Neumann, 1980; Heisler, 1984*a*). Then two different equilibration sequences (Fig. 8) with appropriately chosen P_{CO_2} and temperature values will yield the temperature-dependent buffering matrix required for a model analysis of tissue acid–base regulation during changes in temperature (cf. Heisler, 1984a, 1986a).

Fig. 7. Apparatus for the determination of non-bicarbonate buffer values by the CO_2 titration method. The sample (blood or tissue homogenate) is equilibrated with different P_{CO_2} values by gas exchange at the thin fluid film, produced by intermittent rotation, on the upper cuvette wall. The fluid pH can be monitored continuously by a flat surface, glass electrode with a secondary electrolyte bridge connected to an external, thermostatted reference, or be determined in individual samples taken for total CO_2 measurement. The equilibration process can be accelerated by initially over- or under-shooting P_{CO_2}, and temperature, microprocessor-controlled on the basis of empirical data.

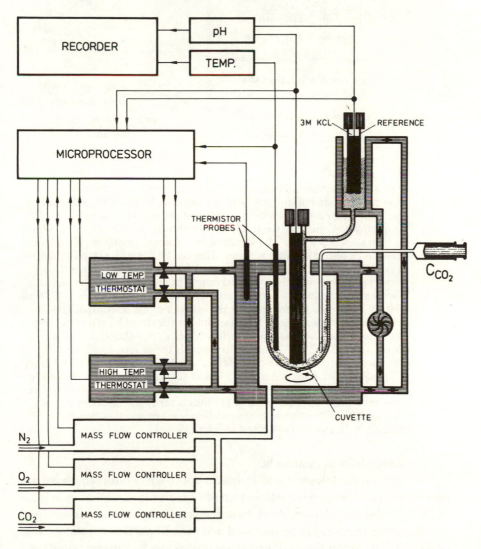

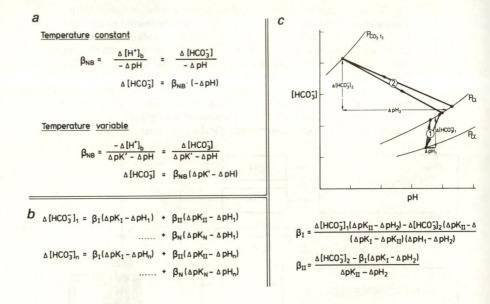

a

Temperature constant

$$\beta_{NB} = \frac{\Delta [H^+]_b}{-\Delta pH} = \frac{\Delta [HCO_3^-]}{-\Delta pH}$$

$$\Delta [HCO_3^-] = \beta_{NB} \cdot (-\Delta pH)$$

Temperature variable

$$\beta_{NB} = \frac{-\Delta [H^+]_b}{\Delta pK' - \Delta pH} = \frac{\Delta [HCO_3^-]}{\Delta pK' - \Delta pH}$$

$$\Delta [HCO_3^-] = \beta_{NB} (\Delta pK' - \Delta pH)$$

b

$$\Delta [HCO_3^-]_1 = \beta_I (\Delta pK_I - \Delta pH_1) + \beta_{II} (\Delta pK_{II} - \Delta pH_1)$$
$$\cdots \cdots + \beta_N (\Delta pK_N - \Delta pH_1)$$

$$\Delta [HCO_3^-]_n = \beta_I (\Delta pK_I - \Delta pH_n) + \beta_{II} (\Delta pK_{II} - \Delta pH_n)$$
$$\cdots \cdots + \beta_N (\Delta pK_N - \Delta pH_n)$$

c

$$\beta_I = \frac{\Delta [HCO_3^-]_1 (\Delta pK_{II} - \Delta pH_2) - \Delta [HCO_3^-]_2 (\Delta pK_{II} - \Delta}{(\Delta pK_I - \Delta pK_{II})(\Delta pH_1 - \Delta pH_2)}$$

$$\beta_{II} = \frac{\Delta [HCO_3^-]_2 - \beta_I (\Delta pK_I - \Delta pH_2)}{\Delta pK_{II} - \Delta pH_2}$$

Fig. 8. Theoretical background of the determination of the non-bicarbonate buffer value. Equilibration with two PCO_2 values at constant temperature results in changes of pH and bicarbonate concentration, yielding the total non-bicarbonate buffer value (*a*). When temperature is variable, the simultaneous changes of buffer pK' values have to be also taken into account. This can be done by establishment of a $\Delta[HCO_3]/\Delta pH/\Delta pK'$ matrix with various assumed $\Delta pK'$ values, evenly scattered over the expected range. In order to solve the matrix the number of equilibration procedures with different changes in pH and [HCO_3] has to be the same as for the number of assumed $\Delta pK'(B)$. It has turned out that usually a ternary buffer system of bicarbonate, and imidazole- and phosphate like buffers ($\Delta pK'_{Im} = -0.021$ U/°C and $\Delta pK'_{Ph} = -0.002$ U/°C) sufficiently models the *in vivo* conditions (Heisler & Neumann, 1980; cf. Heisler, 1986a, 1986c). Then two double CO_2 equilibration procedures (*c*: (1) and (2)) combined with changes in temperature are required to describe the temperature-dependent buffer system. Metabolic production of H^+ ions is taken into account by back and forth equilibration and interpolation (*c*). For more details see Heisler & Neumann (1980) and Heisler (1986a).

Limitations of approaches

The methodological and technical approaches outlined above are suitable for providing an overall pattern of organismic as well as individual tissue acid–base regulation. Acid–base relevant fluxes among tissues, and with the environment, can be followed and their important regulatory role evaluated. The involvement of excretory organs can be differentiated, and protective mechanisms of certain tissues can be better understood.

The available methodology, however, still lacks techniques to differentiate between fluxes of HCO_3^-, OH^-, or of H^+ and NH_4^+ in the opposite direction (not to mention numerous other acid–base relevant ions). This is due to the thermodynamic equilibrium and ready exchange of any tracers with the tremendously large body-water pool and the molecular CO_2 eliminated from the body fluids by diffusion. Certain differentiation into two categories, anions and cations, is possible by monitoring unidirectional and net movements of the main strong ions in the body fluids and the environment (Na^+, Cl^- etc.) which likely act as C_0^- and counterions, transferred simultaneously with acid–base relevant ions. The mass of unidirectional movements of iono- and osmoregulatory background, however, is usually hard to bring into balance with movements of acid–base relevant ions. Approaches like competitive inhibition of transfer mechanisms, or inactivation by the removal of appropriate ions regularly result in a compensatory stimulation of other mechanisms. Distinction between the cause and reaction is then difficult with a complex system of passive ionic movement along the electrochemical gradient together with active carrier-mediated processes, changing potentials and conductances. Accordingly any closer understanding of the more intimate nature of the physiological mechanisms for acid–base regulation can only be sought in simultaneous determination of all relevant parameters. Additionally, evaluation of the sometimes complex acid–base pattern during transitional states may lead to a closer understanding of the general mechanisms delineated by the currently available approaches.

References

Astrup, P. (1956). A simple electrometric technique for the determination of carbon dioxide tension in blood and plasma, total content of carbon dioxide in plasma, and bicarbonate content in 'separated' plasma at a fixed carbon dioxide tension (40 mmHg). *Scandinavian Journal of Clinical and Laboratory Investigation*, **8**, 33–43.

Boron, W. F. (1983). Measurement of intercellular ionic composition and activities in renal tubules. *Annual Review of Physiology*, **45**, 483–96.

Bridges, C. R. & Scheid, P. (1982). Buffering and CO_2 dissociation of body fluids in the pupa of the silkworm moth, *Hyalophora cecropia*. *Respiration Physiology*, **48**, 183–97.

Brønsted, J. N. (1923). Einige Bemerkungen über den Begriff der Säuren und Basen. *Recueil des Traveaux chimiques des Pays-Bas*, **42**, 718–28.

Burton, R. F. (1978). Intracellular buffering. *Respiration Physiology*, **33**, 51–88.

Boutilier, R. G., Heming, T. A. & Iwama, G. K. (1984). Appendix: Physiocochemical parameters for use in fish respiratory physiology. In:

Fish Physiology, vol. XA, ed. W. S. Hoar and D. J. Randall, pp. 403–30. New York and London: Academic Press.

Boutilier, R. G, & Heisler, N. (1987). Blood gases, and extracellular/intracellular acid–base status as a function of temperature in the anuran' amphibians *Xenopus laevis* and *Bufo marinus*. Journal of Experimental Biology **130**, 13–25.

Boutilier, R. G. & Heisler, N. (1988). Acid–base regulation in the anuran amphibian, *Bufo marinus*, during environmental hypercapnia. *The Journal of Experimental Biology* **134**, 79–98.

Cameron, J. N. (1971). Rapid method for determination of total carbon dioxide in small blood samples. *Journal of Applied Physiology*, **31**, 632–4.

Cameron, J. N. & Heisler, N. (1983). Studies of ammonia in the rainbow trout: physico-chemical parameters, acid-base behaviour, and respiratory clearance. *The Journal of Experimental Biology*, **105**, 107–25.

Castellini, M. A. & Somero, G. N. (1981). Buffering capacity of vertebrate muscle: correlations with potential for anaerobic function. *Journal of Comparative Physiology*, **143**, 191–8.

Claiborne, J. B. & Heisler, N. (1984). Acid–base regulation and ion transfers in the carp (*Cyprinus carpio*) during and after exposure to environmental hypercapnia. *The Journal of Experimental Biology*, **108**, 25–43.

Claiborne, J. B. & Heisler, N. (1986). Acid–base regulation and ion transfers in the carp (*Cyprinus carpio*): pH compensation during graded long- and short term environmental hypercapnia and the effect of bicarbonate infusion. *The Journal of Experimental Biology*, **126**, 41–61.

Gertz, K. H., & Loeschke, H. H. (1958). Elektrode zur Bestimmung des CO_2 Drucks. *Naturwissenschaften*, **45**, 160–1.

Hasselbalch, K. A. (1916). Die Berechnung der Wasserstoffzahl des Blutes aus der freien und gebundenen Kohlensäure deselben, und die Sauerstoffbindung des Blutes als Funktion der Wasserstoffzahl. *Biochemische Zeitschrift*, **78**, 112–44.

Heisler, N. (1975). Intracellular pH of isolated rat diaphragm muscle with metabolic and respiratory changes of extracellular pH. *Respiration Physiology*, **23**, 243–55.

Heisler, N. (1982). Intracellular and extracellular acid-base regulation in the tropical fresh-water teleost fish *Synbranchus marmoratus* in response to the transition from water breathing to air breathing. *The Journal of Experimental Biology*, **99**, 9–28.

Heisler, N. (1984a). Role of ion transfer processes in acid-base regulation with temperature changes in fish. *American Journal of Physiology*, **246**, R441–51.

Heisler, N. (1984b). Acid-base regulation in fishes. In: *Fish Physiology*, vol. XA, ed. W. S. Hoar and D. J. Randall, pp. 315–401. New York and London: Academic Press.

Heisler, N. (1986a). Buffering and transmembrane ion transfer processes. In: *Acid–Base Regulation in Animals*, ed. N. Heisler, pp. 3–47. Amsterdam: Elsevier Biomedical Press.

Heisler, N. (1986*b*). Acid–base regulation in fishes. In: *Acid-Base Regulation in Animals*, ed. N. Heisler, pp. 309–56. Amsterdam: Elsevier Biomedical Press.

Heisler, N. (1986*c*). Comparative aspects of acid-base regulation. In: *Acid–Base Regulation in Animals*, ed. N. Heisler, pp. 397–450. Amsterdam: Elsevier Biomedical Press.

Heisler, N. (1988*a*). Acid–base regulation in fishes: I. Mechanisms. In: *Acid Toxicity and Aquatic Animals*, ed. R. Morris, E. W. Taylor, D. J. A. Brown and J. A. Brown. Society of Experimental Biology Seminar Series. Cambridge: University Press.

Heisler, N. (1988*b*). Acid-Base Regulation in Elasmobranch Fishes. In: *Physiology of Elasmobranch Fishes*, ed. T. J. Shuttleworth, pp. 215–52. Heidelberg: Springer.

Heisler, N., Forcht, G., Ultsch, G. F. & Anderson, J. F. (1982). Acid–base regulation in response to environmental hypercapnia in two aquatic salamanders, *Siren lacertina* and *Amphiuma means*. *Respiration Physiology*, **49**, 141–58.

Heisler, N. & Neumann, P. (1980). The role of physico-chemical buffering and of bicarbonate transfer processes in intracellular pH regulation in response to changes of temperature in the larger spotted dogfish (*Scyliorhinus stellaris*). *The Journal of Experimental Biology*, **85**, 99–100.

Heisler, N. & Piiper, J. (1971). The buffer value of rat diaphragm muscle tissue determined by P_{CO_2} equilibration of homogenates. *Respiration Physiology*, **12**, 169–78.

Heisler, N. & Piiper, J. (1972). Determination of intracellular buffering properties in rat diaphragm muscle. *American Journal of Physiology*, **222**, 747–53.

Heisler, N., Weitz, H. & Weitz, A. M. (1976*a*). Hypercapnia and resultant bicarbonate transfer processes in an elasmobranch fish. *Bulletin de Physio-Pathologie respiratoire*, **12**, 77–85.

Heisler, N., Weitz, H. & Weitz, A. M. (1976*b*) Extracellular and intracellular pH with changes of temperature in the dogfish *Scyliorhinus stellaris*. *Respiration Physiology*, **26**, 249–63.

Henderson, L. J. (1909). Das Gleichgewicht zwischen Basen und Säuren im tierischen Organismus. *Ergebnisse der Physiologie*, **8**, 254–325.

Herbst, M. & Piontek, P. (1972). Über den Verlauf des intracellulären pH-Wertes des Skelettmuskels während der Kontraktion. *Pflügers Archiv*, **335**, 213–23.

Holeton, G. F. & Heisler, N. (1983). Contribution of net ion transfer mechanisms to the acid–base regulation after exhausting activity in the larger spotted dogfish (*Scyliorhinus stellaris*). *The Journal of Experimental Biology*, **103**, 31–46.

Holeton, G. F., Neumann, P. & Heisler, N. (1983). Branchial ion exchange and acid–base regulation after strenuous exercise in rainbow trout (*Salmo gairdneri*). *Respiration Physiology*, **51**, 308–18.

Jackson, D. C. & Heisler, N. (1982). Plasma ion balance of submerged anoxic turtles at 3°C: the role of calcium lactate formation. *Respiration Physiology*, **49**, 159–74.

Maffly, R. H. (1968). A conductometric method for measuring micromolar quantities of carbon dioxide. *Analytical Biochemistry*, **23**, 252–62.

Meyerhof, O. & Lohmann, K. (1928). Über die natürlichen Guanidino-phosphorsäuren (Phosphagene) in der quergestreiften Muskulatur. II. Mitteilung: Die physikalisch-chemischen Eigenschaften der Guanidino-phosphorsäuren. *Biochemische Zeitschrift*, **196**, 47–72.

Roos, A. Boron, W. F. (1981). Intracellular pH. *Physiological Reviews*, **61**, 296–434.

Schmidt, P. G. & Smith, E. N. (1983). *In vivo* P–31 NMR measurements of exercise metabolism of the lizard *Anolis carolinensis*. *Biophysical Journal*, **41**, 251a.

Scholander, P. F. (1942). A micro-gas-analyzer. *Review of Scientific Instruments*, **13**, , 264–6.

Scholander, P. F. (1947). Analyzer for accurate estimation of respiratory gases in one-half of cubic centimeter samples. *Journal of Biological Chemistry*, **167**, 235–50.

Severinghaus, J. W. (1965). Blood gas concentrations. Handbook of Physiology, Section 3: Respiration, vol. II. Washington: American Physiological Society.

Siggaard-Andersen, O. (1963, 1974). *The Acid-Base Status of Blood*. Copenhagen: Munksgaard.

Singer, R. B. & Hastings, A. B. (1948). An improved clinical method for the estimation of disturbances of the acid–base balance of human blood. *Medicine* (Baltimore), **27**, 223–42.

Stewart, P. A. (1978). Independent and dependent variables of acid–base control. *Respiration Physiology*, **33**, 9–26.

Stow, R. W., Baer, R. F. & Randall, B. F. (1957). Rapid measurement of the tension of carbon dioxide in blood. *Archives of Physical Medicine and Rehabilitation*, **38**, 646–50.

Thomas, R. C. (1978). *Ion-sensitive microelectrodes. How to make and use them*. New York: Academic.

Van Slyke, D. D. (1917). Studies of acidosis. II. A method for the determination of carbon dioxide and carbonates in solution. *Journal of Biological Chemistry*, **30**, 347–68.

Van Slyke, D. D. (1922). On the measurement of buffer values and on the relationship of buffer value to the reaction constant of the buffer and the concentration and the reaction of the buffer system. *Journal of Biological Chemistry*, **73**, 127–47.

Van Slyke, D. D. & Neill, J. M. (1924). The determination of gases in blood and other solutions by vacuum extraction and manometric measurement. *Journal of Biological Chemistry*, **61**, 523–73.

Waddell, W. J. & Butler, T. C. (1959). Calculation of intracellular pH from the distribution of 5,5–dimethyl 1–2, 4–oxazolidinedione (DMO): Application to skeletal muscle of the dog. *Journal of Clinical Investigation*, **38**, 720–9.

Suppliers of Equipment

Respirometry – methods of and approaches
Abbreviations:
  OMS: oxygen measuring systems
  BOD: specifically for measuring biological oxygen demand
  Zr*S* : zirconium sensor

Ametek Thermox Instruments Division,
150 Freeport Road
Pittsburgh,
PA 15238,
USA
OMS, Zr*S*

Beckman Instruments International S.A.,
PO Box 76,
17 Rue des Pierres-du-Niton,
CH–1211 Geneve 6
Switzerland.
— Inc.,
2500 Harbor Blvd.,
Fullerton,
CA 92634,
USA
OMS, *BOD*

Cyclobios Project Group Ecophysiological Instruments,
Institut für Zoologie,
Universität Innsbruck,
Technikerstrasse 25,
A–6020 Innsbruck,
Austria.
OMS

Delta Analytical,
250 Marcus Blvd.,
Hauppage,
N.Y. 11788
USA
OMS

Fujikura Ltd,
represented in Europe by **Sicovend AG.,**
Widenholfstrasse 6,
CH–8304 Wallisellen/Zürich, Switzerland.
OMS, ZrS

Ingold Electrodes Inc.,
261 Ballardvale Street,
Wilmington,
MA 01887,
USA
Ingold Electrodes Ltd,
Industriezone Nord,
CH–8902 Urdorf/Zürich,
Switzerland.
OMS, special electrodes

Leeds & Northrop Inc.,
Sumneytown Pike,
North Wales, PA 19454, USA.
OMS, *BOD*

Orbisphere Laboratories,
Manoir de Bonvent,
Collonge Bellerive,
CH–1245 Geneve,
Switzerland.
OMS, special electrodes

Plischke & Buhr KG.,
Siemensstrasse 8,
 D–5300 Bonn,
FRG.
Warburg Systems

Radiometer A/S,
Emdrupvej 72,
DK–2400
Copenhagen NV,
Denmark.

Radiometer America Inc.,
811 Sharon Drive, Westlake,
OH 44145,
USA
OMS, special electrodes
StrathKelvin Instruments
15 Lochend Rd,
Bearsden,
Glasgow G61 1DX, UK.
OMS
Techline Instruments,
PO Box 1236,
Fond du Lac,
WI 54935,
USA
OMS, *BOD*
Wissenschaftliche Werkstätten (WTW),
Trifthofstrasse 57a,
D–8120 Weilheim, FRG.
OMS, *BOD*
Yellow Springs Instruments Comp. Inc.
PO Box 279,
Yellow Springs,
OH 45387,
USA
OMS, special electrodes

Doubly-labelled water technique for measuring energy expenditure

Sigma Chemical Co.
Deuterium oxide, 99.8 atom
£35/100 g bottle.
Amersham Intl.
Oxygen–18 enriched water, 20 atom
£133/5 g ampoule,
or **Yeda Res. & Development Co. Ltd,**
PO Box 95,
Rehovot 76100,
Israel.
$7.5/g.

BDH Chemicals.
Guanadine hydrochloride (= guanadinium chloride),
'Aristar'
Riedel-de Haengg,
Seelze, Hannover.
Orthophosphoric Acid, Anal–R: Phosphorous pentoxide
Camlab Ltd
Nuffield Road,
Cambridge CB4 1TH
Vitrex calibrated pipettes:
($10\mu l$, white colour, Type 1260, No impregnation)
Gelman-Hawksley Ltd,
Peter Road,
Lancing,
Sussex.
Cristaseal
Edwards High Vacuum,
Crawley,
W. Sussex
RH10 2LW
Vacuum equipment
VG Medical Systems,
Aston Way,
Middlewich,
Cheshire
CW10 0HT
Mass spectrometry

Metabolic microcalorimetry

ThermoMetric AB
Spjutvagen 5A
S-175 61 JARFALLA/Stockholm
Sweden
Tel 46 8 7959545
Microcalorimeter producer
Thermal Activity Monitor (4-channel)

Respirometer and calorespirometer producer

BioMetric CaloRespirometer system

BioMetric CYCLOBIOS e.V.
Institut für Zoologie
Technikerstr. 25
A-6020 INNSBRUCK
Austria
Cyclobios Twin-Flow respirometer
Cyclobios Oxygraph for mitochondrial and cellular respiration

Telemetry

ACM (Components) Ltd
Mercury House
Calleva Park
Aldermaston
Berks RG7 4QW

All Batteries
Unit 11
Byfleet Industrial Estate
Olds Approach
Watford
Herts WD1 8QY

APC Medical
2 Little Ridge
Ridgeway
Welwyn Garden City
Herts AL7 2BH

F.W.O. Bauch Ltd
Unit 14
Donaldswood Industrial Estate
Donaldswood Rd
Stevenage
Herts SG1 2BH

Corning Ltd (Electrosil Division)
Pallion
Sunderland
Co. Durham SR4 6SU

Goodfellow Metals Ltd
Cambridge Science Park
Milton Rd
Cambridge CB4 4DJ

Hi-tek Electronics Ltd
Ditton Walk
Cambridge CB5 8QD
Lemo (UK) Ltd
12 North St
Worthing
W. Sussex BN11 1DU
A. Short and Sons Ltd
Renata House
116 St John St
London EC1V 4JR
Siliconix Ltd
3 London Rd
Newbury
Berks RG13 1JL
Sprague Electric (UK) Ltd
Airtech 2
Jenner Rd
Flemming Way
Crawley
W. Sussex RH10 2YQ
STC Electronic Services
Edinburgh Way
Harlow
Essex CM20 2DF
Steatite Insulations Ltd
2 The Square
Broad St
Birmingham B15 1AP
Surtech Interconnection Ltd
Intec 2
Wade Rd
Basingstoke
Hampshire RG24 0NL
Tekelec Components Lts
Cumberland House
Baxter Avenue
Southend-on-Sea
Essex SS2 6FA

Unitel Ltd
Unitel House
Fishers Green Rd
Stevenage
Herts SG1 2PT

Methods for measuring blood flow and distribution in intermitently ventilating and diving vertebrates
Electromagnetic flow meters

Zepeda Instruments,
Electromagnetic Flowmeter Research
1937–25th Avenue East,
Seattle,
Washington 98112,
USA
Tel. (206) 3248571
Biotronex Laboratory, Inc.
4225 Howard Avenue,
Kensington,
Maryland 20895,
USA
Tel. (301) 564–1200
Narco Biosystems Inc.
PO Box 12511/7651,
Airport Boulevard,
Houston,
Texas 77017, USA
Tel. (713) 644–7521

Permanent magnet flow transducers

In Vivo Metric,
PO Box 249,
Healdsburg,
California 95448,
USA
Tel. (707) 433–4819

Miniature pulsed Doppler flow meters and transducers

Crystal Biotech, New Englander Industrial Park,
1 Kuniholm Drive,
Holliston,
Massachusetts 01746,
USA
Tel. (617) 429–5977

Titronics Medical Instruments,
PO Box 2202,
Iowa City,
Iowa 52244,
USA
Tel. (309) 338–0836

The Bioengineering Department,
University of Iowa,
Iowa City,
Iowa 52242,
USA
Tel. (319) 335–8644

Microspheres

New England Nuclear,
549 Albany Street,
Boston,
Massachusetts 02118,
USA,
Tel. (617) 482–9595

Miniature vascular occluders

In Vivo Metric,
PO Box 249,
Healdsburg,
California 95488,
USA,
Tel. (707) 433–4819.

APPENDIX 2

Field equipment

Doubly-labelled water technique for measuring energy expenditure

1 Labelled water ($D_2^{18}O$)
2 Residue bottle, for collecting the injectate which remains in the syringe after injection (it can be re-used after purification).
3 Hypodermic needles (Microlance 26G3/8 00.45 × 10).
4 Disposable syringes (1 ml).
5 Absorbent cotton wool.
6 Alcohol.
7 Small plastic box to contain sharps and rubbish.
8 Antiseptic (Germolene).
9 Plastic sample tubes (e.g. tissue culture, 100 × 16 mm).
10 Calibrated pipettes (5μl, Vitrex).
11 Flame torch with extra fine burner (e.g. Soudo 1650, or Tamar LG870).
12 Spare gas cannister.
13 Petrol lighter and matches.
14 Cristaseal for emergency, temporary sealing of capillaries if gas torch fails.
15 Enamel tray for containing hot sharps, etc. during capillary sealing.
16 Data labels (Date, Ring no., Body mass, Site, Time of blood sample, Code to identify capillaries e.g. Bird 1 A/B).
17 Field data sheets.
18 Ringing/marking equipment.
19 Blunt tweezers, for cases where the capillary has to be held too close to the flame to permit use of the fingers.

INDEX